Springer
Berlin
Heidelberg
New York
Barcelona
Hong Kong
London
Milan
Paris
Singapore
Tokyo

Klaus Weihrauch

Computable Analysis
An Introduction

With 44 Figures

 Springer

Author

Prof. Dr. Klaus Weihrauch
FernUniversität Hagen, Fachbereich Informatik
Theoretische Informatik I
Postfach 940, 58084 Hagen
Germany
Klaus.Weihrauch@FernUni-Hagen.de

Series Editors

Prof. Dr. Wilfried Brauer
Institut für Informatik, Technische Universität München
Arcisstraße 21, 80333 München, Germany
Brauer@informatik.tu-muenchen.de

Prof. Dr. Grzegorz Rozenberg
Leiden Institute of Advanced Computer Science
University of Leiden
Niels Bohrweg 1, 2333 CA Leiden, The Netherlands
rozenber@liacs.nl

Prof. Dr. Arto Salomaa
Turku Centre for Computer Science
Lemminkäisenkatu 14 A, 20520 Turku, Finland
asalomaa@utu.fi

Library of Congress Cataloging-in-Publication Data

Weihrauch, K. (Klaus), 1943-
Computable Analysis. An Introduction / Klaus Weihrauch.
p. cm. -- (Texts in theoretical computer science)
Includes bibliographical references and index.
ISBN 3540668179 (alk. paper)
1. Computable functions. 2. Recursion theory. 3. Mathematical analysis. I. Title. II. Series.

ACM Computing Classification (1998): F.1.1, F.1.m, G.1.m

ISBN 3-540-66817-9 Springer-Verlag Berlin Heidelberg New York

Springer-Verlag Berlin Heidelberg New York,
a member of BertelsmannSpringer Science+Business Media GmbH

© Springer-Verlag Berlin Heidelberg 2000
Printed in Germany

The use of general descriptive names, trademarks, etc. in this publication does not imply, even in the absence of a specific statement, that such names are exempt from the relevant protective laws and therefore free for general use.

Cover Design: *design & production* GmbH, Heidelberg
Typesetting: Camera ready by author
SPIN: 10694615   45/3142/PS - 5 4 3 2 1 0 – Printed on acid-free paper

For Susanne

Preface

Computable analysis is a branch of computability theory studying those functions on the real numbers and related sets which can be computed by machines such as digital computers. The increasing demand for reliable software in scientific computation and engineering requires a sound and broad foundation not only of the analytical/numerical but also of the computational aspects of real number computation. Although many researchers have been active in computable analysis, it has never belonged to the main stream of research in computability. Our knowledge of this field is remarkably insufficient and only very few mathematicians or computer scientists know a definition of computable real functions. At present, computable analysis appears as a juxtaposition of several partly independent approaches which are more or less developed. For the interested newcomer this situation is bewildering, since there are not even generally accepted basic definitions; therefore, learning the state of the art from the fragments is a laborious undertaking.

This book is a new attempt to present a coherent basis for computable analysis. It is intended as a textbook suitable for graduate students in computer science or mathematics. Merely elementary knowledge in computability and analysis is assumed as prerequisite. Although many parts of the book offer themselves for extension or generalization, I have tried to concentrate on the most important elementary topics and to remain at a homogeneous moderate "level of abstraction" in order to keep the text short and make it accessible to a broader readership.

The central subject of the book is "Type-2 Theory of Effectivity" (TTE), one of the approaches to effective analysis being discussed today. It is based on definitions of computable real numbers and functions by A. Turing [Tur36] A. Grzegorczyk [Grz55] and D. Lacombe [Lac55]. Basic concepts of TTE are explained informally in Section 1.3. Chapters 2–7 systematically develop foundations of TTE. A framework of "concrete" computability on finite and infinite sequences of symbols is introduced in Chapter 2. Computability on finite and infinite sequences of symbols can be transferred to other sets by using them as names. Computability induced by naming systems is discussed in Chapter 3, where, in particular, the important class of "admissible representations" is introduced. Chapter 4 is devoted to computable real numbers and functions. Computability on spaces of subsets of $\mathbb{R}^n$ and on spaces of real functions are introduced and discussed in Chapters 5 and 6, respectively.

As a refinement of computability, the computational complexity of real functions is introduced in Chapter 7. Computable metric spaces and degrees of discontinuity are extensions of the basic theory which are discussed in Chapter 8. Finally, in Chapter 9 some other approaches to computable analysis are introduced and compared with TTE.

Most sections end with a number of exercises which particulary provide the instructor with material for homework and tests. Unmarked exercises are of medium difficulty. Exercises marked by $\Diamond$ are easy and may be solved in a straightforward manner with a proper understanding of the text. Exercises marked by $\blacklozenge$ are difficult or require a trick to solve. Such a rating of difficulty is, of course, subjective. The reader should make every effort to solve the exercises, at least the easier ones. Similarly, in the text itself, the reader should attempt to prove theorems, whenever possible, without first reading the proof in the text. Many exercises are extensions or generalizations of the material presented in the main text. Throughout the book, the square $\square$ denotes the end of a proof or example.

Since discussion on models of computation and the most adequate concepts and tools for computable analysis is still ongoing, I have put an emphasis on explaining and discussing the central definitions in detail and pointing out their distinctive features.

The origins of this book lie in a number of research papers, and more concretely in Part 3 of a monograph [Wei87], a correspondence course [Wei94] and a technical report [Wei95]. While still writing it, some gaps appeared in the subject which had to be filled, and therefore the book contains material not yet published elsewhere. However, numerous important questions in computable analysis, and even many elementary ones, are still unsettled and await systematic exploration. Although many references are included, the list is far from being complete. I apologize to all those whose work is insufficiently or not mentioned.

I should like to thank the students and collaborators who have contributed to the development of TTE, in particular U. Schreiber, G. Schäfer, C. Kreitz, T. Deil, N. Müller, T. v. Stein, U. Mylatz, M. Schröder, V. Brattka, P. Hertling, X. Zheng and N. Zhong. I have benefited from the discussions with many friends and colleagues, and I am especially grateful to V. Brattka, P. Hertling, M. Schröder, X. Zheng, N. Zhong, J. Zucker and several other people, last but not least the Springer-Verlag copy editors, who have read preliminary versions of the text and made many helpful corrections and suggestions. The research would not have been so efficient without the support of two projects of the DFG (Deutsche Forschungsgemeinschaft) under grants We843/3 and We843/8.

Finally, a big thank to my wife, Susanne, for her patience and encouragement during the numerous phases of writing of this book.

August, 2000   Klaus Weihrauch

Contents

1. Introduction

1.1 The Aim of Computable Analysis

All over the world numerous computers are used for real number computation. They evaluate real functions, find zeroes of functions, determine eigenvalues and integrals and solve differential equations. They perform, or at least are expected to perform, computations on sets like $\mathbb{R}$ (the set of real numbers), $\mathbb{R}^n$, $\mathcal{O}(\mathbb{R})$ (the open subsets of real numbers), $\mathcal{K}(\mathbb{R}^n)$ (the compact subsets of $\mathbb{R}^n$) or $C[0;1]$ (the continuous functions from the real unit interval to the real numbers). The increasing demand for reliable as well as fast software in scientific computation and engineering requires a sound and broad foundation. We agree with L. Blum et al. [BCSS96] (also in [BCSS98], however, see Sect. 9.7):

> Our perspective is to formulate the laws of computation. Thus, we write not from the point of view of the engineer who looks for a good algorithm which solves his problem at hand, or wishes to design a faster computer. The perspective is more like that of a physicist, trying to understand the laws of scientific computation. Idealizations are appropriate, but such idealizations should carry basic truths.
>
> Scientific computation is the domain of computation which is based mainly on the equations of physics. For example, from the equations of fluid mechanics, scientific computation helps to understand better design for airplanes, or assists in weather prediction. The theory underlying this side of computation is called numerical analysis.
>
> There is a substantial conflict between theoretical computer science and numerical analysis. These two subjects with common goals have grown apart. For example, computer scientists are uneasy with calculus, while numerical analysis thrives on it. On the other hand numerical analysts see no use for the Turing machine.
>
> The conflict has its roots in another age-old conflict, that between the continuous and the discrete. Computer science is oriented by the digital nature of machines and by its discrete foundations given by Turing machines. For numerical analysis systems of equations, and differential equations are central and this discipline depends heavily on the continuous nature of the real numbers. [...] Algorithms are primarily a means to solve practical problems. There is not even a formal definition of algorithm in the subject. [...] Thus, we view numerical analysis as an eclectic subject with weak foundations; this certainly in no way denies the great achievements through the centuries.

For a deep understanding and for future development of computation in analysis, a sound theoretical foundation is indispensable. In this book Computable Analysis is developed as the theory of those functions on the real numbers and other sets from analysis, which can be computed by machines. Computable analysis connects the two classical disciplines analysis/numerical analysis and computability/complexity theory. It merges concepts from both of them, in particular, the central concepts of limit and approximation on the one hand and of machine models and discrete computation on the other hand.

1.2 Why a New Introduction?

While analysis and numerical analysis have a very long tradition (mathematicians like Gauß or Lagrange were experts in numerical calculation), it was not until the 1930s that S. Kleene, A. Church, A. Turing and others proposed various definitions of *effectively calculable* functions on the natural numbers, all of which turned out to be equivalent. Meanwhile for functions on the natural numbers or on finite words there exists a well-established and very rich theory of computability and computational complexity [Rog67, HU79, Wei87, Odi89]. Although a number of authors also studied computability on the real numbers, computable analysis is still underdeveloped. In contrast to ordinary computability theory there are several partially non-equivalent suggestions of how to model effectivity in analysis and, in particular, computability of real functions. Even today no theory of computable analysis has been accepted by the majority of mathematicians or computer scientists.

The first author who introduced computable real numbers was A. Turing in his famous article "On computable numbers, with an application to the Entscheidungsproblem" [Tur36, Tur37]. Since that time computable analysis developed continuously. Among the large number of publications there are also some books on computable analysis or closely related topics, for example, R. L. Goodstein[Goo59], D. Klaua [Kla61], S. Mazur [Maz63], O. Aberth [Abe80], B. Kushner [Kuš84], E. Bishop and D. Bridges [BB85], K. Weihrauch [Wei87], M. Pour-El and J. Richards [PER89], J. Traub, G. Wasilkowski and H. Wozniakowski [TWW88], K. Ko [Ko91] and L. Blum, F. Cucker, M. Shub, S. Smale [BCSS98]. While for mathematical branches like linear algebra or recursion theory there are canonical foundations and well established introductions, all these books have important concepts in common, but differ in their contents and technical framework, and each author presents the topic from his individual point of view. This mirrors the fact that computable analysis still has no generally accepted foundation.

This book is a new attempt to present a coherent foundation of computable analysis. It is rooted in a definition of computable real functions via representations introduced by A. Grzegorczyk [Grz55] and later work on the theory of representations by J. Hauck [Hau73, Hau78, Hau80, Hau81, Hau82] and others. To distinguish it from other approaches, it will be called "Type-2 Theory of Effectivity", TTE, for short.

1.3 A Sketch of TTE

Type-2 Theory of Effectivity extends ordinary (Type-1) computability and complexity theory. We note already here that TTE admits two levels of effectivity, continuity and computability as a specialization of continuity. It is applicable to a variety of problems from analysis and provides a common framework for combining approximation, computation and computational complexity. TTE still allows a variety of computability concepts on the real numbers and other sets. We will select the seemingly most important ones and point out their distinctive features. Before we start developing TTE in detail, in this section we explain some essential ideas and concepts informally, in particular, without using a mathematically precise model of computation.

1.3.1 A Model of Computation

Ordinary computability theory first introduces computable partial word functions $f :\subseteq \Sigma^* \to \Sigma^*$ explicitly, for example, by means of Turing machines. For defining computability on other sets M (rational numbers, finite graphs, etc.) words are used as "names" or "codes" of elements of M. While a machine still transfers words to words, the user interprets these words as names of elements from the set M. Equivalently, one can start also with the computable number functions $f :\subseteq \mathbb{N} \to \mathbb{N}$ and use numbers as names. Since the sets Σ^* and $\mathbb{N}$ are only countable, they are not sufficient as sets of names for the uncountable set $\mathbb{R}$ of real numbers.

However, real numbers can be represented by infinite sequences, for example by infinite decimal fractions (example: $3.14159\ldots$ is a name of π). In TTE, infinite sequences are used as names of real numbers, and machines which transfer infinite sequences to infinite sequences are used to compute real functions. Obviously, this method of defining computable functions is not restricted to the real numbers and can be applied to many other sets. Fig. 1.1 shows a machine transforming infinite sequences to infinite sequences.

Fig. 1.1. A machine transforming infinite sequences

On input $(I_0, I_1, I_2, \dots)$, from time to time the machine reads a new sequence element I_k from its input file and from time to time it writes a new sequence element J_m to its output file, where I_{k+1} is read after I_k and J_{m+1} is written after J_m.

1.3.2 A Naming System for Real Numbers

Since infinite decimal fractions are commonly used for representing real numbers, it seems to be natural to take them as inputs and outputs for machines. However, we will prove later that the induced computability concept on the set $\mathbb{R}$ of real numbers is not very interesting. Instead of infinite decimal fractions we will use sequences of nested intervals as names. In this section, a *name* of a real number $x \in \mathbb{R}$ is a sequence $(I_0, I_1, I_2, \dots)$ of closed intervals $[a; b]$ with rational endpoints $a < b$, *rational intervals* for short, such that $I_{n+1} \subseteq I_n$ for all $n \in \mathbb{N}$ and $\{x\} = \bigcap_{n \in \mathbb{N}} I_n$. We assume tacitly that intervals are encoded appropriately, such that, strictly speaking, a name of a real number is an infinite sequence of symbols. Fig. 1.2 shows the first six intervals of a name of a real number x.

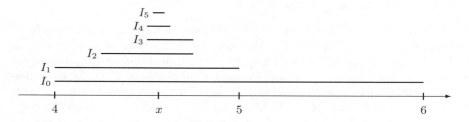

Fig. 1.2. The first elements of a name of x

1.3.3 Computable Real Numbers and Functions

We call a real number *computable*, iff it has a computable name. We illustrate these definitions by examples.

Example 1.3.1 (computable real numbers).
1. Every rational number $r \in \mathbb{Q}$ is computable.
   Define $I_n := [r - 2^{-n}; r + 2^{-n}]$ for all $n \in \mathbb{N}$. Then the sequence $(I_0, I_1, \dots)$ is a computable name of r.
2. $\sqrt{2}$ is computable.
   Define $f : \mathbb{N} \to \mathbb{N}$ by $f(n) := \min\{k \in \mathbb{N} \mid k^2 < 2n^2 \le (k+1)^2\}$, $J_0 := [1; 2]$ and $J_n := [f(n)/n; (f(n) + 2)/n]$ for all $n > 0$. Then the sequence $(J_0, J_1, \dots)$ is computable, $\sqrt{2} \in J_n$ for all n and $\lim_{n \to \infty} \text{length}(J_n) = 0$. But the sequence is not nested in general. Define $I_n := J_0 \cap J_1 \cap \dots \cap J_n$. Then the sequence $(I_0, I_1, I_2, \dots)$ is a computable name of $\sqrt{2}$.

3. $\log_3 5$ is computable.

Define $f(n) := \min\{k \in \mathbb{N} \mid 3^k < 5^n \leq 3^{k+1}\}$ and continue as above.   □

Example 1.3.2 (Specker sequences).   For $A \subseteq \mathbb{N}$

$$x_A := \sum_{i \in A} 2^{-i} \text{ is computable} \iff A \text{ is recursive}$$

Assume that A is recursive. Define $I_n := [s_n; s_n + 2 \cdot 2^{-n}]$ where $s_n := \sum_{i \leq n, i \in A} 2^{-i}$. Then $(I_0, I_1, \ldots)$ is a computable name of x_A.

On the other hand, let $(I_0, I_1, \ldots)$ be a computable name of x_A. If A is finite or co-finite, then A is recursive. Assume that A is neither finite nor co-finite. Then $m/2^n \neq x_A$ for all $m, n \in \mathbb{N}$. For $n = 0, 1, \ldots$ (in this order) decide whether $n \in A$ as follows. Define $A_n := \{i \in A \mid i < n\}$ and $t_n := \sum_{i \in A_n} 2^{-i} + 2^{-n}$. Notice that $0 < \sum_{i \in A, i > n} 2^{-i} < 2^{-n}$. Then $n \in A \implies x_A > t_n$ and $n \notin A \implies x_A < t_n$. Find the smallest $k \in \mathbb{N}$ with $t_n \notin I_k$. If $\max(I_k) < t_n$ then $x_A < t_n$, hence $n \notin A$, if $t_n < \min(I_k)$ then $t_n < x_A$, hence $n \in A$. Therefore, A is recursive.

Let $A \subseteq \mathbb{N}$ be recursively enumerable but not recursive. Then the real number $x_A := \sum_{i \in A} 2^{-i}$ is not computable. From recursion theory we know that $A = \text{range}(f)$ for some computable injective function $f : \mathbb{N} \to \mathbb{N}$. We obtain $x_A = \sum_{i \in \mathbb{N}} 2^{-f(i)}$. Then $(x_0, x_1, \ldots)$ with $x_n = \sum_{i \leq n} 2^{-f(i)}$ is an increasing and bounded computable sequence of rational numbers. Its limit, however, is the non-computable real number x_A. Sometimes such a sequence is called a *Specker sequence* [Spe49].   □

We call a real function $f :\subseteq \mathbb{R} \to \mathbb{R}$ *computable*, iff some machine maps any name of any $x \in \text{dom}(f)$ to a name of $f(x)$. For real functions $f :\subseteq \mathbb{R}^n \to \mathbb{R}$ we consider machines reading n names in parallel. Notice that the machine must behave correctly only for every name of every $x \in \text{dom}(f)$, for other input sequences it may behave arbitrarily.

Example 1.3.3 (computable real functions).
1. Real multiplication $(x, y) \mapsto x \cdot y$ is computable.
   For closed intervals I, J with rational endpoints define the interval $I \cdot J$ by $I \cdot J := \{x \cdot y \mid x \in I, y \in J\}$. There is a machine with two input files which maps the inputs $(I_0, I_1, \ldots)$ and $(J_0, J_1, \ldots)$ to $(I_0 \cdot J_0, I_1 \cdot J_1, \ldots)$. If $(I_0, I_1, \ldots)$ is a name of x and $(J_0, J_1, \ldots)$ is a name of y then $(I_0 \cdot J_0, I_1 \cdot J_1, \ldots)$ is a name of $x \cdot y$. Therefore, multiplication is computable.
2. The square root $\sqrt{\ } :\subseteq \mathbb{R} \to \mathbb{R}$ is computable.
   Since for example $\sqrt{[2; 3]}$ is not a rational interval, we have to modify the above method. We map each input interval I to a rational interval which is slightly longer than $\sqrt{I}$. For any rational numbers a, b with $0 \leq a < b$ there are rational numbers $r, s \geq 0$ such that

$$a - (b - a) < r^2 \leq a < b \leq s^2 < b + (b - a) \, .$$

Therefore, there is a computable function f which for each rational interval I with no negative elements determines (for example by exhaustive search) a rational interval $K = f(I)$ such that $I \subseteq K^2$ and $\text{length}(K^2) < 3 \cdot \text{length}(I)$. For any rational interval $[a;b]$ with $0 \le b$ let $g[a;b] := [\max(0,a);b]$. If $(I_0, I_1, \ldots)$ is a name of $x \ge 0$ then the sequence $(f \circ g(I_0), f \circ g(I_1), \ldots)$ of rational intervals converges to $\sqrt{x}$. This sequence is not necessarily nested. There is a machine M which transforms any name $(I_0, I_1, \ldots)$ of a non-negative real number to the sequence $(J_0, J_1, \ldots)$, where $J_n := f \circ g(I_0) \cap f \circ g(I_1) \cap \ldots \cap f \circ g(I_n)$. Then M computes the square root. $\qquad \square$

We will prove later that elementary real functions like $\exp, \sin, \log$ and $\arcsin$ are computable, that the computable real functions map computable numbers to computable numbers and that the computable real functions are closed under composition. Our definition is essentially the definition of computable real function introduced by A. Grzegorczyk in [Grz55] and studied in more detail in [Grz57], where he proves, in particular, that computable real functions are continuous.

Theorem 1.3.4. Every computable real function is continuous.

Proof: We sketch a proof for the case $f : \mathbb{R} \to \mathbb{R}$. Let M be a machine computing f and let $x \in \text{dom}(f) = \mathbb{R}$. It suffices to show that for every open set $V \subseteq \mathbb{R}$ with $f(x) \in V$, there is some open set $U \subseteq \mathbb{R}$ with $x \in U$ and $f[U] \subseteq V$. Therefore, let V be an open set with $f(x) \in V$. The real number x has a name $([a_0;b_0], [a_1;b_1], \ldots)$ such that $a_i < a_{i+1} < b_{i+1} < b_i$. Let $(J_0, J_1, \ldots)$ be the name of $f(x)$ which M produces on this input. Then $J_n \subseteq V$ for some number n. For producing the initial part $(J_0, J_1, \ldots, J_n)$ of the output, the machine M needs k steps for some k. In k steps the machine M can read at most the first k intervals from the input. We choose $U := (a_k; b_k)$. Obviously, we have $x \in U$. Assume $y \in U$. Then y has a name of the form $([a_0;b_0], [a_1;b_1], \ldots, [a_k;b_k], L_{k+1}, L_{k+2}, \ldots)$. On this input, the machine M writes also $(J_0, J_1, \ldots, J_n)$ within the first k steps which is the initial part of a name of $f(y)$. We obtain $f(y) \in J_n$. This shows $f[U] \subseteq J_n \subseteq V$. Therefore, the function f is continuous at x. $\qquad \square$

In the above proof we have used the essential observation that any finite portion $(J_0, J_1, \ldots, J_n)$ of the output of a computation is determined already by a finite portion $(I_0, I_1, \ldots, I_k)$ of its input. However, we did not use that the transformation from inputs to outputs is computable. As a consequence of the continuity theorem, simple real functions like the step function $s(x) := (0$ if $x < 0$, 1 otherwise$)$ and the Gauß staircase $g(x) := \lfloor x \rfloor$ (the integer part of x) are not computable (Fig 1.3). Obviously, these functions are *easily definable* in our mathematical language, but "easily definable" does not mean "computable" in general. As far as we know neither the step function or the Gauß staircase nor any other non-continuous real function can be computed by physical devices.

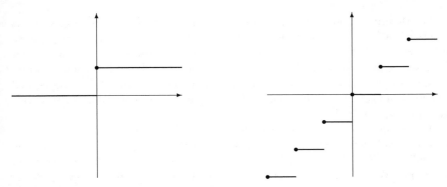

Fig. 1.3. Two non-computable real functions

Some people do not accept the model of computation presented here, since they believe that the above simple functions are or should be called computable. Although the Gauß staircase is not computable according to our present definition, we will see that the framework of TTE admits to define other natural computability concepts such that it is a computable object (Exercise 4.3.2).

1.3.4 Subsets of Real Numbers

A subset $A \subseteq \mathbb{N}$ is *decidable* or *recursive*, iff its characteristic function $\mathrm{cf}_A : \mathbb{N} \to \mathbb{N}$ is computable. Assume for the moment that we call a subset $A \subseteq \mathbb{R}$ recursive, iff its characteristic function $\mathrm{cf}_A : \mathbb{R} \to \mathbb{R}$ is computable. Then $\mathrm{cf}_A : \mathbb{R} \to \mathbb{R}$ is not continuous and hence not computable by the continuity theorem, unless $A = \emptyset$ or $A = \mathbb{R}$. Therefore, this definition is useless.

The *distance function* $d_A : \mathbb{R}^n \to \mathbb{R}$, defined by $d_A(x) := \inf_{y \in A} |y - x|$ is a more useful generalization of the discrete characteristic function. Fig. 1.4 shows the function $1 - \mathrm{cf}_{[0;1]}$ and its continuous counterpart $d_{[0;1]}$. We call a (closed) subset $A \subseteq \mathbb{R}^n$ *recursive*, iff its distance function $d_A : \mathbb{R}^n \to \mathbb{R}$ is computable. It turns out that simple sets like intervals $[a;b] \subseteq \mathbb{R}$ with computable endpoints a and b, the set $\{(x,y) \in \mathbb{R}^2 \mid x \leq y\}$ and the graph $\{(x, f(x)) \mid x \in \mathbb{R}\}$ of any computable function $f : \mathbb{R} \to \mathbb{R}$ are recursive.

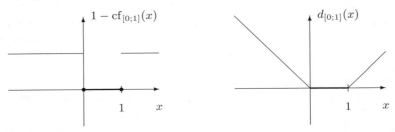

Fig. 1.4. $1 - \mathrm{cf}_{[0;1]}$ and $d_{[0;1]}$

For defining computability on subsets of $\mathbb{R}$, we introduce naming systems of sets of subsets. Since the set $2^{\mathbb{R}}$ of all subsets of $\mathbb{R}$ is too large, we consider only subsets of $2^{\mathbb{R}}$. Here we merely introduce a naming system for the set $\mathcal{O}(\mathbb{R})$ of the open subsets of real numbers. Later we will study various naming systems of the open, the closed and the compact subsets of $\mathbb{R}^n$. As is well known, a subset $U \subseteq \mathbb{R}$ is *open*, iff it is the union of a set of open intervals with rational endpoints.

We define: a *name* of an open subset $U \subseteq \mathbb{R}$ is a sequence $(I_0, I_1, \ldots)$ of open intervals with rational boundaries such that $U = I_0 \cup I_1 \cup \ldots$. In the general case $U \subseteq \mathbb{R}^n$, each I_k is a product of n open intervals with rational endpoints. We call an open set recursively enumerable, iff it has a computable name.

Examples of recursively enumerable open sets are the open intervals $(a; b)$ with computable endpoints a and b, the set $\{(x, y) \subseteq \mathbb{R}^2 \mid x < y\}$ and every interval $(0; x_A)$ where x_A is from Example 1.3.2 with r.e. (recursively enumerable) but non-recursive $A \subseteq \mathbb{N}$. As for real functions, we call a function on the set of open sets computable, iff some machine maps names of arguments to names of the results. The functions *intersection* and *union* on open sets are computable. For example. let M be a machine with two input files which transforms any two names $(I_0, I_1, \ldots)$ and $(J_0, J_1, \ldots)$ of open sets to a sequence $(K_0, K_1, \ldots)$ which is a list of all $I_m \cap J_n$ with $m, n \in \mathbb{N}$. Then M computes the intersection on $\mathcal{O}(\mathbb{R})$.

We define computability for functions of mixed type accordingly. For example, the function $f : \mathbb{R} \to \mathcal{O}(\mathbb{R})$ with $f(x) := \mathbb{R} \setminus \{x\}$ is computable. A machine computing f maps any name of $x \in \mathbb{R}$ to a name of $\mathbb{R} \setminus \{x\}$. Since a computable function maps computable names to computable names, $X \cap Y$ and $X \cup Y$ are recursively enumerable open sets, if X and Y are recursively enumerable open sets, and $\mathbb{R} \setminus \{x\}$ is a recursively enumerable open set, if x is a computable real number.

1.3.5 The Space C[0; 1] of Continuous Functions

As a last example we consider the set $C[0; 1]$ of continuous functions $f : [0; 1] \to \mathbb{R}$. First, we define a naming system. Let us call a function $f : [0; 1] \to \mathbb{R}$ a *rational polygon*, iff its graph is a polygon with finitely many vertices with rational coordinates. We call the set $B(p, r) = \{f \in C[0; 1] \mid d(f, p) < r\}$, where $d(f, p) = \max_{x \in [0;1]} |f(x) - p(x)|$ the *function ball* with *center* $p \in C[0; 1]$ and *radius* $r > 0$. In this section, we generalize our naming system for real numbers as follows. A name of a continuous function $f : [0; 1] \to \mathbb{R}$ is a sequence $(B_0, B_1, \ldots)$, where B_n is a function ball of radius 2^{-n} with $f \in B_n$ the center of which is a rational polygon. Fig. 1.5 shows such a ball.

A name $(B_0, B_1, \ldots)$ of f encloses f arbitrarily narrowly. A function $f : [0; 1] \to \mathbb{R}$ can have a computable name, and it can be computable (by a machine transforming each name of real numbers $x \in [0; 1]$ to a name of $f(x)$). Later, we will prove that these properties are equivalent.

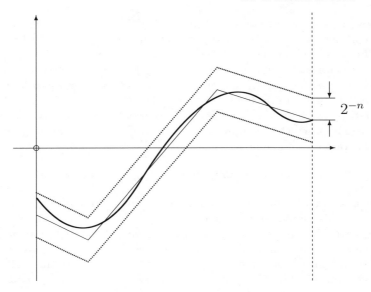

Fig. 1.5. A function f in a function ball with radius 2^{-n} with a rational polygon as center and radius 2^{-n}

Our naming systems of $\mathbb{R}$ and $C[0;1]$ make the evaluation function Apply : $C[0;1] \times [0;1] \to \mathbb{R}$, defined by Apply$(f, x) := f(x)$, computable, that is, there is a machine which transforms any name of any f and any name of any x to a name of $f(x)$. There are many other naming systems of $C[0;1]$, for which the evaluation function becomes computable, however, among all of these our naming system is the "weakest" one (we will explain this later). Thus, our naming system for $C[0;1]$ is tailor-made for the evaluation function.

Integration $f \mapsto \int_0^1 f(x)\mathrm{d}x$ on $C[0;1]$ is an important computable operator, while differentiation $f \mapsto f'$ for continuously differentiable $f \in C[0;1]$ is not computable. By the classical *intermediate value theorem* every continuous function $f : [0;1] \to \mathbb{R}$ with $f(0) < 0 < f(1)$ has a zero, that is, a point x with $f(x) = 0$. Unfortunately, there is no computable operator for zero-finding working correctly for all continuous functions $f : [0;1] \to \mathbb{R}$ with $f(0) < 0 < f(1)$. However, a restricted problem has a computable solution: the operator $Z : f \mapsto$ (the zero of f) for continuous increasing functions with $f(0) < 0 < f(1)$ is computable. Later, we will discuss in detail the problem of zero-finding under various assumptions.

1.3.6 Computational Complexity of Real Functions

Ordinary discrete complexity theory studies resources like time or storage used by machines for computing functions [HU79]. If M is a Turing machine computing a word function $f : \Sigma^* \to \Sigma^*$, then complexity theory considers

$\text{Time}_M(x)$, the number of steps which M makes on input x before halting. This concept cannot be transferred to machines which map infinite sequences to infinite sequences, since these machines do not halt. However, we can consider for each input sequence $s = (I_0, I_1, \ldots)$ and for each number n the number $\text{Time}_M(s)(n)$ of steps which M makes before it prints the first interval J_k of its output $(J_0, J_1, \ldots)$ of length $< 2^{-n}$. For a machine M computing a real function $f : \mathbb{R} \to \mathbb{R}$ we would like to have a complexity $\text{Time}_M(x)(n)$, which depends on the real number x but not on a name of x. For this pupose we define

$$\text{Time}_M(x)(n) := \max\{\text{Time}_M(s)(n) \mid s \text{ is a name of } x\} \ .$$

Since in our naming system we admit arbitrarily redundant names of the real numbers, this maximum does not exist.

However, for the *signed digit representation* of the real numbers, we obtain very natural complexity results. In the signed digit representation, a name of a real number is an infinite binary fraction where the digit $\overline{1} := -1$ can be used in addition to the digits 0 and 1, that is, $a_n \ldots a_0 \bullet a_{-1} a_{-2} \ldots$ with $a_k \in \{0, 1, \overline{1}\}$ is a name of $x := \sum_{k=n}^{-\infty} a_k$. These names encode strongly normalized sequences of nested intervals and induce the same computability on $\mathbb{R}$ as our standard names. The induced computational complexity of real functions is sometimes called "bit complexity". Bit complexity has been studied by several authors, in particular, by Ko [Ko91, Ko98], who applied concepts from discrete complexity theory to prove upper and lower bounds for the complexity of basic numerical operations. It turns out that Turing machines operating on signed digit representations of real numbers can compute addition in time $O(n)$ and multiplication, sin, exp and log in time $O(n^2)$.

1.4 Prerequisites and Notation

We assume that the reader is familiar with the basic concepts of analysis, that is, the theory of real numbers and functions, and of ordinary computability theory at undergraduate level in computer science. For details the reader should consult standard textbooks, e.g. parts of [Die60, HU79].

By $\mathbb{N}$, $\mathbb{Z}$, $\mathbb{Q}$ and $\mathbb{R}$ we denote the set $\{0, 1, \ldots\}$ of natural numbers, of integer numbers, of rational numbers and of real numbers, respectively. For a set A the power set of A, that is, the set of all subsets of A, is denoted by 2^A and the set of all finite subsets of A is denoted by $\text{E}(A)$. If $X = A_1 \times \ldots \times A_n$ is a Cartesian product of sets, then for $n = 0$, $X = \{()\}$ is the set the only element of which is the *empty tuple* "()" .

If $\leq$ is a binary relation on $A \times B$, $x < y$ will abbreviate "$x \leq y$ and $y \not\leq x$".

We will consider multi-valued partial functions and as special cases partial functions and total functions. A *correspondence* or *multi-valued partial function* from A to B is a triple $f = (A, B, R_f)$ such that $R_f \subseteq A \times B$, where A is called the *source*, B the *target* and R_f the *graph* of f. For $X \subseteq A$ we define the *image* of X under f by

$$f[X] := \{b \in B \mid (\exists a \in X)(a, b) \in R_f\} \,.$$

We define the *inverse*, the *domain* and the *range* of f by

$$f^{-1} := (B, A, R_f^{-1}) \text{ such that } R_f^{-1} := \{(b, a) \mid (a, b) \in R_f\} \,,$$

$$\text{range}(f) := f[A] \quad \text{and}$$

$$\text{dom}(f) := f^{-1}[B] \,,$$

respectively. The multi-valued partial function f is completely defined by the family of sets $f[\{a\}]$ $(a \in A)$.

Usually we will denote a correspondence f from A to B by $f :\subseteq A \rightrightarrows B$. A *partial function* $f :\subseteq A \rightarrow B$ from A to B is a multi-valued function $f :\subseteq A \rightrightarrows B$ such that the set $f[\{a\}]$ contains only one element for each $a \in \text{dom}(f)$. A *total function* $f : A \rightarrow B$ from A to B is a partial function $f :\subseteq A \rightarrow B$ such that $\text{dom}(f) = A$. The set of all (total) functions $f : A \rightarrow B$ is denoted by B^A. For a partial function $f :\subseteq A \rightarrow B$, $f(a)$ denotes the single element from $f[\{a\}]$, if $a \in \text{dom}(f)$, and we will write $f(a) = \text{div}$, if $a \notin \text{dom}(f)$. Usually we will call f a *function*, if f is a partial function or a total function.

For multi-valued functions $f_i :\subseteq A \rightrightarrows B_i$ $(i-1 \ldots, k)$ define $(f_1, \ldots, f_k) :\subseteq A \rightrightarrows B_1 \times \ldots \times B_k$ by $(f_1, \ldots, f_k)[\{a\}] := f_1[\{a\}] \times \ldots \times f_k[\{a\}]$. For multi-valued functions $f :\subseteq A \rightrightarrows B$ and $g :\subseteq B \rightrightarrows C$ define the *composition* $g \circ f :\subseteq A \rightrightarrows C$ by $a \in \text{dom}(g \circ f)$, iff $a \in \text{dom}(f)$ and $f[\{a\}] \subseteq \text{dom}(B)$, and $g \circ f[\{a\}] := g[f[\{a\}]]$ for all $a \in \text{dom}(g \circ f)$.

For a multi-valued function $f :\subseteq A \rightarrow B$, $X \subseteq A$ and $Y \subseteq B$, we define the restrictions in the source, the target, the domain and the range as follows:

$$
\begin{array}{llll}
\text{source:} & f|_X :\subseteq X \rightarrow B, & f|_X & := (X, B, R_f \cap X \times B) \\
\text{domain:} & f\rfloor_X :\subseteq A \rightarrow B, & f\rfloor_X & := (A, B, R_f \cap X \times B) \\
\text{target:} & f|^Y :\subseteq A \rightarrow Y, & f|^Y & := (A, Y, R_f \cap A \times Y) \\
\text{range:} & f\rceil^Y :\subseteq A \rightarrow B, & f\rceil^Y & := (A, B, R_f \cap A \times Y) \,.
\end{array}
$$

For combinations of restrictions we use the symbols $|, \rceil, \rfloor$ and $\rceil$, for example $f\rfloor_X^Y$. Usually, $\text{id}_X : X \rightarrow X$ denotes the *identity function* on X.

For any set Σ, Σ^n denotes the set of all words over Σ of length n, $\Sigma^{\leq n}$ the set $\Sigma^0 \cup \ldots \cup \Sigma^n$ and Σ^* the set of all finite words over Σ. We denote the *length* of a word w by $|w|$ and the *empty word* which has length 0 by λ. By Σ^ω we denote the set $\{p \mid p : \mathbb{N} \rightarrow \Sigma\} = \{a_0 a_1 a_2 \ldots \mid a_i \in \Sigma\}$ of all infinite sequences over Σ. Occasionally finite or infinite sequences of symbols will be called *strings*.

We extend concatenation from $\Sigma^* \times \Sigma^*$ to $\Sigma^* \times \Sigma^\omega$. Consider $u, v, w \in \Sigma^*$ and $p \in \Sigma^\omega$. If $x = uvw \in \Sigma^*$ and $q = uvp \in \Sigma^\omega$, then u is a *prefix* of x

and q (abbreviated by $u \sqsubseteq x$ and $u \sqsubseteq q$, respectively), v is a *subword* of x and q (abbreviated by $v \lhd x$ and $v \lhd q$, respectively) and w is a *suffix* of x. A set $A \subseteq \Sigma^*$ is called *prefix-free*, iff $x \not\sqsubseteq y$ for all $x, y \in A$ with $x \neq y$. We define $u^n := uu \ldots u$ (n times) as usual and $u^\omega := uuu \ldots \in \Sigma^\omega$. By $p_{<n}$ we denote the prefix $p(0) \ldots p(n-1)$ of length n of p and by $p_{\leq n}$ the prefix $p(0) \ldots p(n)$. We extend concatenation straightforwardly to sets of finite or infinite sequences: $AB := \{up \mid u \in A, p \in B\}$ for $A \subseteq \Sigma^*$ and $B \subseteq \Sigma^*$ or $B \subseteq \Sigma^\omega$. In particular, $A^n := A \ldots A$ (n factors). (If $A \subseteq \Sigma^*$, it will be clear from the context, whether A^n denotes the Cartesian product or the concatenation product.) The prefix relation can be expressed by means of product: $u \sqsubseteq v \iff v \in u\Sigma^*$ (for $u, v \in \Sigma^*$) and $u \sqsubseteq q \iff q \in u\Sigma^\omega$ (for $u \in \Sigma^*$ and $p \in \Sigma^\omega$).

Later Σ will be a non-empty finite set, an *alphabet*, the symbols of which we will denote in typewriter style (abc...012...#.$ etc.). In emphasizing that an expression denotes a word, we occasionally set it into quotes. For example: if u and v denote words and if 0 and ; are elements of Σ, then "$u0;v$" denotes the word which begins with u, has 0 and ; as the next symbols and ends with the word v.

Ordinary "Type-1" recursion theory considers the computable number functions $f :\subseteq \mathbb{N}^k \to \mathbb{N}$ and the computable word functions $g :\subseteq (\Sigma^*)^k \to \Sigma^*$. Computability can be transferred to other sets M by means of numberings. A *numbering* of a set M is a surjective (that is, "onto") function $\nu :\subseteq \mathbb{N} \to M$. For a given finite alphabet $\Sigma = \{a_1, \ldots, a_n\}$ define the bijective standard numbering $\nu_\Sigma : \mathbb{N} \to \Sigma^*$ of Σ^* as follows:

$$\nu_\Sigma^{-1}(\lambda) := 0; \quad \nu_\Sigma^{-1}(a_{i_k} \ldots a_{i_0}) := i_k \cdot n^k + \ldots + i_0 \cdot n^0 .$$

Then $f :\subseteq \mathbb{N} \to \mathbb{N}$ is computable, iff $\nu_\Sigma \circ f \circ \nu_\Sigma^{-1} :\subseteq \Sigma^* \to \Sigma^*$ is computable (correspondingly for $f :\subseteq \mathbb{N}^m \to \mathbb{N}$). Occasionally, we will extend the definition of computable functions to functions of mixed type ($\mathbb{N}$ and Σ^*) by calling ν_Σ and ν_Σ^{-1} computable and closing under composition. This way we can, for example, define the r.e. subsets $A \subseteq \Sigma^* \times \mathbb{N} \times \Sigma^*$ and the computable functions $f :\subseteq \mathbb{N} \times \Sigma^* \to \mathbb{N}$.

We will need the bijective Cantor pairing function $\langle\ ,\ \rangle : \mathbb{N}^2 \to \mathbb{N}$, defined by $\langle x, y \rangle := y + (x+y)(x+y+1)/2$, which is computable, as well as the projections $\langle\ \rangle_1$ and $\langle\ \rangle_2$ of its inverse. For $n > 2$ arguments, we define inductively

$$\langle x_1, \ldots, x_n \rangle := \langle \langle x_1, \ldots, x_{n-1} \rangle, x_n \rangle$$

as usual.

2. Computability on the Cantor Space

Classically, computability is introduced explicitly for functions $f :\subseteq (\Sigma^*)^n \rightarrow \Sigma^*$ on the set Σ^* of finite words over an arbitrary finite alphabet Σ, for example by means of Turing machines. For computing functions on other sets M such as natural numbers, rational numbers or finite graphs, words are used as "code words" or "names" of elements of M. Under this view a machine transforms words to words without "understanding" the meaning given to them by the user. Equivalently, one can start with computable functions on the natural numbers instead of words and use numbers as "names". Since the set Σ^* of words over a finite alphabet is only countable, this method cannot be applied for introducing computability on uncountable sets M like the set $\mathbb{R}$ of real numbers, the set of open subsets of $\mathbb{R}$ or the set $C[0;1]$ of real continuous functions on the interval $[0;1]$.

We will extend the above concepts by using infinite sequences of symbols as names and by defining computability for functions which transform such infinite sequences. The set Σ^ω of infinite sequences of symbols from a finite alphabet Σ (provided Σ has at least two elements) has the same cardinality as the set of real numbers ("continuum cardinality"), therefore, it can serve as a set of names for every set with at most continuum cardinality. An example is the traditional representation of real numbers by infinite decimal fractions (for example the number π by the sequence $3.14159\ldots$). Also our names of real numbers from Sect. 1.3 are infinite sequences of symbols, if rational intervals are encoded by words. As a first step, we introduce and study computability on finite and infinite sequences of symbols.

In Sect. 2.1 we define the computable functions on finite and infinite sequences of symbols, introduce tupling functions and, among others, prove closure theorems for composition and primitive recursion. In Sect. 2.2 we introduce the Cantor topology on Σ^ω and show that computable functions are continuous. Sect. 2.3 is devoted to naming systems of sets of computable or continuous functions. The reader may skip details in a first reading. Later we will apply essentially the smn- and the utm-theorems (Theorems 2.3.5 and 2.3.13). Finally, in Sect. 2.4 we introduce the recursively enumerable, the decidable and the recursive open subsets. We mention already that two differing concepts "decidable" and "recursive open" must be distinguished on Σ^ω.

Convention 2.0.1. In this chapter we assume that Σ is a fixed finite alphabet containing the symbols 0 and 1. In some examples we will assume that Σ is sufficiently large.

2.1 Type-2 Machines and Computable String Functions

For $k \geq 0$ and $Y_0, Y_1, \ldots, Y_k \in \{\Sigma^*, \Sigma^\omega\}$ we define the computable functions $f :\subseteq Y_1 \times \ldots \times Y_k \to Y_0$ by means of Turing machines with k one-way input tapes, finitely many work tapes and a single one-way output tape. Fig. 2.1 shows such a Turing machine.

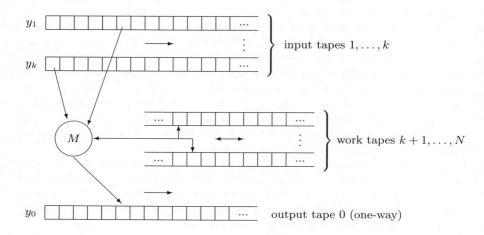

Fig. 2.1. A Turing machine computing $y_0 = f_M(y_1, \ldots, y_k)$

We content ourselves with very short informal definitions. A Turing machine is given by

- an input/output alphabet Σ with a special symbol $B \notin \Sigma$, a work alphabet Γ with $\Sigma \cup \{B\} \subseteq \Gamma$ and a number k of input tapes,
- a flowchart operating on the output tape 0, the k input tapes $1, \ldots, k$ and work tapes $k+1, \ldots, N$ (for some number N). The following statements may occur in the flowchart with the additional restrictions for the input tapes and the output tape given below:
  - the statement "HALT",
  - the statement "i:left" which moves the head on Tape i one position to the left,
  - the statement "i:right" which moves the head on Tape i one position to the right,

- for each $a \in \Gamma$ the statement "i:write(a)" which writes the symbol a on the tape cell scanned by the head of Tape i,
- for each $a \in \Gamma$ the branching statement "i:if(a)" which tests whether a is the symbol scanned by the head of Tape i.

The restrictions are:

- "i:left" and "i:write(a)" are not allowed on an input tape i,
- on the output tape only sequences "0:write(a) ; 0:right" with $a \in \Sigma$ are allowed.

The restrictions guarantee that the input tapes are one-way read-only tapes and that on the output tape only symbols from Σ can be written and no written symbol can be erased (one-way output). We assume that the reader is familiar with the concept of step by step computation for Turing machines. Mathematically exact definitions can be found in introductory text books, for example [HU79, Wei87].

Definition 2.1.1 (Type-2 machine). *A Type-2 machine M is a Turing machine with k input tapes together with a type specification $(Y_1, \ldots, Y_k, Y_0)$ with $Y_i \in \{\Sigma^*, \Sigma^\omega\}$, giving the type for each input tape and the output tape.*

We define the string function $f_M :\subseteq Y_1 \times \ldots \times Y_k \to Y_0$ computed by a Type-2 machine M.

Definition 2.1.2 (computable string function). *The initial tape configuration for input $(y_1, \ldots, y_k) \in Y_1 \times \ldots \times Y_k$ is as follows: for each input tape i the (finite or infinite) sequence $y_i \in Y_i$ is placed on the tape immediately to the right of the head, all other tape cells contain the symbol B; on all the other tapes all tape cells contain the symbol B. For all $y_0 \in Y_0, y_1 \in Y_1, \ldots, y_k \in Y_k$ we define:*

1. **Case $Y_0 = \Sigma^*$:**
   $f_M(y_1, \ldots, y_k) := y_0 \in \Sigma^*$, *iff M halts on input $(y_1, \ldots, y_k)$ with y_0 on the output tape.*
2. **Case $Y_0 = \Sigma^\omega$:**
   $f_M(y_1, \ldots, y_k) := y_0 \in \Sigma^\omega$, *iff M computes forever on input $(y_1, \ldots, y_k)$ and writes y_0 on the output tape.*

We call a string function $f :\subseteq Y_1 \times \ldots \times Y_k \to Y_0$ computable, iff it is computed by a Type-2 machine M.

Notice that $f_M(y_1, \ldots, y_k)$ is undefined, if the machine computes forever but writes only finitely many symbols on the output tape. We do not use the "partial" results of such computations in our definition of semantics. Since Type-2 machines are essentially Turing machines, they are as realistic and powerful as Turing machines. Clearly, infinite inputs or outputs do not exist and infinite computations cannot be finished in reality. But *finite* computations on *finite* initial parts of inputs producing *finite* initial parts of the outputs can be realized on physical devices as long as enough time and memory are available. All doubts about realizability of Type-2 machines have been removed by the recently built machine shown in Fig. 2.2.

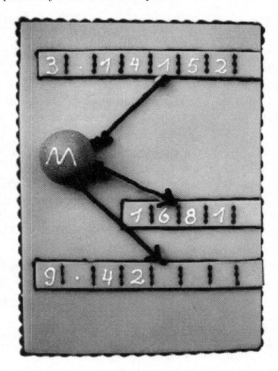

Fig. 2.2. A realization of a Type-2 machine (material: marzipan, year of construction: 1996)

Of course, Type-2 machines can be simulated by digital computers. Therefore, infinite computations of Type-2 machines can be *approximated* by finite physical computations with arbitrary precision. The restriction to one-way output guarantees that any partial output of a finite initial part of a computation cannot be erased in the future and, therefore, is final. For this reason, models of computation with two-way output would not be very useful.

Since it is very cumbersome to specify Turing machines concretely, we will, as usual, only outline algorithms in such a way that constructing concrete Turing machines remains (often rather tedious) routine. If only computability is considered, the reader's favourite algorithmic language may be used instead of Turing machines for formulating algorithms which read from finitely many input files and write on a one-way output file, as long as these algorithms can be translated (at least in principle) to Turing machines. Only in Chap. 7, where we study computational complexity, will we use the special Turing machine model.

Definition 2.1.2 includes the standard definition of computable word functions (choose $Y_0 = Y_1 = \ldots Y_n = \Sigma^*$). We define computable elements of Σ^* and Σ^ω straightforwardly.

Definition 2.1.3 (computable elements).

1. *Every word $w \in \Sigma^*$ is computable.*
2. *A sequence $p \in \Sigma^\omega$ is computable, iff the constant function*
   $f : \{()\} \to \Sigma^\omega$, $f() = p$, *is computable.*
3. *A tuple $(y_1, y_2, \ldots, y_k)$ with $y_i \in \Sigma^*$ or $y_i \in \Sigma^\omega$ is computable, iff each component y_i is computable.*

Obviously, a sequence $p \in \Sigma^\omega$ is computable, iff $p = f(\lambda)$ for some computable function $f : \Sigma^* \to \Sigma^\omega$. We illustrate the definitions by several examples.

Example 2.1.4.

1. A constant function $f : Y_1 \times \ldots \times Y_k \to Y_0$ is computable, iff its value $c \in Y_0$ is computable.
2. Every projection $\mathrm{pr}_i : Y_1 \times \ldots \times Y_k \to Y_i$ is computable.
   There is a Type-2 machine which copies the input tape i to the output tape.
3. Let $\{(u_1, v_1), \ldots, (u_n, v_n)\}$ be a set of pairs of words from Σ^* such that $\{u_1, \ldots, u_n\}$ is prefix-free. Define $f : \Sigma^\omega \to \Sigma^*$ by

   $$f(p) := \begin{cases} v_i & \text{if } u_i \sqsubseteq p \ (1 \leq i \leq n) \\ \lambda & \text{if } u_i \not\sqsubseteq p \text{ for all } 1 \leq i \leq n \ . \end{cases}$$

   Then f is computable. By Theorem 2.2.6 every total computable function $g : \Sigma^\omega \to \Sigma^*$ is of this form.
4. Let $A \subseteq \mathbb{N}$ be a non-empty, recursively enumerable set. Then $A = \{h(0), h(1), \ldots\}$ for some computable function $h : \mathbb{N} \to \mathbb{N}$. Define $f :\subseteq \Sigma^* \to \Sigma^\omega$ by

   $$f(w)(k) := \begin{cases} 1 & \text{if } |w| \notin \{h(0), h(1), \ldots, h(k-1)\} \\ \text{div} & \text{otherwise} \end{cases}$$

   for all $w \in \Sigma^*$ and $k \in \mathbb{N}$. Then the function f is computable. Notice that $\mathrm{dom}(f) = \{w \in \Sigma^* \mid |w| \notin A\}$ is not recursively enumerable, if A is not recursive.
5. Define $f :\subseteq \Sigma^\omega \to \Sigma^*$ by

   $$f(p) := \begin{cases} 1 & \text{if } p \neq 0^\omega \\ \text{div} & \text{otherwise} \ . \end{cases}$$

   There is a Type-2 machine M which reads the input $a_0 a_1 \ldots \in \Sigma^\omega$ and writes 1 and halts, as soon as some i with $a_i \neq 0$ has been found. Then M computes f.

6. The function $f : \Sigma^\omega \to \Sigma^*$ defined by

$$f(p) := \begin{cases} 1 & \text{if } p \neq 0^\omega \\ 0 & \text{otherwise} \end{cases}$$

is not computable.

We assume that some Type-2 machine M computes f. On input $0^\omega = 00\ldots M$ halts with result 0 within t steps for some t. Then M halts with result 0 also on input $q := 0^t 1^\omega \in \Sigma^\omega$. But $f(q) = 1$.

7. Assume $\{0, 1, \ldots, 9, \centerdot\} \subseteq \Sigma$.

There is a Type-2 machine which divides infinite decimal fractions by 3. (Apply the school method.)

No Type-2 machine multiplies infinite decimal fractions by 3.

We assume that some Type-2 machine M computes the real function $x \mapsto 3 \cdot x$. Then it must operate correctly on the name $0\centerdot 333\ldots$ of $1/3$ as input. The output must be $0\centerdot 999\ldots$ or $1\centerdot 000\ldots$. We consider the first case. Let k be the number of steps which M operates before writing the prefix "$0\centerdot$" on the output tape. During this time the machine reads not more than finitely many symbols, say the prefix $0\centerdot w$, from the input tape. Consider now the input sequence $0\centerdot w999\ldots$ which represents a real number $x > 1/3$. Also on this input the machine M will start the output with $0\centerdot$. But since $3 \cdot x > 1$, M does not work correctly on this input (contradiction). In the second case the argument is similar. Therefore, such a machine cannot exist.

A computability concept on the real numbers, under which not even addition and multiplication are computable functions, is not very useful and would hardly be accepted by numerical mathematicians or computer scientists. Notice that in the above proof we have used the one-way output condition for machines. Indeed, machines with two-way output can multiply infinite decimal fractions. However, such machines are useless in practice, since no finite initial part of an infinite computation gives a reliable result, in general.

8. The function $f : \Sigma^* \to \Sigma^\omega$, defined by $f(w) :=$ the expansion of $1/(1 + |w|)$ by an infinite decimal fraction (if there are two fractions choose that with period 0), is computable. (Apply the school method.)

9. Define a function $f :\subseteq \Sigma^\omega \to \Sigma^\omega$ as follows.

$$f(p) := \text{div, if } p(i) = 1 \text{ only for finitely many numbers } i .$$

Otherwise let $i_0 < i_1 < i_2 < \ldots$ be the numbers i with $p(i) = 1$. Define the nth symbol of $f(p)$ by

$$f(p)(n) := \begin{cases} 0 & \text{if } i_n \text{ is an even number} \\ 1 & \text{otherwise .} \end{cases}$$

Again a simple search algorithm shows that f is computable. Notice that $p \in \text{dom}(f)$, iff $p(i) = 1$ infinitely often. $\qquad\square$

The next technical lemma on the initial parts of computations is related to Kleene's T-predicate or Blum's 2nd axiom for complexity measures [Wei87].

Lemma 2.1.5 (partial computations). Let M be a Type-2 machine with k input tapes. Let T_M be the set of all $(u_1, \ldots, u_k, 0^m, v)$ with $u_1, \ldots, u_k, v \in \Sigma^*$ and $m \in \mathbb{N}$ such that on any input $(y_1, \ldots, y_k)$ such that u_i is a prefix of y_i for $i = 1, \ldots, k$ in m steps the machine M reads at most the prefix u_i from the input tape i $(i = 1, \ldots, k)$ and writes v on the output tape. Then the set T_M is recursive.

We omit the elementary proof. By the following lemma, computable functions $f :\subseteq Y \to \Sigma^\omega$ can be represented by computable functions $F :\subseteq Y \times \Sigma^* \to \Sigma^*$.

Lemma 2.1.6 (type conversion). For any function $f :\subseteq Y \to \Sigma^\omega$, $Y = Y_1 \times \ldots \times Y_k$ with $Y_1, \ldots, Y_k \in \{\Sigma^*, \Sigma^\omega\}$, f is computable, iff there is a computable function $F :\subseteq Y \times \Sigma^* \to \Sigma^*$ such that the following two properties hold:

$$[y \in \operatorname{dom}(f) \iff (\forall n \in \mathbb{N})(y, 0^n) \in \operatorname{dom}(F)] \text{ for all } y \in Y ,$$

$$(\forall y \in \operatorname{dom}(f))(\forall n \in \mathbb{N}) F(y, 0^n) = f(y)(n) .$$

Proof: See Exercise 2.1.3.   □

Computability on Σ^* and Σ^ω can be reduced to computability on $\{0, 1\}^*$ and $\{0, 1\}^\omega$, respectively, by encoding the symbols of the alphabet $\Sigma = \{a_1, \ldots, a_n\}$ for example as follows: $a_i \mapsto 1^{i-1}0$ for $i < n$ and $a_n \mapsto 1^n$, see Exercise 2.1.8.

As in ordinary recursion theory, tupling functions are useful tools.

Definition 2.1.7 (tupling functions). *Define the "wrapping function"* $\iota : \Sigma^* \to \Sigma^*$ *by*

$$\iota(a_1 a_2 \ldots a_n) := 110 a_1 0 a_2 0 \ldots a_n 011$$

for all $n \in \mathbb{N}$ and $a_1, a_2, \ldots, a_n \in \Sigma$.
For $x, x_0, x_1, \ldots \in \Sigma^*$, $p, p_0, p_1, \ldots \in \Sigma^\omega$ and $i, j, k \in \mathbb{N}$ with $k \geq 1$ define

$$\langle x_1, \ldots, x_k \rangle := \iota(x_1) \ldots \iota(x_k) \in \Sigma^* ,$$

$$\langle x, p \rangle := \iota(x)p \in \Sigma^\omega ,$$

$$\langle p, x \rangle := \iota(x)p \in \Sigma^\omega ,$$

$$\langle p_1, \ldots, p_k \rangle := p_1(0) \ldots p_k(0)p_1(1) \ldots p_k(1) \ldots \ldots \in \Sigma^\omega ,$$

$$\langle x_0, x_1, \ldots \rangle := \iota(x_0)\iota(x_1) \ldots \in \Sigma^\omega ,$$

$$\langle p_0, p_1, \ldots \rangle \langle i, j \rangle := p_i(j) \quad (\langle p_0, p_1, \ldots \rangle \in \Sigma^\omega) .$$

We use the same notation $\langle\ \rangle$ for each of the above tupling functions. Each of them is injective and even bijective, if the arguments are from Σ^ω. The definition of the wrapping function ι guarantees that subwords $\iota(u)$ and $\iota(v)$ of a word w can overlap only trivially: If $\iota(u)$ is a subword of $\iota(v)$, then $\iota(u) = \iota(v)$, and if x is a suffix of $\iota(u)$ and a prefix of $\iota(v)$, then $\iota(u) = \iota(v) = x$ or $x \in \{11, 1, \lambda\}$.

Theorem 2.1.8 (tupling functions).
1. Each tupling function $\langle\ \rangle : Y_1 \times Y_2 \times \ldots \times Y_k \to Y_0$ is computable, and every projection of its inverse is computable.
2. For both infinite tupling functions $\langle\ \rangle : Y_0 \times Y_1 \ldots \to \Sigma^\omega$, the function $(0^i, q) \mapsto \mathrm{pr}_i \circ \langle\ \rangle^{-1}$ (uniform projection of the inverse) is computable.

The easy proof is left to the reader. As in ordinary recursion theory tupling functions are used to represent functions with several arguments by functions with a single argument (Exercise 2.1.6). By means of infinite tupling we can *define* computable functions $f :\subseteq \Sigma^\omega \times \Sigma^\omega \times \ldots \to Y$, $Y \in \{\Sigma^*, \Sigma^\omega\}$ with infinitely many arguments from Σ^* or Σ^ω:

Definition 2.1.9 (infinitely many arguments). *A function* $f : Y \times Y \times \ldots \to Y_0$ *($Y, Y_0 \in \{\Sigma^*, \Sigma^\omega\}$) is computable, iff*

$$f(p_0, p_1, \ldots) = g \circ \langle p_0, p_1, \ldots \rangle$$

for some computable function $g :\subseteq \Sigma^\omega \to Y_0$.

We continue with two useful characterizations of computable functions $f :\subseteq \Sigma^\omega \to \Sigma^*$ and $f :\subseteq \Sigma^\omega \to \Sigma^\omega$ by means of word functions $f :\subseteq \Sigma^* \to \Sigma^*$.

Definition 2.1.10.
1. *For any partial function $h :\subseteq \Sigma^* \to \Sigma^*$ with prefix-free domain define $h_* :\subseteq \Sigma^\omega \to \Sigma^*$ by*

$$h_*(p) := \begin{cases} h(w) & \text{if } w \sqsubseteq p \text{ and } w \in \mathrm{dom}(h) \\ \mathrm{div} & \text{if } w \not\sqsubseteq p \text{ for all } w \in \mathrm{dom}(h) \end{cases}$$

for all $p \in \Sigma^\omega$ and $w \in \Sigma^*$.
2. *For any monotone total function $h : \Sigma^* \to \Sigma^*$ define $h_\omega :\subseteq \Sigma^\omega \to \Sigma^\omega$ by*

$$p \in \mathrm{dom}(h_\omega) : \Longleftrightarrow h \text{ is unbounded on } p \,,$$

$$h_\omega(p) := \sup_{i \in \mathbb{N}} h(p_{<i}) \text{ for } p \in \mathrm{dom}(h_\omega) \,,$$

for all $p \in \Sigma^\omega$. Here, h is called monotone, iff $u \sqsubseteq v \implies h(u) \sqsubseteq h(v)$, h is called unbounded on p, iff the sequence $(|h(p_{<i})|)_{i \in \mathbb{N}}$ is unbounded, and $\sup_{i \in \mathbb{N}} h(p_{<i})$ is the single sequence $q \in \Sigma^\omega$ with $h(p_{<i}) \sqsubseteq q$ for all i.

Lemma 2.1.11 (computable string functions).
1. A function $f :\subseteq \Sigma^\omega \to \Sigma^*$ is computable, iff $f = h_*$ for some computable function $h :\subseteq \Sigma^* \to \Sigma^*$ with prefix-free domain.
2. A function $f :\subseteq \Sigma^\omega \to \Sigma^\omega$ is computable, iff $f = h_\omega$ for some computable monotone function $h : \Sigma^* \to \Sigma^*$.

Proof: 1. Let M be a Type-2 machine computing $f :\subseteq \Sigma^\omega \to \Sigma^*$. Define $h :\subseteq \Sigma^* \to \Sigma^*$ by

$$h(w) := \begin{cases} f(w0^\omega) & \text{if on input } w0^\omega \text{ machine } M \text{ halts} \\ & \quad \text{after exactly } |w| \text{ steps} \\ \text{div} & \text{otherwise} . \end{cases}$$

Then h is computable, has prefix-free domain and satisfies $f = h_*$. Notice that $\mathrm{dom}(h)$ is even recursive.

On the other hand, let M be a Turing machine computing $h :\subseteq \Sigma^* \to \Sigma^*$ with prefix-free domain. There is a Type-2 machine N which on input $p \in \Sigma^\omega$ searches for the smallest number $k := \langle i, t \rangle$, such that machine M halts on input $p_{<i}$ in exactly t steps, prints the result $h(p_{<i})$ and halts. Since $\mathrm{dom}(h)$ is prefix-free, there is at most one such number k. Obviously, N computes the function h_*.

2. Let M be a Type-2 machine computing $f :\subseteq \Sigma^\omega \to \Sigma^\omega$. Define $h : \Sigma^* \to \Sigma^*$ by: $h(w) :=$ the word $x \in \Sigma^*$ which M on input $w0^\omega$ produces in $|w|$ steps. Then the function h is monotone and computable, and $f = h_\omega$.

On the other hand, let $h : \Sigma^* \to \Sigma^*$ be a computable monotone function. Then for any $p \in \Sigma^\omega$ and $i \in \mathbb{N}$, $h(p_{<i}) \sqsubseteq h(p_{<i+1})$. There is a Type-2 machine N which on input $p \in \Sigma^\omega$ works in stages $n = 0, 1, \ldots$ as follows: In Stage n, N extends the result $h(p_{<n-1})$ from Stage $n-1$ to $h(p_{<n})$. The function $f :\subseteq \Sigma^\omega \to \Sigma^\omega$ computed by the machine N satisfies $f = h_\omega$.   □

The lemma can be generalized easily to functions $f :\subseteq (\Sigma^\omega)^k \to \Sigma^*$ and functions $f :\subseteq (\Sigma^\omega)^k \to \Sigma^\omega$ (Exercise 2.1.12). By the next theorem the computable functions are "essentially" closed under composition.

Theorem 2.1.12 (closure under composition). For $k, n \in \mathbb{N}$ and $X_1, \ldots, X_k, Y_1, \ldots, Y_n, Z \in \{\Sigma^*, \Sigma^\omega\}$ let

$$g_i :\subseteq X_1 \times \ldots \times X_k \to Y_i \quad \text{and} \quad f :\subseteq Y_1 \times \ldots \times Y_n \to Z$$

$(i = 1, \ldots, n)$ be computable functions. Then the composition

$$f \circ (g_1, \ldots, g_n) :\subseteq X_1 \times \ldots \times X_k \to Z$$

- is computable, if $Z = \Sigma^\omega$ or $Y_i = \Sigma^*$ for all i,
- has a computable extension $h :\subseteq X_1 \times \ldots \times X_k \to Z$ with

$$\mathrm{dom}(h) \cap \mathrm{dom}(g_1, \ldots, g_n) = \mathrm{dom}(f \circ (g_1, \ldots, g_n)) \,,$$

if $Z = \Sigma^*$ and $Y_i = \Sigma^\omega$ for some i.

Therefore, the composition $f \circ (g_1, \ldots, g_n)$ is computable, if the final result is infinite or all intermediate results are finite, and it has a computable extension, if the final result is finite and some intermediate result is infinite.

Proof: First, we consider the case $k = n = 1$. From machines M_f for f and M_g for g_1 we construct a machine M for $f \circ g_1$.
Case $Y_1 = \Sigma^*$:
On input $x \in X_1$, the machines M_g and M_f work one after the other. After M_g has stopped, its output $y \in Y_1 = \Sigma^*$ serves as an input for M_f. Obviously, $f_M(x)$ exists, iff $y = g_1(x)$ and $f(y)$ exist. If $f_M(x)$ exists, then $f_M(x) = f \circ g_1(x)$. We obtain $f_M = f \circ g_1$.
Case $Y_1 = \Sigma^\omega$:
For the intermediate result $y \in Y_1 = \Sigma^\omega$, which is the output of M_g and the input for M_f, we use for the moment a special work tape T with two heads, one for writing and one for reading. Now on input $x \in X_1$ the machines M_g and M_f work alternately as follows: M_g works until it writes a symbol on Tape T and interrupts, then M_f makes one step, then M_g continues working until it writes a symbol on Tape T and interrupts, then M_f makes a further step, and so on.

Suppose that $y := g_1(x) \in \Sigma^\omega$ exists. Then $f_M(x)$ exists, iff $f(y)$ exists, iff $f \circ g_1(x)$ exists, and so $\mathrm{dom}(f_M) \cap \mathrm{dom}(g_1) = \mathrm{dom}(f \circ g_1) \cap \mathrm{dom}(g_1) = \mathrm{dom}(f \circ g_1)$. Furthermore, $f_M(x) = f \circ g_1(x)$ if $x \in \mathrm{dom}(f \circ g_1)$, and so f_M extends $f \circ g_1$.

If in addition $Z = \Sigma^\omega$, then $\mathrm{dom}(f_M) \subseteq \mathrm{dom}(g_1)$ by definition of the machine M, and so $\mathrm{dom}(f_M) = \mathrm{dom}(f_M) \cap \mathrm{dom}(g_1) = \mathrm{dom}(f \circ g_1)$, hence $f_M = f \circ g_1$.

It must be mentioned that a Turing tape with 2 heads can be simulated by a Turing tape with a single (two-way) head.

The proof for arbitrary numbers k, n requires only some minor technical extensions. The generalization from $k = 1$ to arbitrarily many input tapes is obvious. For $n = 0$ the theorem is trivial. Consider $n \geq 2$. Now, every input sequence is used by n submachines $M_{g_1}, \ldots, M_{g_n}$. In a first attempt use n one-way read-only heads on each input tape. Every Turing tape with n one-way read-only heads can be simulated by a single one-way read-only head on this tape and n additional work tapes. This ends the proof. □

Notice that the composition of computable functions can be non-computable, only if the final result is finite and some intermediate result is infinite. In

Exercises 2.1.9 and 2.1.10 we give examples of computable functions with non-computable composition. The next important theorem follows from closure under composition.

Theorem 2.1.13. Every computable function maps computable elements to computable elements.

Proof: Let $f :\subseteq Y_1 \times \ldots \times Y_k \to Y_0$ $(Y_0, \ldots, Y_k \in \{\Sigma^*, \Sigma^\omega\})$ be computable and let $y_i \in Y_i$ $(1 \leq i \leq k)$ be computable constants such that $(y_1, \ldots, y_k) \in \mathrm{dom}(f)$. By Definition 2.1.3 the functions $g_i : \{()\} \to Y_i$ such that $g_i() = y_i$ $(1 \leq i \leq k)$ are computable. By Theorem 2.1.12, $f \circ (g_1, \ldots, g_n)$ has a computable extension h, and so

$$f(y_1, \ldots, y_k) = f \circ (g_1, \ldots, g_n)() = h()$$

is a computable element of Y_0 by Definition 2.1.3. $\qquad\square$

The computable functions are also closed under *primitive recursion*.

Theorem 2.1.14 (primitive recursion). Let $f' :\subseteq Y_1 \times \ldots \times Y_k \to Y_0$ be a computable function and for each $a \in \Sigma$ let $f_a :\subseteq \Sigma^* \times Y_0 \times Y_1 \times \ldots \times Y_k \to Y_0$ be a computable function (where $Y_0, Y_1, \ldots, Y_k \in \{\Sigma^*, \Sigma^\omega\}$). Then the function $g :\subseteq \Sigma^* \times Y_1 \times \ldots \times Y_k \to Y_0$, defined by the recursion equations

$$g(\lambda, y_1, \ldots, y_k) = f'(y_1, \ldots, y_k) ,$$
$$g(aw, y_1, \ldots, y_k) = f_a(w, g(w, y_1, \ldots, y_k), y_1, \ldots, y_k)$$

for all $w \in \Sigma^*$, $y_1 \in Y_1, \ldots, y_k \in Y_k$ and $a \in \Sigma$, is computable.

Proof: If $Y_0 = \Sigma^*$ then a Type-2 machine M which computes the function g can be constructed by standard methods, since all intermediate values and computations are finite.

Suppose $Y_0 = \Sigma^\omega$. First, we prove as a special case that there is a computable function $G :\subseteq \Sigma^* \times \Sigma^\omega \to \Sigma^\omega$ with

$$G(\lambda, p) = p ,$$
$$G(aw, p) = F_a(G(w, p))$$

for all $w \in \Sigma^*$, $p \in \Sigma^\omega$ and $a \in \Sigma$, provided $F_a :\subseteq \Sigma^\omega \to \Sigma^\omega$ is computable for all $a \in \Sigma$. (Explicitly written, $G(a_1 \ldots a_m, p) = F_{a_1} \circ \ldots \circ F_{a_m}(p)$.) By Theorem 2.1.11, for each $a \in \Sigma$ there is a computable monotone function $h_a : \Sigma^* \to \Sigma^*$ with $F_a = h_{a,\omega}$. The function $h : \Sigma^* \times \Sigma^* \to \Sigma^*$, defined by

$$h(\lambda, x) = x ,$$
$$h(aw, x) = h_a(h(w, x))$$

for all $w, x \in \Sigma^*$ and $a \in \Sigma$, is computable. The function h can be defined explicitly by $h(a_1 \ldots a_m, x) = h_{a_1} \circ \ldots \circ h_{a_m}(x)$. For each $w \in \Sigma^*$ the function

$h_w : \Sigma^* \to \Sigma^*$, defined by $h_w(x) := h(w, x)$, is monotone. Define $G_w(p) := G(w, p)$. Then for all words w, $h_{w,\omega} = G_w$:

$$h_{\lambda,\omega} = G_\lambda, \quad h_{aw,\omega} = h_{a,\omega} \circ h_{w,\omega} = F_a \circ G_w = G_{a,w} \ .$$

We proceed as in the proof of Theorem 2.1.11. There is a Type-2 machine N which on input $(w, p) \in \Sigma^* \times \Sigma^\omega$ works in stages $n = 0, 1, \ldots$ as follows: In Stage n, N extends the result $h(w, p_{<n-1})$ from Stage $n-1$ to $h(w, p_{<n})$. Then for each word w, $f_N(w, p) = h_{w,\omega}(p) = G_w(p)$, and so $G = f_N$ is computable.

Next, we reduce the general case to the above special case. For $a \in \Sigma$ define $\overline{f}_a :\subseteq \Sigma^\omega \to \Sigma^\omega$ by

$$\overline{f}_a\langle v, r, y \rangle := \langle av, f_a(v, r, y), y \rangle$$

for $v \in \Sigma^*, r \in \Sigma^\omega$ and $y \in Y := Y_1 \times \ldots \times Y_k$, and $\overline{f}_a(z) := \mathrm{div}$, if $z \notin \langle \Sigma^* \times \Sigma^\omega \times Y \rangle$. By Theorems 2.1.8 and 2.1.12, $\overline{f}_a$ is computable. By the above special case, the function $\overline{g} :\subseteq \Sigma^* \times \Sigma^\omega \to \Sigma^\omega$, defined by $\overline{g}(\lambda, p) = p$ and $\overline{g}(aw, p) = \overline{f}_a(\overline{g}(w, p))$, is computable. An induction on w shows $\overline{g}(w, \langle \lambda, f'(y), y \rangle) = \langle w, g(w, y), y \rangle$. Again by Theorems 2.1.8 and 2.1.12, the function g is computable.   □

By specialization we obtain closure under iteration.

Corollary 2.1.15 (iteration).
1. If $f_a :\subseteq \Sigma^\omega \to \Sigma^\omega$ is computable for all $a \in \Sigma$, then $g :\subseteq \Sigma^* \times \Sigma^\omega \to \Sigma^\omega$ with $g(a_1 \ldots a_m, p) = f_{a_1} \circ \ldots \circ f_{a_m}(p)$ is computable.
2. If $f :\subseteq \Sigma^\omega \to \Sigma^\omega$ is a computable function, then $g :\subseteq \Sigma^* \times \Sigma^\omega \to \Sigma^\omega$, defined by $g(w, p) := f^{|w|}(p)$, is computable.

So far we have considered a fixed alphabet Σ. The alphabet can be enlarged without affecting computability of functions (Exercise 2.1.7).

Several other kinds of computable Type-2 functions have been studied, for example enumeration operators $\Phi_z : 2^\mathbb{N} \to 2^\mathbb{N}$, partial recursive operators $f :\subseteq \mathrm{PF} \to \mathrm{PF}$ (where $\mathrm{PF} = \{f \mid f :\subseteq \mathbb{N} \to \mathbb{N}\}$), partial recursive functions $F :\subseteq \mathbb{N}^\mathbb{N} \times \mathbb{N} \to \mathbb{N}$ and $F :\subseteq 2^\mathbb{N} \times \mathbb{N} \to \mathbb{N}$, partial recursive functionals $F :\subseteq \mathrm{PF} \to \mathbb{N} \cup \{*\}$ (Rogers [Rog67] §§ 9.7, 9.8, 15.1/3) and computable functions $F :\subseteq \mathbb{N}^\mathbb{N} \to \mathbb{N}^\mathbb{N}$ and $F :\subseteq \mathbb{N}^\mathbb{N} \to \mathbb{N}$ (Weihrauch [Wei87]). Each of these definitions can be derived from our Type-2 computability and vice versa by means of straightforward encodings. This substantiates the claim that our computability concept is very natural.

By Church's Thesis (also called Church/Turing Thesis)[Rog67, Odi89], a word function $f :\subseteq \Sigma^* \to \Sigma^*$ is computable by a physical machine, iff it is computable by a Turing machine. The above observations might give rise to generalize Church's Thesis as follows: A function $f :\subseteq Y_1 \times \ldots \times Y_k \to$

Y_0 $(Y_0, \ldots, Y_k \in \{\Sigma^*, \Sigma^\omega\})$ is computable by a physical machine, iff it is computable by a Type-2 machine.

We refrain from doing this, since there is no convincing reason to exclude generalized Type-2 machines with two-way output or machines with more sophisticated input/output conventions. The main reason for choosing computability defined by Type-2 machines as a basis for computable analysis is a practical one. We believe that for our purpose this kind of computability is the most useful choice. From Church's Thesis we adopt the conviction that computations on Turing tapes can be realized by physical machines and that every (discrete) computation of a physical machine can be simulated by a Turing tape.

Exercises 2.1.

$\Diamond$ 1. Show that permitting two-way input tapes does not increase the computational power of Type-2 machines.

$\Diamond$ 2. Generalize the Type-2 machine model by admitting two-way output. Show that there is a machine of this kind which multiplies infinite decimal fractions by 3.

3. Prove Lemma 2.1.6.

4. Prove Theorem 2.1.8.

5. Show that there is a computable function $f :\subseteq \Sigma^\omega \to \Sigma^\omega$ with $\mathrm{dom}(f) = \langle \Sigma^* \times \Sigma^\omega \rangle$. Show that there is a computable function $g :\subseteq \Sigma^\omega \to \Sigma^\omega$ with $\mathrm{dom}(g) = \Sigma^\omega \setminus \langle \Sigma^* \times \Sigma^\omega \rangle$.

$\Diamond$ 6. Show that a function $f : \Sigma^* \times \Sigma^\omega \to Y_0$ $(Y_0 \in \{\Sigma^*, \Sigma^\omega\})$ is computable, iff $f(x, p) = g\langle x, p \rangle$ for some computable function $g :\subseteq \Sigma^\omega \to Y_0$.

7. Let Σ and Δ be finite alphabets with $\Sigma \subseteq \Delta$. For $a_0, \ldots, a_k \in \{*, \omega\}$ let $f :\subseteq \Sigma^{a_1} \times \ldots \times \Sigma^{a_k} \to \Sigma^{a_0}$ and $g :\subseteq \Delta^{a_1} \times \ldots \times \Delta^{a_k} \to \Delta^{a_0}$ with $\mathrm{graph}(f) = \mathrm{graph}(g)$. Show that f is computable, iff g is computable. Therefore, extension of the alphabet preserves computability.

8. For $n \geq 2$, computability on Σ^* for $\Sigma = \{a_1, \ldots, a_n\}$ can be reduced to computability on $\{0, 1\}^*$. Define a mapping $c : \Sigma \to \{0, 1\}^*$ by $c(a_i) := 1^{i-1}0$ for $i < n$ and $c(a_n) := 1^n$. Define
   $c_* : \Sigma^* \to \{0, 1\}^*$ by $c_*(b_1 b_2 \ldots b_k) := c(b_1)c(b_2) \ldots c(b_k)$ and
   $c_\omega : \Sigma^\omega \to \{0, 1\}^\omega$ by $c_\omega(b_0 b_1 b_2 \ldots) := c(b_0)c(b_1)c(b_2) \ldots$.
   a) Show that c, c_* are injective and c_ω is bijective.
   b) Show that the target extensions of c_* and c_ω to Σ^* and Σ^ω, respectively, and the inverses of these functions are computable (Convention 2.0.1).
   c) Show that $f :\subseteq \Sigma^{d_1} \times \ldots \times \Sigma^{d_k} \to \Sigma^{d_0}$ $(d_i \in \{*, \omega\})$ is computable, iff $\bar{c} \circ f(\bar{c}^{-1}, \ldots, \bar{c}^{-1})$ is computable, where $\bar{c}(w) := c_*(w)$ and $\bar{c}(p) := c_\omega(p)$ for $w \in \Sigma^*$ and $p \in \Sigma^\omega$.

9. (non-computable composition) Show that there are computable functions $f :\subseteq \Sigma^* \to \Sigma^\omega$ and $g : \Sigma^\omega \to \Sigma^*$, such that $g \circ f :\subseteq \Sigma^* \to \Sigma^*$

is not computable. (Hint: consider f from Example 2.1.4.4 and define $g(p) := \lambda$.)

10. (non-computable composition) Let $f : \Sigma^\omega \to \Sigma^\omega$ be the function from Example 2.1.4.9 and define $g : \Sigma^\omega \to \Sigma^*$ by $g(p) := \lambda$ for all $p \in \Sigma^\omega$. Then f and g are computable. Show that $g \circ f$ cannot be computable.

11. Call a partial function $h :\subseteq \Sigma^* \to \Sigma^*$ *upwards constant*, iff

$$(h(u) \text{ exists and } u \text{ is a prefix of } v) \Longrightarrow h(u) = h(v) .$$

Show that in Definition 2.1.10 and Lemma 2.1.11 functions with prefix-free domain can be replaced by upwards constant functions.

12. Generalize the characterization from Lemma 2.1.11 to functions $f :\subseteq (\Sigma^\omega)^k \to \Sigma^*$ and functions $f :\subseteq (\Sigma^\omega)^k \to \Sigma^\omega$.

13. (Closure under simultaneous recursion, Theorem 2.1.14) For $i = 1, \ldots, l$ let $f_i' :\subseteq Y_1 \times \ldots \times Y_k \to Y_0$ be computable functions and for each $a \in \Sigma$ let $f_{ia} :\subseteq \Sigma^* \times Y_0 \times Y_1 \times \ldots \times Y_k \to Y_0$ be computable functions (where $Y_0, Y_1, \ldots, Y_k \in \{\Sigma^*, \Sigma^\omega\}$). Show that the functions $g_i :\subseteq \Sigma^* \times Y_1 \times \ldots \times Y_k \to Y_0$ $(i = 1, \ldots, l)$, defined by the recursion equations

$$g_i(\lambda, y_1, \ldots, y_k) = f_i'(y_1, \ldots, y_k) ,$$
$$g_i(aw, y_1, \ldots, y_k) = f_{ia}(w, g_1(w, y_1, \ldots, y_k), \ldots,$$
$$g_l(w, y_1, \ldots, y_k), y_1, \ldots, y_k)$$

for all $w \in \Sigma^*$, $y_1 \in Y_1, \ldots, y_k \in Y_k$ and $a \in \Sigma$, are computable.

◆14. The following theorem on *term evaluation* generalizes composition, primitive recursion and iteration.

Assume that the alphabet Σ contains the symbols x and f. Let $\mu : \mathbb{N} \to \mathbb{N}$ be a computable function ($\mu(i)$ will be the arity of the function f_i) and let $n \in \mathbb{N}$.

a) Define the set $\mathrm{Tm} \subseteq \Sigma^*$ of terms in n variables (in Polish notation) inductively as follows:
   - $\mathrm{x}0^k$ is a term for $k = 1, \ldots, n$,
   - if $t_1, \ldots, t_{\mu(i)}$ are terms, then $\mathrm{f}0^i t_1 \ldots t_{\mu(i)}$ is a term (for all $i \in \mathbb{N}$),
   - no other words are terms.

b) For each $i \in \mathbb{N}$ let $f_i :\subseteq (\Sigma^\omega)^{\mu(i)} \to \Sigma^\omega$ be a function. For terms $t \in \mathrm{Tm}$, define the semantics $[\![t]\!] :\subseteq (\Sigma^\omega)^n \to \Sigma^\omega$ recursively as follows:

$$[\![\mathrm{x}0^k]\!](\overline{p}) := p_k ,$$

$$[\![\mathrm{f}0^i t_1 \ldots t_{\mu(i)}]\!](\overline{p}) := f_i([\![t_1]\!](\overline{p}), \ldots, [\![t_{\mu(i)}]\!](\overline{p}))$$

for all $\overline{p} := (p_1, \ldots, p_n) \in (\Sigma^\omega)^n$.

c) Assume that the "universal" function

$$(0^i, \langle p_1, \ldots, p_{\mu(i)} \rangle) \mapsto f_i(p_1, \ldots, p_{\mu(i)})$$

is computable. Prove that the "term evaluation function"

$$(t, p_1, \ldots, p_n) \mapsto [\![\, t\,]\!](p_1, \ldots, p_n) \quad (t \in \mathrm{Tm}, \quad p_1, \ldots, p_n \in \Sigma^\omega)$$

is computable.

d) Derive Theorem 2.1.14 (case $Y_i = \Sigma^\omega$ for all i) and Corollary 2.1.15 as corollaries.

2.2 Computable String Functions are Continuous

Let M be a Type-2 machine such that $f_M :\subseteq \Sigma^\omega \to \Sigma^\omega$. Consider $u \in \Sigma^*$ with $u \sqsubseteq f_M(p)$. Then on input p, the machine M writes the prefix u of the output $f_M(p)$ in t steps for some t. Within t steps, M can read not more than the prefix $w := p_{<t}$ of the input $p \in \Sigma^\omega$. Therefore, the output u depends only on the finite prefix w of p but not on the rest $p(t)p(t+1) \ldots$. We obtain $f_M[w\Sigma^\omega] \subseteq u\Sigma^\omega$. Roughly speaking:

(FP)   Every **finite** portion of the output $f_M(p)$ is already determined by a **finite** portion of the input p.

We have already used this *finiteness property* (FP) repeatedly for proving non-computability of functions (Examples 2.1.4.6/7 and Exercise 2.1.10). We will see that it is equivalent to *continuity* of the function f_M, if we consider the *Cantor topology* on the set Σ^ω.

Since topology is a mathematical theory for studying approximation, it is not surprising that we can use it for describing the approximation of infinite sequences of symbols by finite ones or of real numbers by rational intervals. For readers who are not familiar with the language of topology we summarize briefly some basic definitions below. More details can be found in textbooks on analysis or in any introduction to topology [Die60, Rud64, Eng89].

Definition 2.2.1 (some definitions from topology). A *topology* on a set M is a system $\tau \subseteq 2^M$ of subsets of M, called the *open* sets, with the following three properties:

(T1) $\emptyset$ and M are open.

(T2) If U and V are open, then $U \cap V$ is open.

(T3) If α is a set of open sets, then its union $\bigcup \alpha$ is open.

A subset $A \subseteq M$ is called *closed*, iff its complement $M \setminus A$ is open. The pair (M, τ) is called a *topological space*. An example is the Euclidean space $(\mathbb{R}^n, \tau)$, where τ is the set of open subsets of $\mathbb{R}^n$, which contains $\emptyset$ and $\mathbb{R}^n$ and is closed under finite intersection and arbitrary union. For sets $A, B \subseteq M$, B is called a *neighborhood* of A, iff $A \subseteq U \subseteq B$ for some open set U. B is a neighborhood of a point x, iff it is a neighborhood of $\{x\}$. Then a set $V \subseteq M$ is open, iff V is a neighborhood of every $x \in V$. For a subset $X \subseteq M$ the *closure* $\mathrm{cls}(X)$ is the smallest closed subset of M containing X and the *interior* $\mathrm{int}(X)$ is the greatest open subset of M contained in X.

A *base* of the topology τ is a set $\beta \subseteq \tau$ of open sets, such that for all $U \in \tau$ there is some $\gamma \subseteq \beta$ such that $U = \bigcup \gamma$. A *subbase* of τ is a subset

$\sigma \subseteq \tau$ such that the set $\{\bigcap \gamma \mid \gamma$ is a finite subset of $\sigma\}$ is a base of τ. Let σ be any set of subsets of M, let $\beta := \{\bigcap \gamma \mid \gamma$ is a finite subset of $\sigma\}$ and let $\tau := \{\bigcup \gamma \mid \gamma \subseteq \beta\}$. Then τ is a topology, β is a base of τ, and σ is a subbase of τ. (We use the canonical definitions $\bigcap \emptyset = M$ and $\bigcup \emptyset = \emptyset$.)

Let (M, τ) and (M', τ') be topological spaces. A function $f : M \to M'$ is called *continuous*, iff $f^{-1}[U] \in \tau$ for all $U \in \tau'$. A function $f : M \to M'$ is called *continuous at* $x \in M$, iff for all $V \in \tau'$ with $f(x) \in V$ there is some $U \in \tau$ with $x \in U$ and $f[U] \subseteq V$. (If various topologies are considered, we shall say more precisely (τ, τ')-continuous.) In both cases it suffices to consider $V \in \beta$ for some base β or $V \in \sigma$ for some subbase σ of the topology τ'. A function $f : M \to M'$ is continuous, iff it is continuous at x for all $x \in M$. The composition of two continuous functions is continuous.

Let (M, τ) be a topological space and let $D \subseteq M$. The topology on D *induced* by τ is $\tau|_D := \{D \cap U \mid U \in \tau\}$. We shall call a set $X \subseteq M$ *open in* D, iff $X = D \cap U$ for some open set $U \in \tau$. We shall call a partial function $f :\subseteq M \to M'$ with $\mathrm{dom}(f) = D$ (τ, τ')-continuous (at $x \in D$), iff its restriction $f|_D$ is $(\tau|_D, \tau')$-continuous (at $x \in D$).

A subset $X \subseteq M$ is called a G_δ-set, iff $X = U_0 \cap U_1 \cap \ldots$ for some sequence $(U_0, U_1, \ldots)$ of open sets. We may assume that the sequence of open sets is decreasing (choose $V_n := U_0 \cap \ldots \cap U_n$). For a topological space (M, τ) a set $X \subseteq M$ is *dense*, iff $X \cap U \neq \emptyset$ for all non-empty $U \in \tau$. The *discrete topology* on a set M is the set 2^M of all subsets of M (every subset of M is open).

For any family $(M_i, \tau_i)_{i \in I}$ of topological spaces, the *product space* $(\times_{i \in I} M_i, \bigotimes_{i \in I} \tau_i)$ is defined as follows: $\times_{i \in I} M_i$ is the Cartesian product and the topology $\bigotimes_{i \in I} \tau_i$ has the base
$$\beta := \{\times_{i \in I} U_i \mid U_i \in \tau_i \text{ for all } i \text{ and } U_i \neq M_i \text{ only for finitely many } i\}.$$
On Cartesian products the product topologies are used tacitly.

A *pseudometric space* is a pair (M, d), where M is a set and $d : M \times M \to \mathbb{R}$ is a function (the *distance* or *pseudometric* on M) such that for all $x, y, z \in M$:

(M1) $d(x, y) \geq 0$

(M2) $d(x, x) = 0$

(M3) $d(x, y) = d(y, x)$

(M4) $d(x, z) \leq d(x, y) + d(y, z)$

d is a *metric* and (M, d) is a *metric space*, iff in addition

(M5) $d(x, y) = 0$ implies $x = y$.

For $x \in M$ and $r > 0$, $B(x, r) := \{y \in M \mid d(x, y) < r\}$ is the *open ball with center x and radius r* and $\overline{B}(x, r) := \{y \in M \mid d(x, y) \leq r\}$ is the *closed ball with center x and radius r*. The set of open balls is a base of a topology τ_d on the set M, the topology *generated by the pseudometric* d. Already the set of open balls with rational radius is a base of τ_d. A subset $X \subseteq M$ of a metric space is *compact*, iff every open cover of X has a finite subcover, that is, if $X \subseteq \bigcup T$ for a set T of open sets, then $X \subseteq \bigcup S$ for some finite subset $S \subseteq T$.

We introduce the *discrete topology* and the *Cantor topology* as our standard topologies on Σ^* and Σ^ω, respectively.

Definition 2.2.2 (standard topologies on Σ^* and Σ^ω).
1. $\tau_* := 2^{\Sigma^*} = \{A \mid A \subseteq \Sigma^*\}$ *is called the discrete topology on Σ^*.*
2. $\tau_C := \{A\Sigma^\omega \mid A \subseteq \Sigma^*\}$ *is called the Cantor topology on Σ^ω.*
   (Σ^ω, τ_C) is called the Cantor space (over Σ).
3. *As a canonical base of τ_* we consider the set $\{\{w\} \mid w \in \Sigma^*\}$.*
4. *As a canonical base of τ_C we consider the set $\{w\Sigma^\omega \mid w \in \Sigma^*\}$.*
5. *On $Y := Y_1 \times \ldots \times Y_k$ $(Y_1, \ldots, Y_k \in \{\Sigma^*, \Sigma^\omega\})$ we consider the product topology with the canonical base $\beta_Y := \{y \circ Y \mid y \in (\Sigma^*)^k\}$, where $(y_1, \ldots, y_k) \circ Y := U_1 \times \ldots \times U_k$ with*

$$U_i := \begin{cases} \{y_i\} & \text{if } Y_i = \Sigma^* \\ y_i \Sigma^\omega & \text{if } Y_i = \Sigma^\omega . \end{cases}$$

Obviously, τ_* is a topology on Σ^* and $\{\{w\} \mid w \in \Sigma^*\}$ is a base of τ_*. For any word $w \in \Sigma^*$, $w\Sigma^\omega = \{wq \mid q \in \Sigma^\omega\} = \{p \in \Sigma^\omega \mid w \sqsubseteq p\} \in \tau_C$. For $v, w \in \Sigma^*$ we have $v\Sigma^\omega \cap w\Sigma^\omega = (w\Sigma^\omega$ if $v \sqsubseteq w$, $v\Sigma^\omega$ if $w \sqsubseteq v$, and $\emptyset$ otherwise). Therefore, the set $\beta := \{w\Sigma^\omega \mid w \in \Sigma^*\}$ is closed under finite intersection, and so it is a base of a topology. Since $A\Sigma^\omega = \bigcup \{w\Sigma^\omega \mid w \in A\}$ for each $A \subseteq \Sigma^*$, β is a base of the topology τ_C.

The set Σ^ω of infinite sequences over Σ can be visualized by a tree where every $p \in \Sigma^\omega$ is represented by an infinite descending path from the root. Fig. 2.3 shows this tree for the alphabet $\Sigma = \{0, 1\}$ with two elements.

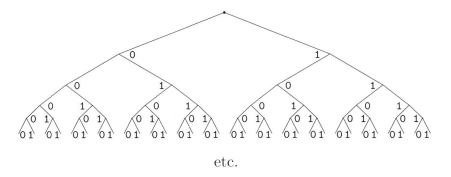

etc.

Fig. 2.3. The tree of infinite paths representing the set $\{0, 1\}^\omega$

A base element $w\Sigma^\omega \in \tau_C$ is represented by a full subtree and an open subset of Σ^ω is represented by a set of full subtrees. Figure 2.4 shows four subtrees representing the open set $001\Sigma^\omega \cup 0110\Sigma^\omega \cup 10\Sigma^\omega \cup 1111\Sigma^\omega$.

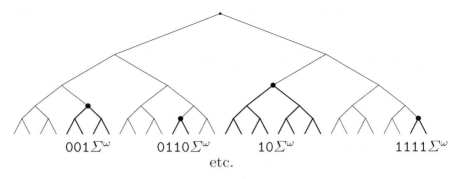

$$001\Sigma^\omega \qquad 0110\Sigma^\omega \qquad 10\Sigma^\omega \qquad 1111\Sigma^\omega$$
etc.

Fig. 2.4. The subtrees representing the set $001\Sigma^\omega \cup 0110\Sigma^\omega \cup 10\Sigma^\omega \cup 1111\Sigma^\omega$

The finiteness property (FP) for the function $f_M : \Sigma^\omega \to \Sigma^\omega$ computed by a machine M can be formulated as follows: For all $p \in \Sigma^\omega$ and $u \in \Sigma^*$ with $f_M(p) \in u\Sigma^\omega$ there is some $w \in \Sigma^\omega$ with $p \in w\Sigma^\omega \subseteq f_M^{-1}[u\Sigma^\omega]$. This means exactly that the function f_M is (τ_C, τ_C)-continuous. Therefore, every computable function $f : \Sigma^\omega \to \Sigma^\omega$ is continuous.

After these preparations we prove our main theorems.

Theorem 2.2.3 (computable $\implies$ continuous). Every computable string function $f :\subseteq Y_1 \times \ldots \times Y_k \to Y_0$ is continuous.

Proof: Let M be a Type-2 machine computing f. Assume $(y_1, \ldots, y_k) \in \text{dom}(f) \subseteq Y := Y_1 \times \ldots \times Y_k$.

Case $Y_0 = \Sigma^*$: It suffices to show that $f^{-1}[\{w\}]$ is open for every base element $\{w\}$ of τ_*. Assume $f(y_1, \ldots, y_k) \in \{w\}$. Since the machine M halts on this input, during its computation from every input tape i with $Y_i = \Sigma^\omega$ it can read only a prefix $u_i \in \Sigma^*$ of $y_i \in \Sigma^\omega$. Define $u_i := y_i$ for $Y_i = \Sigma^*$. Then $y \in (u_1, \ldots, u_k) \circ Y$ and $f[(u_1, \ldots, u_k) \circ Y] \subseteq \{w\}$. The set $(u_1, \ldots, u_k) \circ Y$ is an open neighborhood of $(y_1, \ldots, y_k)$ contained in $f^{-1}[\{w\}]$. Therefore, $f^{-1}[\{w\}]$ is open.

Case $Y_0 = \Sigma^\omega$: It suffices to show that $f^{-1}[w\Sigma^\omega]$ is open for every base element $w\Sigma^\omega$ of τ_C. Assume $f(y_1, \ldots, y_k) \in w\Sigma^\omega$. Since on this input the machine M needs only finitely many computaton steps for producing the prefix w of the result, during this computation from every input tape i with $Y_i = \Sigma^\omega$ it can read only a prefix $u_i \in \Sigma^*$ of $y_i \in \Sigma^\omega$. Define $u_i := y_i$ for $Y_i = \Sigma^*$. Then $y \in (u_1, \ldots, u_k) \circ Y$ and $f[(u_1, \ldots, u_k) \circ Y] \subseteq w\Sigma^\omega$. The set $(u_1, \ldots, u_k) \circ Y$ is an open neighborhood of $(y_1, \ldots, y_k)$ contained in $f^{-1}[w\Sigma^\omega]$. Therefore, $f^{-1}[w\Sigma^\omega]$ is open. $\qquad \square$

Fig. 2.5 illustrates the above proof for the case $f :\subseteq \Sigma^\omega \to \Sigma^\omega$: Any finite portion of the output depends only on a finite portion of the input.

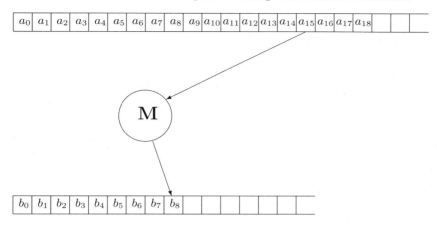

Fig. 2.5. The machine M has computed $b_0 b_1 \ldots b_8$ from the input $a_0 a_1 \ldots a_{15}$

In numerous applications, functions are not computable because they are not even continuous. Continuous but not computable functions can be defined explicitly but occur in practice very rarely. The domains of computable functions have nice topological properties.

Theorem 2.2.4 (domains of computable functions). Assume $Y = Y_1 \times \ldots \times Y_k$ with $Y_1, \ldots, Y_k \in \{\Sigma^*, \Sigma^\omega\}$.
1. If $f :\subseteq Y \to \Sigma^*$ is computable, then $\mathrm{dom}(f)$ is open.
2. If $f :\subseteq Y \to \Sigma^\omega$ is computable, then $\mathrm{dom}(f)$ is a G_δ-set.

Proof: 1. In the proof of Theorem 2.2.3 we have shown that every element $(y_1, \ldots, y_k) \in \mathrm{dom}(f)$ has an open neighborhood $(u_1, \ldots, u_k) \circ Y \subseteq \mathrm{dom}(f)$. Therefore, $\mathrm{dom}(f)$ is open.

2. Let M be a Type-2 machine which computes f. Then, for each $n \in \mathbb{N}$ there is a machine M_n which on input $y \in Y$ halts, if M on input y writes at least n symbols on the output tape, and does not halt otherwise. By Property 1, $\mathrm{dom}(f_{M_n})$ is open for each n, therefore,
$\mathrm{dom}(f_M) = \mathrm{dom}(f_{M_0}) \cap \mathrm{dom}(f_{M_1}) \cap \ldots$ is a G_δ-set. □

The Cantor topology can be generated from a *metric*.

Lemma 2.2.5 (metric on Cantor space). Define $d : \Sigma^\omega \times \Sigma^\omega \to \mathbb{R}$ by $d(p, p) := 0$ and $d(p, q) := 2^{-n}$ where n is the smallest number with $p(n) \neq q(n)$ for $p \neq q$.
Then $M := (\Sigma^\omega, d)$ is a metric space, τ_C (the Cantor topology) is the topology generated by d, and Σ^ω is a compact set .

The proof is left to the reader. The sets $w\Sigma^\omega$ which form a base of the Cantor topology are the open balls as well as the closed balls in the metric space (Σ^ω, d). As a closed subset of a compact set, every ball is compact. In the following a set $A \subseteq \Sigma^\omega$ is called *clopen*, iff it is closed and open, and *decidable*, iff its characteristic function is computable (Sect. 2.4). As a consequence of compactness, total computable functions $f : \Sigma^\omega \to \Sigma^*$, clopen subsets and decidable subsets $X \subseteq \Sigma^\omega$ are very simple.

Theorem 2.2.6 (decidable set, total continuous function).
1. A subset $X \subseteq \Sigma^\omega$ is clopen, iff it is decidable, iff $X = A\Sigma^\omega$ for some finite set $A \subseteq \Sigma^*$.
2. A (*total*) function $f : \Sigma^\omega \to \Sigma^*$ is continuous, iff it is computable, iff there are pairs $(u_1, v_1), \ldots, (u_k, v_k) \in \Sigma^* \times \Sigma^*$ such that $\{u_1, \ldots, u_k\}$ is prefix-free, $\Sigma^\omega = u_1\Sigma^\omega \cup \ldots \cup u_k\Sigma^\omega$ and $f[u_i\Sigma^\omega] = \{v_i\}$ for all i.

Proof: Let $f : \Sigma^\omega \to \Sigma^*$ be a continuous function. Then for each $p \in \Sigma^\omega$ there are words u_p, v_p with $p \in u_p\Sigma^\omega$ and $f[u_p\Sigma^\omega] = \{v_p\}$. Since $\Sigma^\omega = \bigcup_{p \in \Sigma^\omega} u_p\Sigma^\omega$ and Σ^ω is compact, there are finitely many pairs $(u_i, v_i) \in \Sigma^* \times \Sigma^*$, $i \in I$, such that $\Sigma^\omega = \bigcup_{i \in I} u_i\Sigma^\omega$ and $f[u_i\Sigma^\omega] = \{v_i\}$ for all i. If $u_i \sqsubseteq u_j$ then $u_j\Sigma^\omega \subseteq u_i\Sigma^\omega$ and $v_i = v_j$. Therefore, the pair (u_j, v_j) is unnecessary and can be omitted. After finitely many omissions we obtain a sequence $(u_1, v_1), \ldots, (u_k, v_k) \in \Sigma^* \times \Sigma^*$ of pairs with the properties stated in Part 2. The remaining statements of the theorem can be proved easily. $\square$

In this book we will develop a theory of computability and simultaneously a formally similar "theory of continuity".

Exercises 2.2.

$\Diamond$ 1. Define continuous functions $f :\subseteq \Sigma^\omega \to \Sigma^*$ and $g : \Sigma^\omega \to \Sigma^\omega$ which are not computable.

2. Show that the characteristic function of the set $\{0^\omega\} \subseteq \Sigma^\omega$ is not continuous (Example 2.1.4.6).

$\Diamond$ 3. Show that $\Sigma^\omega \setminus \{0^\omega\}$ is open but not closed.

4. a) Show that $G := \{p \in \Sigma^\omega \mid p(i) = 1 \text{ infinitely often}\}$ is a G_δ-set but neither open nor closed. (Its complement is not a G_δ-set, Exercise 2.3.2.)
   b) Conclude that the composition $g \circ f$ is not computable, where f is the function from Example 2.1.4.9 and $g : \Sigma^\omega \to \Sigma^*$ is any computable function.

5. Let $D := \{0, 1, \ldots, 9\}^*\{\bullet\}\{0, 1, \ldots, 9\}^\omega \subseteq \Sigma^\omega$ be the set of all infinite decimal fractions. Show that the set D is
   a) not open,
   b) not closed,
   c) the intersection of an open and a closed subset of Σ^ω.

6. Prove Lemma 2.2.5.
7. Complete the proof of Theorem 2.2.6.
8. Let $(M_i, \tau_i) := (\{0, 1\}, 2^{\{0,1\}})$ for all $i \in \mathbb{N}$. Identify $\times_{i \in \mathbb{N}} M_i$ with $\{0, 1\}^\omega$. Show that τ_C is the product topology.
9. Define $\mathbb{B} := \mathbb{N}^\omega = \{p \mid p : \mathbb{N} \to \mathbb{N}\}$ and a topology $\tau_\mathbb{B}$ on $\mathbb{B}$ by the basic open basic sets $w\mathbb{N}^\omega$ with $w \in \mathbb{N}^*$. $\mathbb{B}$ (or $(\mathbb{B}, \tau_\mathbb{B})$) is called Baire space. In [Wei85, KW85, Wei87] the Baire space is used in TTE instead of the Cantor space. Define $\beta : \mathbb{B} \to \Sigma^\omega$ by $\beta(a_0 a_1 a_2 \ldots) := 0^{a_0} 1 0^{a_1} 1 0^{a_2} 1 \ldots$. Show that
   a) range(β) is a G_δ-subset of Σ^ω,
   b) β and its inverse β^{-1} are continuous,
   A function $f :\subseteq \mathbb{B} \to \mathbb{B}$ is computable according to [Wei85, KW85, Wei87], iff $\beta \circ f \circ \beta^{-1}$ is computable according to Definition 2.1.2. In particular, every computable function $f :\subseteq \mathbb{B} \to \mathbb{B}$ is continuous.
10. Let $f :\subseteq \Sigma^\omega \to \Sigma^*$ be a computable function with compact domain. Show that f has a computable extension $g : \Sigma^\omega \to \Sigma^*$.
♦11. For $X \subseteq \Sigma^*$, $X \in \Pi_2$, iff there is a recursive set $T \subseteq \Sigma^* \times \mathbb{N} \times \mathbb{N}$ such that $X = \{w \mid (\forall i)(\exists j) T(w, i, j\}$ [Rog67, Odi89]. Show that $X \subseteq \Pi_2$, iff $X = \mathrm{dom}(f)$ for some computable function $f :\subseteq \Sigma^* \to \Sigma^\omega$.

2.3 Standard Representations
of Sets of Continuous String Functions

In ordinary (Type-1) recursion theory the concept of an "effective Gödel numbering" or "admissible Gödel numbering" $\varphi : \mathbb{N} \to \mathrm{P}^{(1)}$ of the set $\mathrm{P}^{(1)}$ of partial recursive functions (that is, computable number functions) $f :\subseteq \mathbb{N} \to \mathbb{N}$ is fundamental [Rog67, Wei87, Odi89]. This numbering φ is defined uniquely up to equivalence by the *universal Turing machine theorem*, utm-theorem for short, and the *smn-theorem*. In this section we generalize the numbering φ to notations of the computable functions $f :\subseteq \Sigma^a \to \Sigma^b$ by means of Σ^* and to representations of certain continuous functions $f :\subseteq \Sigma^a \to \Sigma^b$ by means of Σ^ω, where $a, b \in \{*, \omega\}$.

First, we introduce general *notations and representations*.

Definition 2.3.1 (notation, representation).
  1. *A notation of a set M is a surjective function $\nu :\subseteq \Sigma^* \to M$.*
  2. *A representation of a set M is a surjective function $\delta :\subseteq \Sigma^\omega \to M$.*

A "naming system" is a notation or a representation.
Occasionally we abbreviate $\nu(w)$ by ν_w and $\delta(p)$ by δ_p.

Sometimes we will say that $p \in Y$ is a γ-*name* of $x \in M$, if $\gamma :\subseteq Y \to M$ is a naming system and $\gamma(p) = x$. Notice that we do not consider functions $\gamma :\subseteq \Sigma^* \cup \Sigma^\omega \to M$ with finite and infinite sequences as names. We compare naming systems by continuous and by computable *reduction* or *translation*.

Definition 2.3.2 (reduction and equivalence). *For arbitrary functions* $\gamma :\subseteq Y \to M$ *and* $\gamma' :\subseteq Y' \to M'$ *with* $Y, Y' \in \{\Sigma^*, \Sigma^\omega\}$ *we define:*

1. *The function* $f :\subseteq Y \to Y'$ *translates or reduces* γ *to* γ', *iff* $\gamma(y) = \gamma' f(y)$ *for all* $y \in \mathrm{dom}(\gamma)$.
2. $\gamma \leq \gamma'$, *iff some computable function translates* γ *to* γ'.
3. $\gamma \leq_t \gamma'$, *iff some continuous function translates* γ *to* γ'.
4. $\gamma \equiv \gamma' :\Longleftrightarrow \gamma \leq \gamma'$ *and* $\gamma' \leq \gamma$.
5. $\gamma \equiv_t \gamma' :\Longleftrightarrow \gamma \leq_t \gamma'$ *and* $\gamma' \leq_t \gamma$.

(Pronunciations: $\leq$ *"reducible to",* $\leq_t$ *"continuously reducible to",* $\equiv$ *"equivalent to",* $\equiv_t$ *"continuously equivalent to").*

Occasionally, we extend the definitions of reduction and equivalence to numberings by calling the standard numbering ν_Σ and its inverse ν_Σ^{-1} computable and closing the computable functions under composition (Sect. 1.4). This extension allows $Y = \mathbb{N}$ or $Y' = \mathbb{N}$.

Notice that we do not require $M = M'$, $Y = Y'$, $\mathrm{range}(\gamma) = M$ or $\mathrm{range}(\gamma') = M'$. As usual, $\gamma < \gamma'$ means ($\gamma \leq \gamma'$ and not $\gamma' \leq \gamma$), etc.. Since the identity on Y is computable and continuous, the relations $\leq$ and $\leq_t$ are reflexive. Since continuous functions are closed under composition and computable functions closed under composition up to extension (Theorem 2.1.12), the relations $\leq$ and $\leq_t$ are transitive. Therefore, on the class of naming systems, $\leq$ and $\leq_t$ are *pre-orders* and $\equiv$ and $\equiv_t$ are *equivalence relations*.

Consider the "information" about elements of M which can be obtained from names of a naming system computationally. Since intuitively information cannot be increased by translation, we may call γ' *poorer* than γ or γ *richer* than γ', if $\gamma < \gamma'$ or $\gamma <_t \gamma'$ or .

In elementary computability theory, Turing machines computing word functions $f :\subseteq \Sigma^* \to \Sigma^*$ are encoded canonically by words from Σ^*. If $\psi_w :\subseteq \Sigma^* \to \Sigma^*$ is the word function computed by the Turing machine with code w, then the partial notation $w \mapsto \psi_w$ of the computable word functions satisfies the utm-theorem and the smn-theorem:

utm(ψ): The function $(w, x) \mapsto \psi_w(x)$ is computable.

smn(ψ): For every computable function $f :\subseteq \Sigma^* \times \Sigma^* \to \Sigma^*$ there is a total computable function $r : \Sigma^* \to \Sigma^*$ with $r(x) \in \mathrm{dom}(\psi)$ for all $x \in \Sigma^*$ and $f(x, y) = \psi_{r(x)}(y)$ for all $x, y \in \Sigma^*$.

We generalize these properties as follows.

Definition 2.3.3 (general utm- and smn-property). *Let* $a, b, c \in \{*, \omega\}$, *let* G^{ab} *be a set of functions* $g :\subseteq \Sigma^a \to \Sigma^b$ *and let* $\zeta :\subseteq \Sigma^c \to G^{ab}$ *be a naming system of* G^{ab}. *Define:*

utm(ζ): *There is a computable ("universal") function* $u :\subseteq \Sigma^c \times \Sigma^a \to \Sigma^b$ *with* $\zeta_x(y) = u(x, y)$ *for all* $x \in \mathrm{dom}(\zeta)$ *and* $y \in \Sigma^a$.

smn(ζ): *For every computable function* $f :\subseteq \Sigma^c \times \Sigma^a \to \Sigma^b$ *there is a total computable function* $s : \Sigma^c \to \Sigma^c$ *such that* $s(x) \in \mathrm{dom}(\zeta)$ *for all* $x \in \Sigma^c$ *and* $f(x,y) = \zeta_{s(x)}(y)$ *for all* $x \in \Sigma^c$ *and* $y \in \Sigma^a$.

If ζ is a total function, then the universal function is defined uniquely. First we define the standard notations of the computable functions $f :\subseteq \Sigma^a \to \Sigma^b$.

Definition 2.3.4 (standard notations of computable functions).
Consider a canonical encoding of Type-2 machines with one input tape by words $w \in \Sigma^*$, *such that the set* TC *of code words is recursive. For all* $a, b \in \{*, \omega\}$ *define*
1. $\mathrm{P}^{ab} := \{ f :\subseteq \Sigma^a \to \Sigma^b \mid f \text{ is computable} \}$,
2. *a notation* $\xi^{ab} : \Sigma^* \to \mathrm{P}^{ab}$ *of the set* P^{ab} *by*

$$\xi^{ab}(w) := \begin{cases} \textit{the nowhere defined function, if } w \notin \mathrm{TC}, \\ \textit{the function } f \in \mathrm{P}^{ab} \textit{ computed by the} \\ \textit{Type-2 machine with code } w \textit{ otherwise} . \end{cases}$$

Since canonical encodings of Turing machines are defined in every introduction to computability theory [HU79], we use them here without further explanation. The following generalized utm- and the smn-properties can be proved in exactly the same way as the "classical" theorems. Therefore, we omit the proofs.

Theorem 2.3.5 (utm- and smn-property for ξ^{ab}).   For all $a, b \in \{*, \omega\}$ we have $\mathrm{utm}(\xi^{ab})$ and $\mathrm{smn}(\xi^{ab})$.

For obtaining a powerful and elegant theory we introduce representations of sets of continuous functions $f :\subseteq \Sigma^a \to \Sigma^b$.
Consider the case $a = *$: Since every subset of Σ^* is open, every function $f :\subseteq \Sigma^* \to \Sigma^b$ is continuous. The set of all functions $f :\subseteq \Sigma^* \to \Sigma^b$ has the same cardinality as Σ^ω or $\mathbb{R}$ ("continuum cardinality") and, therefore, has a representation.
Consider the case $a = \omega$: Let $f : \Sigma^\omega \to \Sigma^b$ be a constant function. Then f as well as all its restrictions are continuous. The set of all these restrictions has the same cardinality as the power set of Σ^ω which is bigger than Σ^ω. Therefore, the set of all continuous functions $f :\subseteq \Sigma^\omega \to \Sigma^b$ has no representation. We solve the problem by considering only continuous functions with "natural" domains which in some sense represent all continuous functions.

Definition 2.3.6 (classes of continuous functions).

$\mathrm{F}^{**} := \{ f \mid f :\subseteq \Sigma^* \to \Sigma^* \}$,

$\mathrm{F}^{*\omega} := \{ f \mid f :\subseteq \Sigma^* \to \Sigma^\omega \}$,

$\mathrm{F}^{\omega *} := \{ f \mid f :\subseteq \Sigma^\omega \to \Sigma^* \text{ is continuous and } \mathrm{dom}(f) \text{ is open} \}$,

$\mathrm{F}^{\omega\omega} := \{ f \mid f :\subseteq \Sigma^\omega \to \Sigma^\omega \text{ is continuous and } \mathrm{dom}(f) \text{ is a } \mathrm{G}_\delta\text{-set} \}$.

Remember that by Theorem 2.2.4 the domain of every computable function from $F^{\omega*}$ is open and the domain of every computable function from $F^{\omega\omega}$ is a G_δ-set. The following two theorems may serve as further justifications for considering the above sets $F^{\omega*}$ and $F^{\omega\omega}$. The first theorem is a continuous version of Lemma 2.1.11.

Theorem 2.3.7 (characterization of $F^{\omega*}$ and $F^{\omega\omega}$).
1. $F^{\omega*} = \{h_* \mid h :\subseteq \Sigma^* \to \Sigma^* \text{ has prefix-free domain}\}$.
2. $F^{\omega\omega} = \{h_\omega \mid h : \Sigma^* \to \Sigma^* \text{ is monotone}\}$.

By the next theorem, the sets $F^{\omega*}$ and $F^{\omega\omega}$ represent essentially all partial continuous functions $f :\subseteq \Sigma^\omega \to \Sigma^*$ and $f :\subseteq \Sigma^\omega \to \Sigma^\omega$, respectively. The second part is a special case of a well-known extension theorem for continuous functions on metric spaces [Kur66].

Theorem 2.3.8 (continuous extension).
1. Every continuous partial function $f :\subseteq \Sigma^\omega \to \Sigma^*$ has an extension in $F^{\omega*}$.
2. Every continuous partial function $f :\subseteq \Sigma^\omega \to \Sigma^\omega$ has an extension in $F^{\omega\omega}$.

Proof of Theorems 2.3.7 and 2.3.8:
1. $h_* \in F^{\omega*}$ for every $h :\subseteq \Sigma^* \to \Sigma^*$ with prefix-free domain:
Let $f := h_*$. If $f(p) = y$ then there is a unique $x \sqsubseteq p$ with $h(x) = y$. We obtain $f^{-1}[\{y\}] = \bigcup_{h(x)=y} x\Sigma^\omega$ (which is an open set) for every $y \in \Sigma^*$. Since the set of all $\{y\}$ is a base of the discrete topology on Σ^*, f is continuous. Since $f^{-1}[\Sigma^*] = \bigcup_{y \in \Sigma^*} f^{-1}[\{y\}]$, $\text{dom}(f)$ is open.
2. If $f :\subseteq \Sigma^\omega \to \Sigma^*$ is continuous, then h_* extends f for some $h :\subseteq \Sigma^* \to \Sigma^*$ with prefix-free domain:
Define $h :\subseteq \Sigma^* \to \Sigma^*$ by

$$h(x) = y :\Longleftrightarrow (f[x\Sigma^\omega] = \{y\} \text{ and } f[x'\Sigma^\omega] = \{y\} \text{ for no } x' \sqsubset x) .$$

The function h is well-defined and has prefix-free domain. Suppose, $f(p) = y$. Since the set $\{y\}$ is open and f is continuous, there is some shortest word $x \in \Sigma^*$ such that $p \in x\Sigma^\omega$ and $f[x\Sigma^\omega] = \{y\}$. We conclude $h(x) = y$ and $h_*(p) = y$. Therefore, h_* extends f.
3. If $V \subseteq \text{dom}(h_*)$ is open, then the restriction of h_* to V is equal to g_* for some function $g :\subseteq \Sigma^* \to \Sigma^*$:
There is some prefix-free set $B \subseteq \Sigma^*$ with $V = B\Sigma^\omega$. Define $C := \{\max(u, v) \mid u \in \text{dom}(h), v \in B\}$, where $\max(u, v) := (u, \text{ if } v \sqsubseteq u, v, \text{ if } u \sqsubseteq v, \text{ div otherwise})$. C is prefix-free, since $\text{dom}(h)$ and B are prefix-free. We obtain $C\Sigma^\omega = \bigcup\{\max(u, v)\Sigma^\omega \mid u \in \text{dom}(h), v \in B\} = \bigcup\{u\Sigma^\omega \cap v\Sigma^\omega \mid u \in \text{dom}(h), v \in B\} = \text{dom}(h)\Sigma^\omega \cap B\Sigma^\omega = V$. Define $g :\subseteq \Sigma^* \to \Sigma^*$ by $\text{dom}(g) = C$ and $g(w) := h(u)$, where u is the (uniquely determined) word from $\text{dom}(h)$ with $u \sqsubseteq w$. Consider $p \in V$ and $h_*(p) = y$. Then there are

words $u \in \mathrm{dom}(h)$ and $v \in B$ with $u \sqsubseteq p$, $v \sqsubseteq p$ and $h(u) = y$. We obtain $w := \max(u, v) \in C$ and $g(w) = h(u) = y$, that is, $g_*(p) = y$. If, on the other hand, $g_*(p)$ exists, then $g(w)$ exists for some $w \in C$ with $w \sqsubseteq p$. Then there are words $u \in \mathrm{dom}(h)$ and $v \in B$ with $w := \max(u, v)$, hence $v \sqsubseteq p$ and $p \in B\Sigma^\omega = V$. Therefore, g_* is the restriction of h_* to V.

The above three properties imply Theorems 2.3.7.1 and 2.3.8.1.

4. $h_\omega \in \mathrm{F}^{\omega\omega}$ for every monotone function $h : \Sigma^* \to \Sigma^*$:

Assume $h_\omega(p) = q$ and $q \in v\Sigma^\omega$. Then $v \sqsubseteq h(u)$ for some $u \sqsubseteq p$. We obtain $p \in u\Sigma^\omega$ and $h_\omega[u\Sigma^\omega] \subseteq h(u)\Sigma^\omega \subseteq v\Sigma^\omega$. Therefore, h_ω is continuous. For $n \in \mathbb{N}$ define $B_n := \{u \in \Sigma^* \mid |h(u)| \geq n\}$. The sets $B_n\Sigma^\omega$ are open. Then $p \in \mathrm{dom}(h_\omega)$, iff $p \in B_n\Sigma^\omega$ for all n, that is, $\mathrm{dom}(h_\omega)$ is the G_δ-set $\bigcap_n B_n\Sigma^\omega$.

5. If $f :\subseteq \Sigma^\omega \to \Sigma^\omega$ is continuous, then h_ω extends f for some monotone function $h : \Sigma^* \to \Sigma^*$:

We define $h : \Sigma^* \to \Sigma^*$ inductively as follows:

$$h(\lambda) := \lambda \,,$$

$$h(xa) := \begin{cases} h(x) & \text{if } f[xa\Sigma^\omega] \subseteq h(x)b\Sigma^\omega \text{ for no } b \in \Sigma \,, \\ h(x)b & \text{for the unique } b \in \Sigma \text{ with} \\ & f[xa\Sigma^\omega] \subseteq h(x)b\Sigma^\omega, \text{ otherwise} \end{cases}$$

for all $x \in \Sigma^*$ and $a \in \Sigma$. Obviously, the function h is monotone. Suppose $f(p) = q$. An easy induction proof shows $x \sqsubset p \Longrightarrow h(x) \sqsubset q$. It remains to show that the sequence $|h(p_{<n})|_n$ is unbounded. Assume $h(p_{<n}) = u$ for all $n \geq N$. For some $b \in \Sigma$, $q \in ub\Sigma^\omega$. By continuity of f, $f[p_{<n+1}\Sigma^\omega] \subseteq ub\Sigma^\omega$ for some $n \geq N$. Then by definition, $h(p_{<n+1}) = h(p_{<n}p_n) = h(p_{<n})b = ub$ (contradiction). We conclude $h_\omega(p) = q$. Therefore, h_ω extends f.

6. If $G \subseteq \mathrm{dom}(h_\omega)$ is a G_δ-set, then the restriction of h_ω to G is equal to g_ω for some monotone function $g : \Sigma^* \to \Sigma^*$:

There is a decreasing sequence $(V_n)_{n \in \mathbb{N}}$ of open sets such that $G = \bigcap_{n \in \mathbb{N}} V_n$. We may assume $V_0 = \Sigma^\omega$. Define $d : \Sigma^* \to \Sigma^*$ by $d(x) := y$ where $y \sqsubseteq x$ and $|y|$ is the greatest number $k \leq |x|$ with $x\Sigma^\omega \subseteq V_k$. Then $\mathrm{dom}(d_\omega) = G$ and $d_\omega(p) = p$ for all $p \in G$. Choose $g := h \circ d$.

Properties 4, 5 and 6 imply Theorems 2.3.7.2 and 2.3.8.2. $\square$

The functions from the sets F^{ab}, $a, b \in \{*, \omega\}$, are essentially closed under composition. The following theorem is a continuous counterpart to Theorem 2.1.12. We leave the proof to the reader.

Theorem 2.3.9 (composition). Consider $g \in \mathrm{F}^{ab}$ and $f \in \mathrm{F}^{bc}$ where $a, b, c \in \{*, \omega\}$. If $b = \omega$ and $c = *$, then $f \circ g$ has an extension $d \in \mathrm{F}^{a*}$ with $\mathrm{dom}(d) \cap \mathrm{dom}(g) = \mathrm{dom}(f \circ g)$, otherwise $f \circ g \in \mathrm{F}^{ac}$.

Now we introduce standard representations of the sets F^{ab}. We derive them from our standard notations ξ^{ab} of the computable functions.

Definition 2.3.10 (standard representations of the sets F^{ab}). *For all $a, b \in \{*, \omega\}$ define the standard representation $\eta^{ab} : \Sigma^\omega \to \mathrm{F}^{ab}$ of F^{ab} by*

$$\eta^{ab}(\langle x, p \rangle)(y) := \xi_x^{\omega b} \langle p, y \rangle$$

for all $x \in \Sigma^*$, $p \in \Sigma^\omega$ and $y \in \Sigma^a$,

$$\eta^{ab}(q) := \text{ the nowhere defined function,}$$

if for no $x \in \Sigma^*$, $\iota(x)$ is a prefix of q. We will abbreviate $\eta^{ab}(q)$ by η_q^{ab}.

Therefore, roughly speaking, $\eta_{\langle x, p \rangle}^{ab}(y)$ is the result of the Type-2 machine M with code x applied to input $\langle p, y \rangle$. We may regard M as an *oracle machine* [Rog67, Odi89, HU79, Wei87], which for any fixed oracle $p \in \Sigma^\omega$ computes the continuous function $\eta_{\langle x, p \rangle}^{ab}$. Occasionally we will call q a "program" of the function η_q^{ab}. First, we must show that the representations η^{ab} are well-defined.

Lemma 2.3.11. For all $a, b \in \{*, \omega\}$, η^{ab} is a representation of F^{ab}.

Proof: Let us say that $p \in \Sigma^\omega$ is a list of $Q \subseteq \Sigma^*$, iff $Q = \{x \in \Sigma^* \mid \iota(x) \triangleleft p\}$ (Definition 2.1.7).
Case $a = b = *$:
Obviously, $\eta_q^{**} \in \mathrm{F}^{**}$ for every $q \in \Sigma^\omega$.

There is a Type-2 machine M with one input tape which on input $\langle p, y \rangle$, $p \in \Sigma^\omega$ and $y \in \Sigma^*$, works as follows: the machine M searches for a subword $\iota\langle y, z \rangle$, $z \in \Sigma^*$ of p. It prints z on the output tape and halts, if it has found such a subword, and diverges otherwise. Consider $f \in \mathrm{F}^{**}$. Let p be a list of $\{\langle y, z \rangle \mid f(y) = z\}$. Then $f_M\langle p, y \rangle = f(y)$ for all $y \in \Sigma^*$. Let x be a code of the machine M (that is, $f_M = \xi_x^{\omega *}$). Then $\eta_{\langle x, p \rangle}^{**}(y) = \xi_x^{\omega *} \langle p, y \rangle = f_M\langle p, y \rangle = f(y)$ for all $y \in \Sigma^*$, and so $\eta_{\langle x, p \rangle}^{**} = f$. Therefore, $\mathrm{F}^{**} = \mathrm{range}(\eta^{**})$.
Case $a = *, b = \omega$:
Obviously, $\eta_q^{*\omega} \in \mathrm{F}^{*\omega}$ for every $q \in \Sigma^\omega$.

There is a Type-2 machine M with one input tape which on input $\langle p, y \rangle$, $p \in \Sigma^\omega$ and $y \in \Sigma^*$, works in stages $n = 0, 1, \ldots$ as follows: Let z_n be the word on the output tape before Stage n. Then in Stage n the machine M searches for a subword $\iota\langle y, z \rangle$, $z \in \Sigma^*$, of p with $z_n \sqsubseteq z$. It extends z_n on the output tape to z and finishes Stage n, if it has found such a subword (and remains in Stage n forever without further writing otherwise). Consider $f \in \mathrm{F}^{*\omega}$. Let p be a list of all $\langle y, z \rangle$ with $y \in \mathrm{dom}(f)$ and $z \sqsubseteq f(y)$. Then $f_M\langle p, y \rangle = f(y)$ for all $y \in \Sigma^*$. As above we obtain $\eta_{\langle x, p \rangle}^{*\omega} = f$, where x is a code of the machine M. Therefore, $\mathrm{F}^{*\omega} = \mathrm{range}(\eta^{*\omega})$.
Case $a = \omega, b = *$:
For fixed $x \in \Sigma^*$ and $p \in \Sigma^\omega$, $\eta_{\langle x, p \rangle}^{\omega *}(q) = \xi_x^{\omega *} \langle p, q \rangle = \xi_x^{\omega *} \circ g(q)$ where $g(q) := \langle p, q \rangle$. By Theorems 2.2.3 and 2.2.4 and Definition 2.3.4, $\xi_x^{\omega *} \in \mathrm{F}^{\omega *}$.

Since $g \in \mathrm{F}^{\omega\omega}$ and $\mathrm{dom}(g) = \Sigma^\omega$, $\eta^{\omega*}_{\langle x,p \rangle} \in \mathrm{F}^{\omega*}$ by Theorem 2.3.9. Therefore, $\mathrm{range}(\eta^{\omega*}) \subseteq \mathrm{F}^{\omega*}$.

There is a Type-2 machine M with one input tape which on input $\langle p, q \rangle$, $p, q \in \Sigma^\omega$, works as follows: M searches for a subword $\iota\langle y, z \rangle$, $y, z \in \Sigma^*$ of p such that $y \sqsubseteq q$. It prints z on the output tape and halts, if it has found such a subword, and diverges otherwise.

Consider $f \in \mathrm{F}^{\omega*}$. Then $f = h_*$ for some function $h :\subseteq \Sigma^* \to \Sigma^*$ with prefix-free domain by Theorem 2.3.7, and there is a list $p \in \Sigma^\omega$ of $\{\langle y, z \rangle \mid h(y) = z\}$. Then $f_M\langle p, q \rangle = h_*(q)$ for all $q \in \Sigma^\omega$. If x is a code of the machine M, $\eta^{\omega*}_{\langle x,p \rangle}(q) = \xi^{\omega\omega}_x\langle p, q \rangle = f_M\langle p, q \rangle = h_*(q) = f(q)$ for all $q \in \Sigma^\omega$, and so $\eta^{\omega*}_{\langle x,p \rangle} = f$. Therefore, $\mathrm{F}^{\omega*} = \mathrm{range}(\eta^{\omega*})$.

Case $a = b = \omega$:
The property $\mathrm{range}(\eta^{\omega\omega}) \subseteq \mathrm{F}^{\omega\omega}$ can be proved as in the last case.

There is a Type-2 machine M with one input tape which on input $\langle p, q \rangle$, $p, q \in \Sigma^\omega$, works in stages $n = 0, 1, \ldots$ as follows: Let z_n be the word on the output tape before Stage n. Then in Stage n the machine M searches for a subword $\iota\langle y, z \rangle$, $y, z \in \Sigma^*$, of p with $y \sqsubset q$ and $z_n \sqsubset z$. It extends z_n on the output tape to z and finishes Stage n, if it has found such a subword.

Now, the property $\mathrm{F}^{\omega\omega} \subseteq \mathrm{range}(\eta^{\omega\omega})$ can be proved as in the last case. Therefore, $\mathrm{F}^{\omega\omega} = \mathrm{range}(\eta^{\omega\omega})$. $\square$

Computable functions have computable names:

Lemma 2.3.12. A function $f :\subseteq \Sigma^a \to \Sigma^b$ is computable, iff $f = \eta^{ab}_p$ for some computable $p \in \Sigma^\omega$.

Proof: If f is computable, then there is some Turing machine with code x such that $f(y) = \xi^{\omega b}_x \langle 0^\omega, y \rangle$. Then $f(y) = \eta^{ab}_{\langle x, 0^\omega \rangle}(y)$. Choose $p := \langle x, 0^\omega \rangle$. If, on the other hand, $f = \eta^{ab}_{\langle x,p \rangle}$ for some computable $p \in \Sigma^\omega$ then $f(y) = \xi^{\omega b}_x \langle p, y \rangle$, hence f is computable. $\square$

Theorem 2.3.13 (utm- and smn-theorem for η^{ab}).   For each $a, b \in \{*, \omega\}$, we have $\mathrm{utm}(\eta^{ab})$ and $\mathrm{smn}(\eta^{ab})$.

Proof: Let $v :\subseteq \Sigma^* \times \Sigma^\omega \to \Sigma^b$ be a computable universal function of $\xi^{\omega b}$. Define $u :\subseteq \Sigma^\omega \times \Sigma^a \to \Sigma^b$ by

$$u(\langle x, p \rangle, y) := v(x, \langle p, y \rangle)$$

for all $x \in \Sigma^*$, $p \in \Sigma^\omega$ and $y \in \Sigma^a$,

$$u(q, y) := \mathrm{div} ,$$

if for no $x \in \Sigma^*$, $\iota(x)$ is a prefix of q. The function u is computable. Since

$$\eta^{ab}_{\langle x,p \rangle}(y) = \xi^{\omega b}_x \langle p, y \rangle = v(x, \langle p, y \rangle) = u(\langle x, p \rangle, y) ,$$

the representation η^{ab} has the utm-property (Definition 2.3.3).

Let $g :\subseteq \Sigma^\omega \times \Sigma^a \to \Sigma^b$ be computable. Then there is some $x \in \mathrm{TC}$ (Definition 2.3.4) with $g(p, y) = \xi_x^{\omega b}\langle p, y \rangle$ for all $p \in \Sigma^\omega$ and $y \in \Sigma^a$. Define $s : \Sigma^\omega \to \Sigma^\omega$ by $s(p) := \langle x, p \rangle$. Then s is computable and

$$\eta_{s(p)}^{ab}(y) = \eta_{\langle x,p \rangle}^{ab}(y) = \xi_x^{\omega b}\langle p, y \rangle = g(p, y)$$

for all $p \in \Sigma^\omega$ and $y \in \Sigma^a$. Therefore, the representation η^{ab} has the smn-property (Definition 2.3.3). □

H. Rogers [Rog67] has proved that the numbering $\varphi : \mathbb{N} \to \mathrm{P}^{(1)}$ of the set $\mathrm{P}^{(1)}$ of partial recursive functions $f :\subseteq \mathbb{N} \to \mathbb{N}$ is, up to equivalence, the only numbering of $\mathrm{P}^{(1)}$ satisfying the utm-theorem and the smn-theorem. This equivalence theorem can be generalized easily to our naming systems of function spaces. In the following we apply Definitions 2.3.3 and 2.3.4.

Theorem 2.3.14. For notations β, γ and δ of a set G^{ab} of functions $f :\subseteq \Sigma^a \to \Sigma^b$ with utm(δ) and smn(δ)we have:
1. $[\beta \leq \gamma$ and utm(γ)$] \Longrightarrow$ utm(β)
2. $[\mathrm{smn}(\beta)$ and $\beta \leq \gamma] \Longrightarrow \mathrm{smn}(\gamma)$
3. $[\mathrm{utm}(\beta)$ and smn(γ)$] \Longrightarrow \beta \leq \gamma$
4. utm(β) $\Longleftrightarrow \beta \leq \delta$
5. smn(β) $\Longleftrightarrow \delta \leq \beta$
6. $[\mathrm{utm}(\gamma)$ and smn(γ)$] \Longleftrightarrow \gamma \equiv \delta$

The statements hold accordingly for representations β, γ and δ.

Proof: See Exercise 2.3.4. □

Properties 1 and 2 describe heredity of the smn- and the utm-property under reduction. Property 6 is a general form of the equivalence theorem by Rogers. It applies, in particular, to our notations ξ^{ab} and to our representations η^{ab} substituted for γ: A notation (representation) γ is equivalent to ξ^{ab} (η^{ab}), iff it has the utm-property and the smn-property. Property 4 deserves special attention. For a representation δ it can be interpreted as follows:

- δ is *complete* in the set of all representations of G^{ab} which have the utm-property.
- δ is, up to equivalence, the *poorest* representation of G^{ab} with the utm-property.

Therefore, the equivalence class of δ can be defined axiomatically by the utm-property and a "maximality" or "universality" property. Fig. 2.6 illustrates Property 4 for the case $\delta = \eta^{\omega\omega}$ (ζ is below ζ', if $\zeta \leq \zeta'$). Later we will characterize many other important naming systems similarly.

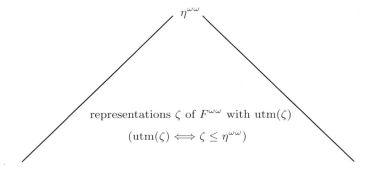

Fig. 2.6. Characterization of $\eta^{\omega\omega}$

Example 2.3.15. Let $a, b \in \{*, \omega\}$. Let M be the machine considered in the proof of Lemma 2.3.11, Case (a, b). Define a function $\delta^{ab} : \Sigma^\omega \to \mathrm{F}^{ab}$ by $\delta_p^{ab}(y) := f_M\langle p, y\rangle$ for all $p \in \Sigma^\omega$ and $y \in \Sigma^a$. Then $\delta^{ab} \equiv \eta^{ab}$. □

For representations, Theorems 2.3.13 and 2.3.14 have purely topological versions.

Theorem 2.3.16 (topological utm- and smn-property for η^{ab}).
For each $a, b \in \{*, \omega\}$, we have $\mathrm{tutm}(\eta^{ab})$ and $\mathrm{tsmn}(\eta^{ab})$, where for a representation δ of F^{ab}:
1. $\mathrm{tutm}(\delta)$: There is a function $u \in \mathrm{F}^{\omega b}$ such that $u\langle p, y\rangle = \delta_p(y)$ for all $p \in \mathrm{dom}(\delta)$ and $y \in \Sigma^a$.
2. $\mathrm{tsmn}(\delta)$: For every function $f \in \mathrm{F}^{\omega b}$ there is a continuous function $S : \Sigma^\omega \to \Sigma^\omega$ with $S(p) \in \mathrm{dom}(\delta)$ and $f\langle p, y\rangle = \delta_{S(p)}(y)$ for all $p \in \Sigma^\omega$ and $y \in \Sigma^a$.

Proof: Define a function $u :\subseteq \Sigma^\omega \to \Sigma^a$ by $u(r) := v(p, y)$, if $r = \langle p, y\rangle$ with $p \in \Sigma^\omega$ and $y \in \Sigma^a$, $u(r) :=$ div otherwise, where v is the universal function of η^{ab}. By Theorem 2.3.13, u is computable, hence $u \in \mathrm{F}^{\omega b}$.

Assume $f \in \mathrm{F}^{\omega b}$. Then $f = \eta_r^{\omega b}$ for some $r \in \Sigma^\omega$. Define $v :\subseteq \Sigma^\omega \times \Sigma^a \to \Sigma^b$ by $v(\langle p, q\rangle, y) := u(q, \langle p, y\rangle)$ for all $p, q \in \Sigma^\omega$ and $y \in \Sigma^a$, where u is the universal function of $\eta^{\omega b}$, which is computable by Theorem 2.3.13. Since v is computable, $v(\langle p, q\rangle, y) = \eta_{R\langle p, q\rangle}^{ab}(y)$ for some computable function $R : \Sigma^\omega \to \Sigma^\omega$ by Theorem 2.3.13. Define $S(p) := R\langle p, r\rangle$. □

We transfer the last two statements of Theorem 2.3.14 from the computable to the continuous case.

Corollary 2.3.17. For representations γ and δ of F^{ab} with $\mathrm{tutm}(\delta)$ and $\mathrm{tsmn}(\delta)$ we have:
1. $[\mathrm{tutm}(\gamma)$ and $\mathrm{tsmn}(\gamma)] \iff \gamma \equiv_t \delta$
2. $\forall \alpha[\mathrm{tutm}(\alpha) \iff \alpha \leq_t \delta]$

The second statement can be expressed informally as follows:

- η^{ab} is *continuously complete* in the set of all representations of F^{ab} which have a continuous universal function.
- η^{ab} is, up to continuous equivalence, the *poorest* representation of F^{ab} with a continuous universal function.

Therefore, the continuous equivalence class of η^{ab} can be defined axiomatically by the continuous utm-property and a "maximality" or "universality" property.

As in Type-1 computability theory the utm-theorem and the smn-theorem can be used to prove computability of numerous operations on computable or continuous Type-2 functions, for example composition (cf. Theorem 2.1.12).

Lemma 2.3.18 (computable composition). There is a computable function $f : \Sigma^\omega \times \Sigma^\omega \to \Sigma^\omega$ such that for all $a, b, c \in \{*, \omega\}$

$$\eta^{ac}_{f(p,q)} \begin{cases} = \eta^{bc}_q \circ \eta^{ab}_p & \text{if } b = * \text{ or } c = \omega \\ \text{extends } \eta^{bc}_q \circ \eta^{ab}_p & \text{if } b = \omega \text{ and } c = * . \end{cases}$$

Proof : Define $h : \Sigma^\omega \times \Sigma^a \to \Sigma^c$ by $h(\langle p, q\rangle, r) := \eta^{bc}_q \circ \eta^{ab}_p(r)$ $(p, q \in \Sigma^\omega)$. Then $h(\langle p, q\rangle, r) = u^{bc}(q, u^{ab}(p, r))$, where u^{ab} and u^{bc} are the computable universal functions of η^{ab} and η^{bc}, respectively (Theorem 2.3.13). By Theorem 2.1.12 there is a computable function g such that $g = h$, if $b = *$ or $c = \omega$, and g extends h, otherwise (!!). By the smn-theorem for η^{ac} there is a computable function $s : \Sigma^\omega \to \Sigma^\omega$ such that $\eta^{ac}_{s(t)}(r) = h(t, r)$ for all $t \in \Sigma^\omega$ and $r \in \Sigma^a$. Define $f(p, q) := s\langle p, q\rangle$. $\qquad\square$

Exercises 2.3.

1. Define $f :\subseteq \Sigma^\omega \to \Sigma^*$ by $f(0^i aq) := 0^i$ for all $i \in \mathbb{N}$, $a \in \Sigma$, $a \neq 0$, and $q \in \Sigma^\omega$ and $f(0^\omega) = \mathrm{div}$. Then f is a computable function which has no proper continuous extension. Show that the function $f :\subseteq \Sigma^\omega \to \Sigma^\omega$ from Example 2.1.4.9 has a G_δ-domain and no proper continuous extension. Therefore, the extension Theorem 2.3.8 is sharp.

2. Show that the set $F := \{p \in \Sigma^\omega \mid p(i) = 0 \text{ for almost all } i\}$ is not a G_δ-set (cf. Exercise 2.2.4). For a proof assume that F is a G_δ-set. Then $F = A_0 \cap A_1 \cap \dots$ for open sets A_n. Now $0^\omega \in F$, therefore, $0^\omega \in A_0$. Since A_0 is open, there is n_0 such that $0^{n_0}\Sigma^\omega \subseteq A_0$. Now $0^{n_0}10^\omega \in F$, therefore, $0^{n_0}10^\omega \in A_1$. Since A_1 is open, there is n_1 such that $0^{n_0}10^{n_1}\Sigma^\omega \subseteq A_1$.

By repetition we obtain a sequence $p = 0^{n_0} 1 0^{n_1} 1 \ldots \in A_0 \cap A_1 \cap \ldots = F$, contradiction.

3. Prove the composition theorem 2.3.9.

4. Prove Theorem 2.3.14.

5. Prove $\delta^{ab} \equiv \eta^{ab}$ for the representations δ^{ab} from Example 2.3.15.

6. Prove $\xi^{ab} \equiv \zeta^{ab}$ for all $a, b \in \{*, \omega\}$, where $\zeta_w^{ab} := \eta_{\langle w, 0^\omega \rangle}^{ab}$.

7. Prove $\xi^{ab} \leq \eta^{ab}$ for all $a, b \in \{*, \omega\}$.

8. Show that for all $Y_0, \ldots, Y_k \in \{\Sigma^*, \Sigma^\omega\}$ there is a computable function $g :\subseteq \Sigma^\omega \times Y_1 \times \ldots \times Y_k \to Y_0$ such that a function $f :\subseteq Y_1 \times \ldots \times Y_k \to Y_0$ is continuous, iff there is a sequence $p \in \Sigma^\omega$ such that $f(y_1, \ldots, y_k) = g(p, y_1, \ldots, y_k)$ for all $(y_1, \ldots, y_k) \in \mathrm{dom}(f)$.

9. Prove Lemma 2.3.18 with ξ replacing η.

10. Show that there is a computable function $f : \Sigma^\omega \times \Sigma^\omega \to \Sigma^\omega$ such that $\eta_{f(p,q)}^{*\omega}(x) = \eta_p^{**}(x)\eta_q^{*\omega}(x)$ for all $p, q \in \Sigma^\omega$ and $x \in \Sigma^*$.

11. Show that there is a computable function $f : \Sigma^\omega \to \Sigma^\omega$ such that $\eta_{f(p)}^{**} = (\eta_p^{**})^{-1}$ whenever η_p^{**} is injective.

♦12. Computability on Σ^ω can be derived from *domain*-computability (Sect. 9.5):

Define a partially ordered space $(\Sigma^{\leq \omega}, \sqsubseteq)$ by $\Sigma^{\leq \omega} := \Sigma^* \cup \Sigma^\omega$ and $x \sqsubseteq y$, iff $x = y$ or x is a prefix of y. For every increasing sequence $x_0 \sqsubseteq x_1 \sqsubseteq \ldots$ the least upper bound $\sup_{i \to \infty} x_i$ exists. Call a function $f : \Sigma^{\leq \omega} \to \Sigma^{\leq \omega}$ *continuous*, iff it is monotone and $f(\sup_i x_i) = \sup_i f(x_i)$ for every increasing sequence $x_0 \sqsubseteq x_1 \sqsubseteq \ldots$. Call f *computable*, iff it is continuous and $\{(u, v) \in \Sigma^* \times \Sigma^* \mid v \sqsubseteq f(u)\}$ is r.e. Show:

a) $\mathrm{F}^{\omega\omega} = \{f|_{\Sigma^\omega}^{\Sigma^\omega} \mid f : \Sigma^{\leq \omega} \to \Sigma^{\leq \omega} \text{ is continuous}\}$,

b) $g :\subseteq \Sigma^\omega \to \Sigma^\omega$ is computable, iff $g = f|_{\Sigma^\omega}^{\Sigma^\omega}$ for some computable $f : \Sigma^{\leq \omega} \to \Sigma^{\leq \omega}$.

For $\mathbb{N}^\omega$ instead of Σ^ω (cf. Exercise 2.2.9) a proof can be found in [Wei87].

2.4 Effective Subsets

A subset $A \subseteq \Sigma^*$ is called *recursive* or *decidable*, iff its characteristic function is computable, and it is called *recursively enumerable*, r.e. for short, iff it is the domain of a computable function $f :\subseteq \Sigma^* \to \Sigma^*$. The set A is recursive, iff it is r.e. and its complement is r.e. [Rog67]. We generalize these concepts as follows, where we use the term *decidable*. *Recursive* subsets will be defined later (Definition 2.4.9).

Definition 2.4.1 (r.e. open in Z, decidable in Z). *Consider* $X \subseteq Z \subseteq Y := Y_1 \times \ldots \times Y_k$, *where* $k \geq 1$ *and* $Y_1, \ldots Y_k \in \{\Sigma^*, \Sigma^\omega\}$.

1. *X is called recursively enumerable open (r.e. open) in Z, iff* $X = \mathrm{dom}(f) \cap Z$ *for some computable function* $f :\subseteq Y \to \Sigma^*$.

2. *X is called decidable in Z, iff X and $Z \setminus X$ are r.e. open in Z.*

If $Z = Y$, we omit "in Z".

For $k = 1$ and $Z = Y_1 = \Sigma^*$ we obtain the ordinary definition of r.e. and decidable sets. There is a striking similarity to topological concepts, where "r.e. open" corresponds to "open": By definition,

1. X is *open* in Z, iff $X = U \cap Z$ for some open set $U \subseteq Y$,
2. X is *open and closed* in Z, iff X and $Z \setminus X$ are open in Z.

As usual, we abbreviate "open and closed" by "clopen". The following corollary justifies the term "r.e. open". Later we will also introduce r.e. closed sets (Sect. 5.1).

Corollary 2.4.2. If X is r.e. open (decidable) in Z, then X is open (clopen) in Z.

This follows immediately from Theorem 2.2.4. Decidable sets can also be defined by computable characteristic functions. Notice that the decision function below is not total in general but may behave arbitrarily for $z \notin Z$.

Theorem 2.4.3 (characterization of decidable and clopen sets).
1. X is clopen in Z, iff there is a continuous function $f :\subseteq Y \to \Sigma^*$ with $f(z) = 1$, if $z \in X$, and $f(z) = 0$, if $z \in Z \setminus X$.
2. X is decidable in Z, iff there is a computable function $f :\subseteq Y \to \Sigma^*$ with $f(z) = 1$, if $z \in X$, and $f(z) = 0$, if $z \in Z \setminus X$.

Proof: 1. Let X and $Z \setminus X$ be open in Z. Then there are open sets $U, V \subseteq Y$ with $X = U \cap Z$ and $Z \setminus X = V \cap Z$. Define $f :\subseteq Y \to \Sigma^*$ by $\mathrm{dom}(f) := Z$, $f(z) := 1$, if $z \in U \cap Z$, and $f(z) := 0$, if $z \in V \cap Z$. The function f is continuous, since $f^{-1}[\{1\}] = U \cap \mathrm{dom}(f)$ and $f^{-1}[\{0\}] = V \cap \mathrm{dom}(f)$. Furthermore, $f(z) = 1$, if $z \in X$, and $f(z) = 0$, if $z \in Z \setminus X$.
On the other hand, let f be continuous with $f(z) = 1$, if $z \in X$, and $f(z) = 0$, if $z \in Z \setminus X$. Then $f^{-1}[\{1\}] = U \cap \mathrm{dom}(f)$ for some open set U. We obtain $X = f^{-1}[\{1\}] \cap Z = U \cap \mathrm{dom}(f) \cap Z = U \cap Z$. Therefore, X is open in Z. For a similar reason $Z \setminus X$ is open in Z.
2. Let X and $Z \setminus X$ be r.e. open in Z. Then there are Type-2 machines M and N with $X = \mathrm{dom}(f_M) \cap Z$ and $Z \setminus X = \mathrm{dom}(f_N) \cap Z$. Let L be a machine which on input x simulates the computation of M on input x in parallel with the computation of N on input z. As soon as the simulation of M on input z halts, L writes the output 1 and halts. As soon as the simulation of N on input z halts, L writes the output 0 and halts. Obviously, the machine L decides, whether $z \in X$ or not for all $z \in Z$.
On the other hand, let L be a machine which decides, whether $z \in X$ or not for all $z \in Z$. From L we construct a machine M which on input $z \in Y$ halts, iff the machine L halts on input z with result 1. Then $X = \mathrm{dom}(f_M) \cap Z$, that is, X is r.e. open in Z. The case "$Z \setminus X$" can be proved accordingly. $\square$

Remember that by Theorem 2.2.6 each decidable set $X \subseteq \Sigma^\omega$ has the form $X = A\Sigma^\omega$ for some finite $A \subseteq \Sigma^*$.

Example 2.4.4.

1. Z is decidable in Z. If X is r.e. open (decidable), then X is r.e. open (decidable) in Z, whenever $X \subseteq Z$.
2. The set $X = \{p \in \Sigma^\omega \mid p \neq 0^\omega\}$ is r.e. open. Its complement $\{0^\omega\}$ is not open, since it does not contain a set $w\Sigma^\omega$ (Definition 2.2.2). Therefore, X is not closed, hence not clopen and not decidable.
3. Let $\langle\ \rangle$ be the tupling function for $Y_1 \times \ldots \times Y_k$ (Definition 2.1.7). Then $X \subseteq Y_1 \times \ldots \times Y_k$ is r.e. open (decidable), iff $\langle X \rangle$ is r.e. open (decidable) in $\langle Y_1 \times \ldots \times Y_k \rangle$. □

The r.e. open and decidable sets have the expected closure properties.

Theorem 2.4.5 (union, intersection, complement). Assume $X_1, X_2, Z \subseteq Y_1 \times \ldots \times Y_k$ with $X_1, X_2 \subseteq Z$, assume $U, W \subseteq Y_0$ with $U \subseteq W$ and $f :\subseteq Y_1 \times \ldots \times Y_k \to Y_0$, where $Y_0, \ldots, Y_k \in \{\Sigma^*, \Sigma^\omega\}$.

1. If X_1 and X_2 are r.e. open in Z, then $X_1 \cap X_2$ and $X_1 \cup X_2$ are r.e. open in Z.
2. If X_1 and X_2 are decidable in Z, then $X_1 \cap X_2$, $X_1 \cup X_2$ and $Z \setminus X_1$ are decidable in Z.
3. If f is computable and U is r.e. open (decidable) in W, then $f^{-1}[U]$ is r.e. open (decidable) in $f^{-1}[W]$.

The statements hold accordingly with "open", "clopen" and "continuous" instead of "r.e. open", "decidable" and "computable", respectively.

Proof: 1. There are Type-2 machines M_i such that $X_i = \mathrm{dom}(f_{M_i}) \cap Z$ $(i = 1, 2)$. Let M be a machine which simulates M_1 and M_2 in parallel and halts, if M_1 and M_2 halt (case intersection) or if M_1 halts or M_2 halts (case union).

2. See Exercise 2.4.1.

3. Let U be r.e. open in W. Then $U = \mathrm{dom}(g) \cap W$ for some computable function $g :\subseteq Y_0 \to \Sigma^*$. By Theorem 2.1.12, $g \circ f$ has a computable extension h with $\mathrm{dom}(h) \cap \mathrm{dom}(f) = \mathrm{dom}(g \circ f)$. Consider $y \in f^{-1}[W]$. Then $y \in \mathrm{dom}(f)$, $f(y) \in W$ and $y \in f^{-1}[U] \iff f(y) \in U \iff g \circ f(y)$ exists $\iff y \in \mathrm{dom}(g \circ f) \iff y \in \mathrm{dom}(h)$. Therefore, $f^{-1}[U] = \mathrm{dom}(h) \cap f^{-1}[W]$, hence $f^{-1}[U]$ is r.e. open in $f^{-1}[W]$. If U is decidable in W, then U as well as $W \setminus U$ are r.e. open in W. As we have shown above, $f^{-1}[U]$ as well as $f^{-1}[W \setminus U] = f^{-1}[W] \setminus f^{-1}[U]$ are r.e. open in $f^{-1}[W]$. Therefore, $f^{-1}[U]$ is decidable in $f^{-1}[W]$.

The proof of the topological version is trivial. □

> **Corollary 2.4.6 (pre-image).** Consider f, U and W from Theorem 2.4.5.
> 1. If f is computable, U is r.e. open and $f^{-1}[W]$ is r.e. open, then $f^{-1}[U]$ is r.e. open.
> 2. If f is a total computable function and U is r.e. open (decidable), then $f^{-1}[U]$ is r.e. open (decidable).

In the following we characterize the recursively enumerable open subsets of $Y := Y_1 \times \ldots \times Y_k$ ($Y_1, \ldots, Y_k \in \{\Sigma^*, \Sigma^\omega\}$) and define the co-r.e. open and the recursive open subsets of Y. We use the notation $y \circ Y$ introduced in Definition 2.2.2.

> **Theorem 2.4.7 (r.e. open sets).** For $X \subseteq Y := Y_1 \times \ldots \times Y_k$ ($Y_1, \ldots, Y_k \in \{\Sigma^*, \Sigma^\omega\}$) the following properties are equivalent:
> 1. X is r.e. open,
> 2. $X = A \circ Y$ for some r.e. set $A \subseteq (\Sigma^*)^k$,
> 3. X is open and $\{y \in (\Sigma^*)^k \mid y \circ Y \subseteq X\}$ is r.e.

Proof: For the sake of simplicity we consider the case $Y = \Sigma^* \times \Sigma^\omega$. The general case is proved correspondingly.

$1 \Longrightarrow 2$: Let X be r.e. open. Then $X = \mathrm{dom}(f_M)$ for some Type-2 machine M. Define a computable function $g :\subseteq \Sigma^* \times \Sigma^* \to \Sigma^*$ by

$$g(x_1, x_2) := \begin{cases} 1 & \text{if } M \text{ halts on input } (x_1, x_2 0^\omega) \text{ after reading at most} \\ & \quad \text{the first } |x_2| \text{ symbols from the second input tape,} \\ \mathrm{div} & \text{otherwise.} \end{cases}$$

The set $A := g^{-1}[\{1\}]$ is r.e.

Assume $(x_1, x_2) \in A$. Then the machine M halts on input $(x_1, x_2 q)$ for every $q \in \Sigma^\omega$, and so $(x_1, x_2) \circ Y \subseteq X$. Therefore, $A \circ Y \subseteq X$.

On the other hand assume $(x_1, p) \in X$. Then for some prefix x_2 of p the machine M halts on input (x_1, p) after reading at most the first $|x_2|$ symbols from the second input tape. Therefore, $(x_1, x_2) \in A$, and $(x, p) \in A \circ Y$. We obtain $X \subseteq A \circ Y$.

$2 \Longrightarrow 1$: Let M be a Turing machine with $A = \mathrm{dom}(f_M)$. Let N be a Type-2 machine which on input $(x, p) \in \Sigma^* \times \Sigma^\omega$ searches the smallest number $n = \langle k, m \rangle$ such that M on input $(x, p_{<k})$ halts in m steps and halts as soon as it has found such a number n. Then $A \circ Y = \mathrm{dom}(f_N)$.

$2 \Longrightarrow 3$: Suppose, $y \circ Y \subseteq A \circ Y$. Since $y \circ Y$ is a compact subset of Y (Exercise 6), there is a finite subset $C \subseteq A$ such that $y \circ Y \subseteq C \circ Y$. Since this property is decidable in y and C, the set $\{y \in (\Sigma^*)^k \mid y \circ Y \subseteq A \circ Y\}$ is r.e., if A is r.e. Clearly, X is open, if $X = A \circ Y$.

$3 \Longrightarrow 2$: If X is open and $B := \{y \in (\Sigma^*)^k \mid y \circ Y \subseteq X\}$, then $X = B \circ Y$. $\square$

For the special case $Y_1 = \ldots = Y_k = \Sigma^*$, Theorem 2.4.7 becomes trivial, since $A \circ Y = A$ for all $A \subseteq (\Sigma^*)^k$. Properties 2.4.7.2 and 2.4.7.3 could be strengthened for defining recursive open subsets or Y. The first alternative is useless:

Example 2.4.8. For every r.e. open subset $X \subseteq \Sigma^\omega$ there is a recursive subset $B \subseteq \Sigma^*$ such that $X = B\Sigma^\omega$. For a proof assume $X = \text{dom}(f_M)$ for some Type-2 machine M. Define a computable function $g : \Sigma^* \to \Sigma^*$ by

$$g(x) := \begin{cases} 1 & \text{if } M \text{ on input } x0^\omega \text{ halts in exactly } |x| \text{ steps,} \\ 0 & \text{otherwise} \end{cases}$$

and $B := g^{-1}[\{1\}]$. Then for any $p \in \Sigma^\omega$,

$$\begin{aligned}
p \in B\Sigma^\omega &\iff (\exists\, n)\ p_{<n} \in B \\
&\iff (\exists\, n)\ g(p_{<n}) = 1 \\
&\iff (\exists\, n)\ M \text{ on input } p_{<n}0^\omega \text{ halts in } n \text{ steps} \\
&\iff (\exists\, n)\ M \text{ on input } p \text{ halts in } n \text{ steps} \\
&\iff p \in \text{dom}(f_M)
\end{aligned}$$

$\square$

The second alternative leads to the following important definition:

Definition 2.4.9 (r.e., co-r.e. and recursive open sets). *For any open subset* $X \subseteq Y := Y_1 \times \ldots \times Y_k$ *(*$Y_1, \ldots, Y_k \in \{\Sigma^*, \Sigma^\omega\}$*) define:*

$$X \text{ is r.e.} \iff \{y \in (\Sigma^*)^k \mid y \circ Y \subseteq X\} \text{ is r.e.}, \tag{2.1}$$

$$X \text{ is co-r.e.} \iff \{y \in (\Sigma^*)^k \mid y \circ Y \not\subseteq X\} \text{ is r.e.}, \tag{2.2}$$

$$X \text{ is recursive} \iff \{y \in (\Sigma^*)^k \mid y \circ Y \subseteq X\} \text{ is decidable.} \tag{2.3}$$

By this definition, an open set $X \subseteq Y$ is recursive, iff it is r.e. and co-r.e. Formula (2.1) is consistent with Definition 2.4.1 by Theorem 2.4.7. Notice that for $Y = (\Sigma^*)^k$, X is r.e. open, iff X is r.e. in the usual sense, X is co-r.e. open, iff its complement $Y \setminus X$ is r.e., and X is recursive open, iff X is recursive in the usual sense. Later we will define r.e. open subsets and recursive open subsets of the Euclidean space $\mathbb{R}^n$ similarly (Sect. 5.1). Definition 2.4.9 is particularly simple, since every base set $y \circ Y$ $(y \in (\Sigma^*)^k)$ is not only open but also compact.

By Theorem 2.4.5, the r.e. open subsets of Y are closed under union and intersection. The recursive open subsets of Σ^ω are closed under intersection (proof straightforward) but:

Lemma 2.4.10. *The recursive open subsets of* $Y := Y_1 \times \ldots \times Y_k$ *(*$Y_1, \ldots, Y_k \in \{\Sigma^*, \Sigma^\omega\}$*) are* **not** *closed under union.*

Proof: See Exercise 2.4.7. $\square$

Similarly, the recursive open subsets of the real numbers are not closed under union, cf. Corollary 5.1.18 and Exercise 5.1.15. As in ordinary recursion theory, the projection theorem is a useful tool.

Theorem 2.4.11 (projection theorem). For $X \subseteq Y := Y_1 \times \ldots \times Y_k$
$(Y_1, \ldots, Y_k \in \{\Sigma^*, \Sigma^\omega\})$ the following properties are equivalent:
1. X is r.e. open.
2. There is some decidable set $V \subseteq Y \times \Sigma^*$ such that
   $X = \{y \in Y \mid (\exists u \in \Sigma^*)(y, u) \in V\}$.
3. There is some r.e. open set $V \subseteq Y \times \Sigma^*$ such that
   $X = \{y \in Y \mid (\exists u \in \Sigma^*)(y, u) \in V\}$.

Proof: The proof is similar to that of the classical special case $k = 1$ and $Y_1 = \Sigma^*$ (Exercise 2.4.2). □

Lemma 2.4.12. Every decidable set $X \subseteq Y_1 \times \ldots \times Y_k$ $(Y_i \in \{\Sigma^*, \Sigma^\omega\})$ is recursive. There are recursive sets which are not decidable.

Proof: For the first part the proof of Theorem 2.2.6 can be generalized (as Exercise 2.4.9). The set $\Sigma^\omega \setminus \{0^\omega\} = \bigcup \{0^n 1 \Sigma^\omega \mid n \in \mathbb{N}\}$ is recursive but not decidable by Theorem 2.2.6. □

Exercises 2.4.

1. Prove Theorem 2.4.5.2
2. Prove the projection theorem.
3. Prove the continuous version of the projection theorem, where "r.e. open" and "decidable" are replaced by "open" and "clopen", respectively.
4. Prove the statement from Example 2.4.4.3.
♦ 5. For $A, B \subseteq \Sigma^\omega$ define $A \leq B$ ($A \leq_w B$, "Wadge reducible"), iff $A = f^{-1}[B]$ for some computable (continuous) function $f : \Sigma^\omega \to \Sigma^\omega$. Show that
   a) A is r.e. open, iff $A \leq (\Sigma^\omega \setminus \{0^\omega\})$,
   b) A is open, iff $A \leq_w (\Sigma^\omega \setminus \{0^\omega\})$,
   c) A is a G_δ-set, iff $A \leq_w \{p \in \Sigma^\omega \mid p(i) = 0 \text{ infinitely often}\}$.
6. Show that $y \circ Y$ is a compact subset of Y for every $y \in (\Sigma^*)^k$ (Definition 2.2.2).
7. Consider $\Sigma = \{0, 1\}$. Let $f : \mathbb{N} \to \mathbb{N}$ be a computable function such that $K := \text{range}(f)$ is not recursive. Define

$$X := \bigcup \{0^i 10^m 1 \Sigma^\omega \mid i, m \in \mathbb{N}\} \quad \text{and} \quad Y := \bigcup \{0^i 10^m \Sigma^\omega \mid i = f(m)\} .$$

Show that X and Y are recursive open sets and that $X \cup Y$ is not recursive open.
(Remark: $i \in K \iff 0^i 1 \Sigma^\omega \subseteq X \cup Y$)

8. Consider Theorem 2.4.7. Call $B \subseteq (\Sigma^*)^k$ prefix-free, iff $(y {\circ} Y \subseteq y' {\circ} Y \Longrightarrow y = y')$ for all $y, y' \in B$. Show that $X \subseteq Y$ is r.e. open, iff $X = B \circ Y$ for some prefix-free recursive set $B \subseteq (\Sigma^*)^k$.

9. Show that every decidable subset $X \subseteq Y_1 \times \ldots \times Y_k$ $(Y_i \in \{\Sigma^*, \Sigma^\omega\})$ is recursive open.

◊10. Show that the open sets $\Sigma^\omega \setminus \{0^\omega\}$ and $\Sigma^\omega \setminus \langle \Sigma^* \times \Sigma^\omega \rangle$ are recursive but not decidable.

◊11. Find an r.e. open set $X \subseteq \Sigma^\omega$ which is not recursive open.

♦12. For every subset $X \subseteq \Sigma^\omega$ define $W_X := \{w \in \Sigma^* \mid w\Sigma^\omega \cap X \neq \emptyset\}$. Call X *H-recursive* or *H-r.e.*, iff W_X is recursive or r.e., respectively ([Hau73], Definition 1). Show that

a) X is H-r.e. iff $\Sigma^\omega \setminus \mathrm{cls}(X)$ is r.e. open,

b) X is H-recursive iff $\Sigma^\omega \setminus \mathrm{cls}(X)$ is recursive open.

3. Naming Systems

So far we have defined computability on the sets Σ^* of finite words and Σ^ω of infinite sequences explicitly by Type-2 machines (Sect. 2.1). In TTE we introduce computability on other sets M by using finite or infinite words as "names". Machines, therefore, still transform "concrete" sequences of symbols. Only the user of the machine interprets theses sequences as finite or infinite names of "abstract objects". Although there are several other suggestions to define computability on sets or structures, in this book we will confine ourselves exclusively to computability concepts induced by naming systems. As we have seen, some concepts from computability theory have formally similar topological counterparts. In the following we will continue to develop computability theory, considering also the weaker topological aspects whenever advisable.

Sect. 3.1 is devoted to computability and the topological concepts induced on sets by naming systems. We introduce some standard naming systems and transfer the theorems on composition and primitive recursion to functions on named sets. In Sect. 3.2 we introduce and discuss the important class of admissible representations. Most representations which we will use later are admissible. In Sect. 3.3 we construct new naming systems from given ones and investigate their topological and computational properties. The most important ones are representations of products and of spaces of continuous functions. The reader may skip conjunction and disjunction in a first reading.

3.1 Continuity and Computability Induced by Naming Systems

In this section we transfer topological concepts and computability from Σ^* and Σ^ω to sets M by means of naming systems and illustrate the definitions by examples.

We have already introduced notations $\nu :\subseteq \Sigma^* \to M$ and representations $\delta :\subseteq \Sigma^\omega \to M$, which we called naming systems (Definition 2.3.1). We will not consider naming systems where both, finite and infinite sequences of symbols, are used as names simultaneously. (They can be modelled by representations if necessary.) Let $\gamma :\subseteq Y \to M$ be a naming system. Then every element of

M has a name since a naming system is surjective. Not every element of Σ^* or Σ^ω must be a name of an element of M, since a naming system may be a partial function. Furthermore, an element of M may have several names, since a naming system is not one-one in general. We have also defined computable and continuous reducibility and equivalence for comparing naming systems (Definition 2.3.2).

For later use we first introduce some concrete naming systems.

Convention 3.1.1. We still assume that Σ is a fixed finite alphabet containing the symbols 0 and 1 and consider sufficiently large alphabets in examples.

Definition 3.1.2 (some standard notations and representations).
1. *The identity* $\mathrm{id}_{\Sigma^*} : \Sigma^* \to \Sigma^*$ *is a notation of* Σ^* *and the identity* $\mathrm{id}_{\Sigma^\omega} : \Sigma^\omega \to \Sigma^\omega$ *is a representation of* Σ^ω.
2. *Define the binary notation* $\nu_{\mathbb{N}} :\subseteq \Sigma^* \to \mathbb{N}$ *of the natural numbers by:* $\mathrm{dom}(\nu_{\mathbb{N}}) := \{0\} \cup 1\{0,1\}^*$ *and* $\nu_{\mathbb{N}}(a_k \ldots a_0) = \sum_{i=0}^{k} a_i \cdot 2^i$ $(a_0, \ldots, a_k \in \{0,1\})$.
3. *Define a notation* $\nu_{\mathbb{Z}} :\subseteq \Sigma^* \to \mathbb{Z}$ *of the integers by* $\mathrm{dom}(\nu_{\mathbb{Z}}) := \{0\} \cup 1\{0,1\}^* \cup \text{-}1\{0,1\}^*$, $\nu_{\mathbb{Z}}(w) = \nu_{\mathbb{N}}(w)$ *and* $\nu_{\mathbb{Z}}(\text{-}w) = -\nu_{\mathbb{N}}(w)$ *for* $w \in \mathrm{dom}(\nu_{\mathbb{N}}) \setminus \{0\}$.
4. *Define a notation* $\nu_{\mathbb{Q}} :\subseteq \Sigma^* \to \mathbb{Q}$ *by* $\mathrm{dom}(\nu_{\mathbb{Q}}) := \{u/v \mid u \in \mathrm{dom}(\nu_{\mathbb{Z}}), v \in \mathrm{dom}(\nu_{\mathbb{N}}), \nu_{\mathbb{N}}(v) \neq 0\}$ *and* $\nu_{\mathbb{Q}}(u/v) := \nu_{\mathbb{Z}}(u)/\nu_{\mathbb{N}}(v)$. *Later we will abbreviate* $\nu_{\mathbb{Q}}(w)$ *by* $\overline{w}$.
5. *Define the enumeration representation* $\mathrm{En} : \Sigma^\omega \to 2^{\mathbb{N}}$ *by* $\mathrm{En}(p) := \{n \in \mathbb{N} \mid 110^{n+1}11 \lhd p\}$.
6. *Define the characteristic function representation* $\mathrm{Cf} : \Sigma^\omega \to 2^{\mathbb{N}}$ *by* $\mathrm{Cf}(p) := \{i \mid p(i) = 1\}$.
7. *For* $\Delta \subseteq \Sigma$ *define the enumeration representation* $\mathrm{En}_\Delta : \Sigma^\omega \to 2^{\Delta^*}$ *by* $\mathrm{En}_\Delta(p) := \{w \in \Delta^* \mid \iota(w) \lhd p\}$ *(Definition 2.1.7)*.
8. *Define a representation* $\delta_{\mathbb{B}}$ *of the Baire space* $\mathbb{B} = \mathbb{N}^\omega$ *by* $\mathrm{dom}(\delta_{\mathbb{B}}) := \{p \in \{0,1\}^\omega \mid p(n) = 1 \text{ infinitely often}\}$ *and* $\delta_{\mathbb{B}}(0^{n_0}10^{n_1}10^{n_2}1\ldots) := (n_0 n_1 n_2 \ldots)$.

In the following definition we introduce effectivity of sets and functions. This is also the place where *multi-valued functions* (Sect. 1.4) naturally come into play.

Definition 3.1.3 (effectivity induced by naming systems).   *Let* $\gamma :\subseteq Y \to M$ *and* $\gamma_0 :\subseteq Y_0 \to M_0$ *be naming systems. Assume* $x \in M$, $X \subseteq M$, $f :\subseteq M \to M_0$ *and* $F :\subseteq M \rightrightarrows M_0$.
1. *The element* x *is called* γ-*computable, iff* $x = \gamma(y)$ *for some computable* $y \in Y$.
2. *The set* X *is called* γ-*open* (γ-*clopen*, γ-*r.e.*, γ-*decidable*), *iff* $\gamma^{-1}[X]$ *is open (clopen, r.e. open, decidable) in* $\mathrm{dom}(\gamma)$. *The set* τ_γ *of the* γ-*open subsets of* M *is called the final topology of* γ *on* M.

3. *A function $g :\subseteq Y \to Y_0$ is called a (γ, γ_0)-realization of the function f, iff $f \circ \gamma(y) = \gamma_0 \circ g(y)$ for all $y \in Y$ such that $f \circ \gamma(y)$ exists. The function f is called (γ, γ_0)-continuous (-computable), iff it has a continuous (computable) (γ, γ_0)-realization.*

4. *A function $g :\subseteq Y \to Y_0$ is called a (γ, γ_0)-realization of the multi-valued function F, iff $\gamma_0 \circ g(y) \in F[\{\gamma(y)\}]$ for all $y \in Y$ such that $\gamma(y) \in \mathrm{dom}(F)$. The multi-valued function F is called (γ, γ_0)-continuous (-computable), iff it has a continuous (computable) (γ, γ_0)-realization.*

5. *A choice function for the multi-valued function F is a function $h :\subseteq M \to M_0$ such that $h(y) \in F[\{y\}]$ for all $y \in \mathrm{dom}(F)$.*

By substituting $(\gamma_1(y_1), \ldots, \gamma_k(y_k))$ for $\gamma(y)$, $\mathrm{dom}(\gamma_1) \times \ldots \times \mathrm{dom}(\gamma_k)$ for $\mathrm{dom}(\gamma)$ etc. we generalize the definitions straightforwardly to the case $M = M_1 \times \ldots \times M_k$ with naming systems $\gamma_i :\subseteq Y_i \to M_i$ $(i = 1, \ldots, k)$ and thus, define $(\gamma_1, \ldots, \gamma_k)$-computable points, $(\gamma_1, \ldots, \gamma_k)$-open subsets, $(\gamma_1, \ldots, \gamma_k, \gamma_0)$-computable functions etc. .

As long as we consider various naming systems for sets we will use the somewhat unwieldy prefixes "$(\gamma_1, \ldots, \gamma_k, \gamma_0)$–" etc. . They can be omitted, if each set has a naming system fixed in advance. The set $X \subseteq M$ is γ-open, if the set of its γ-names is open in $\mathrm{dom}(\gamma)$. The set τ_γ is indeed a topology on M since $\gamma^{-1}[\ ]$ commutes with union and intersection. For any representation γ of M the final topology τ_γ is the finest (that is, greatest as a subset of 2^M) topology τ on M for which $\gamma :\subseteq \Sigma^\omega \to M$ is (τ_C, τ)-continuous. For any notation γ of M, τ_γ is the discrete topology 2^M. In Fig. 3.1, the "concrete" function $g :\subseteq Y \to Y_0$ is a (γ, γ_0)-realization of the "abstract" function $f :\subseteq M \to M_0$.

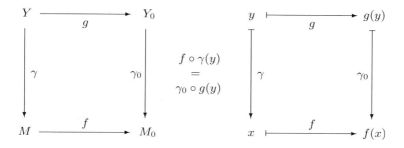

Fig. 3.1. $g :\subseteq Y \to Y_0$ is a (γ, γ_0)-realization of $f :\subseteq M \to M_0$

The condition

$$f \circ \gamma(y) = \gamma_0 \circ g(y) \quad \text{for all} \quad y \in Y \quad \text{such that} \quad f \circ \gamma(y) \quad \text{exists}$$

means that g maps any γ-name y of some $x \in \text{dom}(f)$ to some γ_0-name of $f(x)$. Notice that g may behave arbitrarily on all other arguments $y \in Y \setminus \text{dom}(f \circ \gamma)$. Consequently, g is a (γ, γ_0)-realization also of every restriction of the function f. In contrast to partial recursive functions and functions from F^{ab}, a (γ, γ_0)-computable function f has no "natural" domain (Exercise 3.1.5). Definition 3.1.3.3 can be considered as the special case of Definition 3.1.3.4, where $F(y)$ has exactly one element for all $y \in \text{dom}(F)$.

A (γ, γ_0)-realization g of a function f must be "(γ, γ_0)-extensional" on $\text{dom}(f \circ \gamma)$, i.e,

$$\gamma(y) = \gamma(y') \in \text{dom}(f) \implies \gamma_0 \circ g(y) = \gamma_0 \circ g(y') . \qquad (3.1)$$

Many-valued partial functions can be used as models, if it suffices to determine only a single element from a set of possible ones. For a (γ, γ_0)-realization g of a multi-valued function F, instead of (3.1) only the weaker property

$$\gamma(y) = \gamma(y') = x \in \text{dom}(f) \implies \gamma_0 \circ g(y), \gamma_0 \circ g(y') \in F[\{x\}]$$

is necessary (the function g may map equivalent names to non-equivalent ones).

By the axiom of choice, every multi-valued function has a choice function. In our context, we are interested in (γ, γ_0)-continuous or (γ, γ_0)-computable choice functions. If a multi-valued function has a (γ, γ_0)-computable choice function, then it is (γ, γ_0)-computable, but we will see that it may be (γ, γ_0)-computable without having a (γ, γ_0)-continuous choice function.

Example 3.1.4. Let Σ be a fixed sufficiently large alphabet.
1. The real number $\pi \in \mathbb{R}$ is $\rho_{b,10}$-computable, where $\rho_{b,10} :\subseteq \Sigma^\omega \to \mathbb{R}$ is the representation of the real numbers by infinite decimal fractions. The sequence $\mathbf{3.14159}\ldots \in \Sigma^\omega$ is the computable unique $\rho_{b,10}$-name of $\pi \in \mathbb{R}$.
2. A tuple $(x_1, \ldots, x_k)$ is $(\gamma_1, \ldots, \gamma_k)$-computable, iff x_i is γ_i-computable for $1 \leq i \leq k$.
3. A sequence on M is a function $f : \mathbb{N} \to M$ usually written as $(f_n)_{n \in \mathbb{N}}$. A sequence $(f_n)_{n \in \mathbb{N}}$ will be called γ-computable, iff it is a $(\nu_\mathbb{N}, \gamma)$-computable function.
4. Let ν and ν_1 be notations of M and let γ be a naming system of M'. Then
   a) every element of M is ν-computable (Definition 2.1.3);
   b) every subset $X \subseteq M$ is ν-clopen, in particular, the final topology of ν is the discrete topology on M;
   c) $\nu \equiv_t \nu'$;
   d) every function $f :\subseteq M \to M$ is (ν, γ)-continuous.

5. The sets $\{0\}$, $\mathbb{N}$, $\mathbb{Z}$ and $\{a \in \mathbb{Q} \mid a > 0\}$ are $\nu_{\mathbb{Q}}$-decidable, the set $\{(a,b) \in \mathbb{Q}^2 \mid a < b\}$ is $(\nu_{\mathbb{Q}}, \nu_{\mathbb{Q}})$-decidable.

6. A set $A \subseteq \mathbb{N}$ is En-computable, iff it is r.e. , and Cf-computable, iff it is recursive.

7. $\nu_{\mathbb{N}} \leq \nu_{\mathbb{Z}} \leq \nu_{\mathbb{Q}}$: The identity function id : $\Sigma^* \to \Sigma^*$ is computable and translates (Definition 2.3.2) $\nu_{\mathbb{N}}$ to $\nu_{\mathbb{Z}}$. The function $g : \Sigma^* \to \Sigma^*$ defined by $g(w) := $ "$w/1$" is computable and translates $\nu_{\mathbb{Z}}$ to $\nu_{\mathbb{Q}}$.

8. Addition and multiplication are $(\nu_{\mathbb{N}}, \nu_{\mathbb{N}}, \nu_{\mathbb{N}})$-computable, $(\nu_{\mathbb{Z}}, \nu_{\mathbb{Z}}, \nu_{\mathbb{Z}})$-computable and $(\nu_{\mathbb{Q}}, \nu_{\mathbb{Q}}, \nu_{\mathbb{Q}})$-computable. Subtraction is $(\nu_{\mathbb{Z}}, \nu_{\mathbb{Z}}, \nu_{\mathbb{Z}})$-computable and $(\nu_{\mathbb{Q}}, \nu_{\mathbb{Q}}, \nu_{\mathbb{Q}})$-computable. Inversion Inv :$\subseteq \mathbb{Q} \to \mathbb{Q}$ is $(\nu_{\mathbb{Q}}, \nu_{\mathbb{Q}})$-computable:

   There is a Turing machine M which adds natural numbers in binary notation. This means, $+(\nu_{\mathbb{N}}(u), \nu_{\mathbb{N}}(v)) = \nu_{\mathbb{N}} \circ f_M(u, v)$ for all "binary numbers" $u, v \in \mathrm{dom}(\nu_{\mathbb{N}})$, that is, the function f_M is a $(\nu_{\mathbb{N}}, \nu_{\mathbb{N}}, \nu_{\mathbb{N}})$-realization of addition on $\mathbb{N}$.

   There is a computable function $g : \Sigma^* \to \Sigma^*$ with $g($"u/v"$) = $ "v/u" and $g($"$-u/v$"$) = $ "$-v/u$"for all $u, v \in \mathrm{dom}(\nu_{\mathbb{N}})$. Assume that Inv $\circ \nu_{\mathbb{Q}}(w)$ exists. Then Inv $\circ \nu_{\mathbb{Q}}(w) = \nu_{\mathbb{Q}} \circ g(w)$ for all $w \in \nu_{\mathbb{Q}}$ with $\nu_{\mathbb{Q}}(w) \neq 0$. Therefore, g is a $(\nu_{\mathbb{Q}}, \nu_{\mathbb{Q}})$-realization of the function Inv.

   The remaining proofs are left as Exercise 3.1.12.

9. Cf $\leq$ En but En $\not\leq_t$ Cf. (Of course, this implies Cf $\leq_t$ En and En $\not\leq$ Cf): There is a Type-2 machine M which on input $p \in \Sigma^\omega$ for each $n = 0, 1, \ldots$ reads the input symbol $p(n)$ and writes $110^{n+1}11$ on the output tape, if $p(n) = 1$, and writes 1 otherwise. Obviously, f_M translates Cf to En. Assume that some continuous function $f : \Sigma^\omega \to \Sigma^\omega$ translates En to Cf. Then $f(111\ldots) = 000\ldots$. By continuity, there is some n with $f[1^n \Sigma^\omega] \subseteq 0\Sigma^\omega$. But $f(1^n 1101^\omega) = 1000\ldots$ (contradiction).

10. Union and intersection on $2^{\mathbb{N}}$ are both $(\mathrm{En}, \mathrm{En}, \mathrm{En})$-computable and $(\mathrm{Cf}, \mathrm{Cf}, \mathrm{Cf})$-computable. Complementation on $2^{\mathbb{N}}$ is $(\mathrm{Cf}, \mathrm{Cf})$-computable but not $(\mathrm{En}, \mathrm{En})$-continuous:

   There is a Type-2 machine M with 2 input tapes which reads the input tapes and for each $n \in \mathbb{N}$ writes $110^n 11$ on the output tape as soon as it has found $110^n 11$ as a subword on one of the input tapes. If $\mathrm{En}(p) = A$ and $\mathrm{En}(q) = B$ then $\mathrm{En} \circ f_M(p, q) = A \cup B$, that is, $\cup(\mathrm{En}(p), \mathrm{En}(q)) = \mathrm{En} \circ f_M(p, q)$ for all $p, q \in \Sigma^\omega$. Therefore, f_M is an $(\mathrm{En}, \mathrm{En}, \mathrm{En})$-realization of union.

   Suppose that complementation has a continuous $(\mathrm{En}, \mathrm{En})$-realization $h : \Sigma^\omega \to \Sigma^\omega$. Then $\mathbb{N} \setminus \mathrm{En}(p) = \mathrm{En} \circ h(p)$ for all $p \in \Sigma^\omega$. Since $0 \in \mathbb{N} \setminus \mathrm{En}(111\ldots)$, $11011 \lhd h(111\ldots)$. Therefore, there are $n \in \mathbb{N}$ and $u \in \Sigma^*$ with $h[1^n \Sigma^\omega] \subseteq u11011\Sigma^\omega$. But $h(1^n 110111^\omega) \not\subseteq u11011\Sigma^\omega$ (contradiction).

   The remaining proofs are left as Exercise 3.1.14.

11. The Scott topology τ_S on $2^{\mathbb{N}}$ is the topology generated by the base $\beta :=$ $\{O_E \mid E \subseteq \mathbb{N} \text{ finite}\}$, where $O_E := \{A \subseteq \mathbb{N} \mid E \subseteq A\}$. The set β is

closed under intersection, since $O_E \cap O_F = O_{E \cup F}$. Therefore, it is a base of a topology. The set $\sigma := \{O_{\{n\}} \mid n \in \mathbb{N}\}$ is a subbase of τ_S. We show that τ_S is the final topology $\tau_{\mathrm{En}} = \{U \subseteq 2^{\mathbb{N}} \mid \mathrm{En}^{-1}[U] \text{ is open}\}$ of the representation En. For each subbase element $O_{\{n\}}$ of τ_S we have $\mathrm{En}^{-1}[O_{\{n\}}] = \{p \in \Sigma^\omega \mid 110^{n+1}11 \lhd p\}$, an open subset of Σ^ω. Therefore, $\tau_S \subseteq \tau_{\mathrm{En}}$. On the other hand for $w \in \Sigma^*$, $\mathrm{En}[w\Sigma^\omega] = O_E \in \tau_S$, where $E := \{n \in \mathbb{N} \mid 110^{n+1}11 \lhd w\}$. Assume $U \in \tau_{\mathrm{En}}$. Then $\mathrm{En}^{-1}[U] = A\Sigma^\omega$ for some $A \subseteq \Sigma^*$, hence $U = \mathrm{En}[A\Sigma^\omega] = \bigcup_{w \in A} \mathrm{En}[w\Sigma^\omega] \in \tau_S$.

12. The set $\{(A, n) \in 2^{\mathbb{N}} \times \mathbb{N} \mid n \in A\}$ is $(\mathrm{En}, \nu_{\mathbb{N}})$-r.e. , $(\mathrm{Cf}, \nu_{\mathbb{N}})$-decidable but not $(\mathrm{En}, \nu_{\mathbb{N}})$-clopen.

13. Is there a "method" to find some $n \in A$ for each non-empty subset A of $\mathbb{N}$? Our terminology admits precise answers. Consider the multi-valued function $F :\subseteq 2^{\mathbb{N}} \rightrightarrows \mathbb{N}$ such that $\mathrm{R}_F = \{(A, n) \mid n \in A\}$.
   a) F is $(\mathrm{En}, \nu_{\mathbb{N}})$-computable and $(\mathrm{Cf}, \nu_{\mathbb{N}})$-computable,
   b) F has no $(\mathrm{En}, \nu_{\mathbb{N}})$-continuous choice function,
   c) F has a $(\mathrm{Cf}, \nu_{\mathbb{N}})$-computable choice function.
   13a: There is a Type-2 machine which on input $p \in \Sigma^\omega$ prints a binary name u of n and halts, as soon as it has found a subword $1100^n11 \lhd p$. Then $\nu_{\mathbb{N}} \circ f_M(p) \in \mathrm{En}(p)$ whenever $\mathrm{En}(p) \in \mathrm{dom}(F)$. Therefore, f_M is a computable $(\mathrm{En}, \nu_{\mathbb{N}})$-realization of F. (The argument is similar for Cf.)
   13b: Assume that $f :\subseteq 2^{\mathbb{N}} \to \mathbb{N}$ is an $(\mathrm{En}, \nu_{\mathbb{N}})$-continuous choice function of F with a continuous realization $h :\subseteq \Sigma^\omega \to \Sigma^*$. Then $h(110111^\omega) = u$ with $\nu_{\mathbb{N}}(u) = 0$ and $h(1100111^\omega) = v$ with $\nu_{\mathbb{N}}(v) = 1$. By continuity of h there are numbers i, j such that $h[110111^i \Sigma^\omega] = \{u\}$ and $h[1100111^j \Sigma^\omega] = \{v\}$. The sequences $p := 110111^i 1100111^\omega$ and $q := 1100111^j 110111^\omega$ are En-names of $\{0, 1\}$. Then $h(p) = u$ and $h(q) = v$, therefore, $f(\{0, 1\}) = f \circ \mathrm{En}(p) = \nu_{\mathbb{N}} \circ h(p) = 0$ and $f(\{0, 1\}) = f \circ \mathrm{En}(q) = \nu_{\mathbb{N}} \circ h(q) = 1$ (contradiction).
   13c: The function $f :\subseteq 2^{\mathbb{N}} \to \mathbb{N}$, defined by $f(\emptyset) := \mathrm{div}$ and $f(A) := \min(A)$, if $A \neq \emptyset$, is a $(\mathrm{Cf}, \nu_{\mathbb{N}})$-computable choice function of F. $\quad\square$

The next lemma expresses reducibility by induced computability and vice versa.

Lemma 3.1.5 (reducibility versus computability). Consider naming systems $\gamma :\subseteq Y \to M$ and $\gamma_0 :\subseteq Y_0 \to M_0$.
   1. $\gamma \leq \gamma_0$ ($\gamma \leq_t \gamma_0$), iff $M \subseteq M_0$ and the identical embedding id_{MM_0} is (γ, γ_0)-computable (-continuous).
   2. A function $f :\subseteq M \to M_0$ is (γ, γ_0)-computable (-continuous), iff $f \circ \gamma \leq \gamma_0$ ($f \circ \gamma \leq_t \gamma_0$).

The easy proof is left as Exercise 3.1.1. Computable functions map computable elements to computable elements and are closed under composition.

Theorem 3.1.6 (composition).   Let $f :\subseteq M_1 \times \ldots \times M_n \to M_0$, $g_i :\subseteq N_1 \times \ldots \times N_k \to M_i$ $(i = 1, \ldots, n)$ be functions, and let $\gamma_i :\subseteq Y_i \to M_i$ and $\gamma'_j :\subseteq Y'_j \to N_j$ be naming systems.
1. If f is $(\gamma_1, \ldots, \gamma_n, \gamma_0)$-computable and $x \in \mathrm{dom}(f)$ is $(\gamma_1, \ldots, \gamma_n)$-computable, then $f(x)$ is γ_0-computable.
2. If f is $(\gamma_1, \ldots, \gamma_n, \gamma_0)$-computable and g_i is $(\gamma'_1, \ldots, \gamma'_k, \gamma_i)$-computable $(i = 1, \ldots, n)$, then $f \circ (g_1, \ldots, g_k)$ is $(\gamma'_1, \ldots, \gamma'_k, \gamma_0)$-computable.

Part 2 holds accordingly for multi-valued functions.

The proof is left as Exercise 3.1.2. The computable functions are also closed under primitive recursion.

Theorem 3.1.7 (primitive recursion, iteration).   Let $\gamma :\subseteq Y \to M$ and $\gamma' :\subseteq Y \to M'$ be naming systems.
1. Let $f :\subseteq M \to M'$ be (γ, γ')-computable and let $f_a :\subseteq \Sigma^* \times M' \times M \to M'$ be $(\mathrm{id}_{\Sigma^*}, \gamma', \gamma, \gamma')$-computable for each $a \in \Sigma$. Define $g :\subseteq \Sigma^* \times M \to M'$ by

$$g(\lambda, x) = f(x) , \tag{3.2}$$
$$g(aw, x) = f_a(w, g(w, x), x) \tag{3.3}$$

for all $x \in M$, $a \in \Sigma$ and $w \in \Sigma^*$.
Then the function g is $(\mathrm{id}_{\Sigma^*}, \gamma, \gamma')$-computable.
2. Let $f :\subseteq M \to M'$ be (γ, γ')-computable and let $f' :\subseteq \mathbb{N} \times M' \times M \to M'$ be $(\nu_\mathbb{N}, \gamma', \gamma, \gamma')$-computable. Define $g' :\subseteq \mathbb{N} \times M \to M'$ by

$$g'(0, x) = f(x) ,$$
$$g'(n + 1, x) = f'(n, g'(n, x), x)$$

for all $x \in M$ and $n \in \mathbb{N}$.
Then the function g is $(\nu_\mathbb{N}, \gamma, \gamma')$-computable.
3. Let $h :\subseteq M \to M$ be (γ, γ)-computable. Then $H :\subseteq \mathbb{N} \times M \to M$, defined by

$$H(0, x) = x, \quad H(n + 1, x) = h \circ H(n, x)$$

(that is, $H(n, x) := h^n(x)$) is $(\nu_\mathbb{N}, \gamma, \gamma)$-computable.

Proof: 1. Let $\bar{f}$ be a computable realization of f and let $\bar{f}_a$ be a computable realization of f_a for all $a \in \Sigma$. By Theorem 2.1.14 there is a computable function $\bar{g}$ with

$$\bar{g}(\lambda, x) = \bar{f}(x) ,$$
$$\bar{g}(aw, x) = \bar{f}_a(w, \bar{g}(w, x), x) .$$

A simple induction shows that $\bar{g}$ realizes g.

2. Define $\nu(w) := |w|$. Since $\nu \equiv \nu_\mathbb{N}$, it suffices to prove the statement for ν instead of $\nu_\mathbb{N}$. For each $a \in \Sigma$ define $f_a(w, y, x) := f'(|w|, y, x)$ and define g from f and the f_a by primitive recursion according to 1 above. Since f' is $(\nu, \gamma', \gamma, \gamma')$-computable, each f_a is $(\mathrm{id}_{\Sigma^*}, \gamma', \gamma, \gamma')$-computable. Therefore, g is $(\mathrm{id}_{\Sigma^*}, \gamma, \gamma')$-computable by Property 1. An inductive proof shows $g'(|w|, x) = g(w, x)$ for all w, x. Since g is $(\mathrm{id}_{\Sigma^*}, \gamma, \gamma')$-computable, g' is (ν, γ, γ')-computable.

3. Consider the following special case of Part 2: $f(x) := x$, $f'(w, z, x) := h(z)$. Then $g'(n, x) = h^n(x) = H(n, x)$, and so H is $(\nu_\mathbb{N}, \gamma, \gamma)$-computable. $\square$

The next theorem describes heredity of the induced effectivity concepts under reduction.

Theorem 3.1.8 (heredity under reduction).   Let $\gamma :\subseteq Y \to M$, $\gamma' :\subseteq Y' \to M$, $\gamma_0 :\subseteq Y_0 \to M_0$ and $\gamma_0' :\subseteq Y_0' \to M_0$ be naming systems. Assume $x \in M$, $X \subseteq M$, $f :\subseteq M \to M_0$ and $F :\subseteq M \rightrightarrows M_0$.

1. If x is γ'-computable and $\gamma' \leq \gamma$, then x is γ-computable.
2. If X is γ-open (-clopen) and $\gamma' \leq_t \gamma$, then X is γ'-open (clopen), therefore, $\gamma' \leq_t \gamma \implies \tau_\gamma \subseteq \tau_{\gamma'}$.
3. If X is γ-r.e. (-decidable) and $\gamma' \leq \gamma$, then X is γ'-r.e. (-decidable).
4. If the function f is (γ, γ_0)-continuous, $\gamma' \leq_t \gamma$ and $\gamma_0 \leq_t \gamma_0'$, then f is (γ', γ_0')-continuous. (Accordingly for the function F)
5. If the function f is (γ, γ_0)-computable, $\gamma' \leq \gamma$ and $\gamma_0 \leq \gamma_0'$, then f is (γ', γ_0')-computable. (Accordingly for the function F)

The properties hold correspondingly for the general case $M = M_1 \times \ldots \times M_k$ with naming systems $\gamma_i :\subseteq Y_i \to M_i$ and $\gamma_i' :\subseteq Y_i' \to M_i$ $(i = 1, \ldots, k)$.

Proof: 1. Let $f :\subseteq Y' \to Y$ be a computable translation from γ' to γ. Let y be a computable γ'-name of x. Then $f(y)$ is a computable γ-name of x.

2. Let X be γ-open and let $f :\subseteq Y' \to Y$ be a continuous translation from γ' to γ. Then $\gamma^{-1}[X]$ is open in $\mathrm{dom}(\gamma)$. By Theorem 2.4.5 $f^{-1}[\gamma^{-1}[X]]$ is open in $f^{-1}[\mathrm{dom}(\gamma)]$. Then $f^{-1}[\gamma^{-1}[X]] \cap \mathrm{dom}(\gamma') = \gamma'^{-1}[X]$ is open in $f^{-1}[\mathrm{dom}(\gamma)] \cap \mathrm{dom}(\gamma') = \mathrm{dom}(\gamma')$. Therefore, X is γ'-open. Let X be γ-clopen. Then X and $M \setminus X$ are γ-open and hence γ'-open. Therefore, X is γ'-clopen.

3. Replace in the proof of Property 2 "open", "clopen" and "continuous" by "r.e.", "decidable" and "computable", respectively.

4. Let h translate γ' to γ, let g be a (γ, γ_0)-realization of f and let h_0 translate γ_0 to γ_0'. Then $h_0 \circ g \circ h$ is a (γ', γ_0')-realization of f. If h, g, h_0 are continuous, then $h_0 \circ g \circ h$ is continuous. For the multi-valued function F the argument applies as well.

5. The proof resembles that of Property 4. In this case notice that $h_0 \circ g \circ h$ has a computable extension, if h, g, h_0 are computable. $\square$

From Property 2 it follows immediately that naming systems γ and γ' have the same final topology, $\tau_\gamma = \tau_{\gamma'}$, if they are continuously equivalent, that is, $\gamma \equiv_t \gamma'$. We conclude with an important corollary characterizing the variety of computability concepts which can be defined on a set by means of naming systems.

Corollary 3.1.9.
1. Naming systems $\gamma :\subseteq Y \to M$ and $\gamma' :\subseteq Y' \to M$ induce the same computability concepts on the set M, iff they are equivalent.
2. Naming systems $\gamma :\subseteq Y \to M$ and $\gamma' :\subseteq Y' \to M$ induce the same topological concepts on the set M, iff they are continuously equivalent.

Proof: If $\gamma \equiv \gamma'$, then by Theorem 3.1.8 γ and γ' induce the same computability concept on M. Suppose $\gamma \not\leq \gamma'$. Then the identity on M is (γ, γ)-computable but not (γ, γ')-computable, that is, γ and γ' induce different computability concepts on M. The topological case can be proved correspondingly.   □

We have reduced computability on countable sets M to computability on the set Σ^* of words by means of notations. Equivalently it can be reduced to computability on the set $\mathbb{N}$ of natural numbers by means of numberings. The connection is given by the bijection $\nu \mapsto \nu \circ \nu_\Sigma$ from notations to numberings (Exercise 3.1.18). The theory of numberings has been investigated in detail [Mal71, Erš73, Erš75, Erš77].

Exercises 3.1.

◇ 1. Prove Lemma 3.1.5.
◇ 2. Prove Theorem 3.1.6.
◇ 3. Prove Example 3.1.4.4.
   4. Show that for any naming system $\gamma :\subseteq Y \to M$ a set $X \subseteq M$ is γ-decidable, iff its characteristic function $\mathrm{cf}_X : M \to \mathbb{N}$ is $(\gamma, \nu_\mathbb{N})$-computable.
   5. There is a more restricted, stronger kind of realization of functions or multi-valued functions f. A (γ, γ_0)-realization g of a function f is called strong, iff $g(y) = \mathrm{div}$ for all $y \in \mathrm{dom}(\gamma)$ with $\gamma(y) \notin \mathrm{dom}(f)$. This "strong" (γ, γ_0)-computability and -continuity has been considered, for example in [KW85, Wei87, Wei93]. Prove Theorem 3.1.8.4 and 5 for strong computability in the case that γ and γ_0 are representations.
   6. Consider Definition 3.1.3. Call a function $g :\subseteq Y \to Y_0$ (γ, γ_0)-extensional in $x \in M$, iff

$$\gamma(p) = \gamma(q) = x \implies \gamma_0 \circ g(p) = \gamma_0 \circ g(q) \in M_0 .$$

Define $f :\subseteq M \to M_0$ such that $\mathrm{dom}(f)$ is the set of all $x \in M$ in which the function g is (γ, γ_0)-extensional and $f(x) := \gamma_0 \circ g(p)$ for some

$p \in \gamma^{-1}[\{x\}]$.

Show that the function g is a (γ, γ_0)-realization of the function f, and g is a (γ, γ_0)-realization of a function f', iff f' is a restriction of f.

Not much is known about such "maximal" functions f for naming systems γ and γ_0 realized by functions $g \in F^{ab}$ in general.

7. Sometimes it is reasonable to use a finite or infinite sequence of symbols as a name of *many* elements of a set M. We formalize this situation by a *set-valued* naming system $\gamma : \Sigma^a \to 2^M$ (notice that $\text{dom}(\gamma) = \Sigma^a$). Informally, $p \in \Sigma^a$ serves as a name of every $x \in \gamma(p)$. If $\text{card}(\gamma(p)) \leq 1$ for every $p \in \Sigma^a$, then γ can be identified with an ordinary naming system $\bar{\gamma} :\subseteq \Sigma^a \to M$ defined by $\bar{\gamma}(p) = x \iff x \in \gamma(p)$. Continuous and computable reduction and equivalence are defined by means of translation according to Definition 2.3.2. A translation from $\gamma : \Sigma^a \to 2^M$ to $\gamma' : \Sigma^b \to 2^{M'}$ is a function $f :\subseteq \Sigma^a \to \Sigma^b$ such that $x \in \gamma(p) \implies x \in \gamma' \circ f(p)$ for all p, x, that is, $\gamma(p) \subseteq \gamma' \circ f(p)$ for all p, which is true by convention, if $\gamma(p) = \emptyset$ and $f(p) = \text{div}$. Relative computability and continuity of functions is defined via realizations according to Definition 3.1.3.3. A function $g :\subseteq \Sigma^a \to \Sigma^b$ is called a (γ, γ')-realization of a function $f :\subseteq M \to M'$, iff $x \in \gamma(p) \cap \text{dom}(f) \implies f(x) \in \gamma' \circ g(p)$ for all p, x, that is, $f[\gamma(p)] \subseteq \gamma' \circ g(p)$ for all p. (Set-valued representations are used in [Wei93]).

As an example consider composition $g \circ f$ for $f \in F^{*\omega}$ and $g \in F^{\omega*}$. If g and f are computable, then $g \circ f$ has a computable extension (Theorem 2.1.12) which is not computable in general (Exercise 2.1.9), and so computability of composition cannot be expressed by means of Definition 3.1.3. For $a, b \in \{*, \omega\}$ define a set-valued representation by

$$\alpha^{ab}(p) := \{f :\subseteq \Sigma^a \to \Sigma^b \mid \eta_p^{ab} \text{ extends } f\} .$$

Show that composition $f, g \mapsto g \circ f$ for *continuous* $f :\subseteq \Sigma^a \to \Sigma^b$ and *continuous* $g :\subseteq \Sigma^b \to \Sigma^c$ is $(\alpha^{ab}, \alpha^{bc}, \alpha^{ac})$-computable. (This means that there is a computable function $h :\subseteq \Sigma^\omega \to \Sigma^\omega$ such that $\eta_{h(p,q)}^{ac}$ extends $g \circ f$, if η_p^{ab} extends f and $\eta_q^{c,d}$ extends g.)

◇ 8. Show that division by 3 is $(\rho_{b,10}, \rho_{b,10})$-computable but multiplication by 3 is not even $(\rho_{b,10}, \rho_{b,10})$-continuous, where $\rho_{b,10} :\subseteq \Sigma^\omega \to \mathbb{R}$ is the representation of the real numbers by infinite decimal fractions.

9. Show that
   ◇ a) every notation has an equivalent representation,
   ♦ b) there is a total representation of $\{0, 1\}$ which has no equivalent notation. (Hint: cardinality)

10. Define $\nu : \Sigma^* \to \mathbb{N}$ by $\nu(w) := |w|$. Show: $\nu \equiv \nu_\mathbb{N}$.

11. $\nu_\mathbb{Q}|^\mathbb{N} \leq \nu_\mathbb{Z}|^\mathbb{N} \leq \nu_\mathbb{N}$ (Sect. 1.4).

12. Prove the remaining statements from Example 3.1.4.8 .

13. Prove the remaining statements from Example 3.1.4.10 .

14. Define a representation $\delta : \{0,1\}^\omega \to \Sigma^\omega$ such that for any representations $\gamma :\subseteq \Sigma^\omega \to M$ and $\gamma_0 :\subseteq \Sigma^\omega \to M_0$ a function $f :\subseteq M \to M_0$ is (γ, γ_0)-computable, iff it is $(\gamma \circ \delta, \gamma_0 \circ \delta)$-computable.

15. Define a representation $\mathrm{En}^c : \Sigma^\omega \to 2^{\mathbb{N}}$ by $\mathrm{En}^c(p) := \mathbb{N} \setminus \mathrm{En}(p)$. Show that for every representation δ of $2^{\mathbb{N}}$:

$$(\delta \le \mathrm{En} \text{ and } \delta \le \mathrm{En}^c) \iff \delta \le \mathrm{Cf} .$$

$$(\delta \le_t \mathrm{En} \text{ and } \delta \le_t \mathrm{En}^c) \iff \delta \le_t \mathrm{Cf} .$$

Therefore, Cf is the greatest lower bound of En and En^c w.r.t. computable and to continuous reducibility.

16. Show that $\sigma := \{C_n \mid n \in \mathbb{N}\} \cup \{D_n \mid n \in \mathbb{N}\}$ is a subbase of the final topology of the representation $\mathrm{Cf} : \Sigma^\omega \to 2^{\mathbb{N}}$, where $C_n := \{A \subseteq \mathbb{N} \mid n \in A\}$ and $D_n := \{A \subseteq \mathbb{N} \mid n \notin A\}$.

17. Let $\rho_{\mathrm{b},10} :\subseteq \Sigma^\omega \to \mathbb{R}$ be the decimal representation of the set of real numbers. Show: $X \subseteq \mathbb{R}$ is open, iff it is $\rho_{\mathrm{b},10}$-open.

18. A numbering $\nu :\subseteq \mathbb{N} \to M$ induces a computability theory on the set M (modify Definition 3.1.3 accordingly). Let $\nu_\Sigma : \mathbb{N} \to \Sigma^*$ be the bijective standard numbering of Σ^* (Sect. 1.4). Show that for any notation $\nu :\subseteq \Sigma^* \to M$, ν and the associated numbering $\nu \circ \nu_\Sigma$ induce the same computability concept on the set M.

19. Show that every partial $(\mathrm{En}, \mathrm{En})$-computable function $f :\subseteq 2^{\mathbb{N}} \to 2^{\mathbb{N}}$ has a total $(\mathrm{En}, \mathrm{En})$-computable extension $f' : 2^{\mathbb{N}} \to 2^{\mathbb{N}}$.

20. (See [Wei87] Theorem 3.3.17)
    a) There is a representation δ with $\delta \equiv \mathrm{En}$ such that $\delta^{-1}[\emptyset]$ has one element, $\delta^{-1}[A]$ is countable, if A is finite, and $\delta^{-1}[A]$ is uncountable, if A is infinite.
 ◆ b) For every representation δ with $\delta \equiv \mathrm{En}$, $\delta^{-1}[A]$ is at least countably infinite, if A is non-empty and finite, and $\delta^{-1}[A]$ is uncountable, if A is infinite.

◇21. Let f be a $(\delta_1, \delta_2, \delta_3)$-computable function and let c be a δ_2-computable point. Then the function g, defined by $g(x) := f(x,c)$, is (δ_1, δ_3)-computable.

22. Let $\delta :\subseteq \Sigma^\omega \to M$ be a representation. Show that the set M must be countable, if the diagonal $\{(x,x) \mid x \in M\}$ is (δ, δ)-open (cf. Theorem 4.1.16).

23. Consider Definition 3.1.3. Show that a *total* function $f : M \to M_0$ is $(\tau_\gamma, \tau_{\gamma_0})$-continuous, if it is (γ, γ_0)-continuous [Wie80] (also [Eng89] Exercise 2.4.F).

24. Define a multi-valued function $F :\subseteq \mathrm{P}^{**} \rightrightarrows \Sigma^*$ by
    $\mathrm{R}_F := \{(f,w) \mid w \in \mathrm{dom}(f)\}$.
    a) Show that F is $(\xi^{**}, \mathrm{id}_{\Sigma^*})$-computable.
 ◆ b) Show that F has no $(\xi^{**}, \mathrm{id}_{\Sigma^*})$-computable choice function.

3.2 Admissible Naming Systems

Let M be a countable set, for example the set of integer numbers, the set of polynomials with rational coefficients, the set of finite graphs with non-negative integer vertices or the set $P^{\omega\omega}$ of computable functions $f :\subseteq \Sigma^\omega \to \Sigma^\omega$. Usually, computability on such a set is defined via a notation $\nu :\subseteq \Sigma^* \to M$. As we have shown (Corollary 3.1.9), two notations $\nu, \nu' :\subseteq \Sigma^* \to M$ induce the same kind of computability on the set M, iff $\nu \equiv \nu'$. Since the set M has more than countably many total notations (provided it has at least two elements) and since there are only countably many computable word functions for reduction, there are still more than countably many different kinds of computability on the set M that can be defined via notations.

For commonly used sets M only a single computability concept is of practical interest, which is defined by a so-called "effective" notation. Our notations $\nu_\mathbb{N}, \nu_\mathbb{Z}$ and $\nu_\mathbb{Q}$ (Definition 3.1.2) of the natural, the integer and the rational numbers, respectively, and the notations ξ^{ab} of the sets P^{ab} (Definition 2.3.4) are examples of such "effective" notations. In most applications, "ad hoc" definitions of notations of a set M turn out to be equivalent, and the meaning of the word "effective" remains unexplained. Therefore, an independent characterization of the computability theory (that is, the equivalence class of the "effective" notation) by "natural" axioms derived from a structure on the set M is useful. We will give some examples in Exercises 3.2.2, 3.2.3 and 3.2.4.

An interesting example is the notation ξ^{**} of the set P^{**} of the computable functions $f :\subseteq \Sigma^* \to \Sigma^*$. The equivalence theorem by Rogers [Rog67] states that a notation ν is equivalent to ξ^{**}, iff it satisfies the utm- and the smn-theorem (Theorem 2.3.14). Therefore, the "effectivity" of ξ^{**} and the induced computability can be characterized by the utm- and the smn-property. Since ξ^{**} is $\leq$-complete in the set of all notations ν of P^{**} which satisfy the utm-theorem (Theorem 2.3.14), the utm-property and the completeness property characterize the intended computability theory on the set P^{**}. The set $\{A \subseteq \Sigma^* \mid A \text{ is recursive}\}$ is an example of a set which has at least two non-equivalent important ("effective") notations: $U(w) = A :\iff \xi_w^{**}$ is the characteristic function of A and $V(w) := \text{dom}(\xi_w^{**})$.

For representations $\delta :\subseteq \Sigma^\omega \to M$ similar arguments show that the set of $\equiv_t$-classes as well as the set of $\equiv$-classes (and therefore, the sets of continuity theories and of computability theories) on M are larger than the continuum (provided the set M has at least two elements). Also in this case only very few of them are of practical interest. Indeed, many "ad hoc" definitions of representations have proved to be not very useful. An example is the decimal representation $\rho_{b,10}$ of the real numbers (which traditionally is considered to be very natural) since by Example 2.1.4.7 not even multiplication by 3 is $(\rho_{b,10}, \rho_{b,10})$-computable. For this reason we will discuss each representation we will use in detail and put emphasis on characterizing the induced computability theory.

We will now introduce a class of very natural representations and computability theories. In Sect. 1.3.1 we have considered a machine model, where infinite sequences $(I_0, I_1, \ldots)$ of pieces of finite information are used as names of "abstract objects" and machines transform such infinite sequences to infinite sequences (Fig. 3.2).

Fig. 3.2. A machine transforming infinite sequences

For the case of real numbers we have used intervals $I = [a; b]$ with rational endpoints as pieces of finite information. We have assumed tacitly that these intervals are encoded appropriately, such that the concrete machine can read them. We generalize this concept as follows.

Definition 3.2.1 (effective/computable topological space).

1. *An effective topological space is a triple* $\mathbf{S} = (M, \sigma, \nu)$ *where* M *is a non-empty set,* $\sigma \subseteq 2^M$ *is a countable system of subsets of* M *such that*

$$x = y \quad if \quad \{A \in \sigma \mid x \in A\} = \{A \in \sigma \mid y \in A\}, \qquad (3.4)$$

   and $\nu :\subseteq \Sigma^* \to \sigma$ *is a notation of* σ.
2. *Let* $\tau_{\mathbf{S}}$ *be the topology on* M *generated by* σ *as a subbase .*
3. *A computable topological space is an effective topological space for which the equivalence problem*

$$\{(u, v) \mid u, v \in \mathrm{dom}(\nu) \text{ and } \nu(u) = \nu(v)\} \quad is \quad r.e. . \qquad (3.5)$$

We will call the elements $A \in \sigma$ *atomic properties*. Notice that for a computable topological space the numbering ν must have an r.e. domain. By (3.4) any two different elements of M can be distinguished by their atomic properties. In other words, every element of the set M can be identified by its atomic properties. The set β of finite intersections $I_1 \cap \ldots \cap I_k$ of atomic properties is a countable base of the topology $\tau_{\mathbf{S}}$, that is, $\tau_{\mathbf{S}} = \{\bigcup \alpha \mid \alpha \subseteq \beta\}$ where $\beta = \{\bigcap \gamma \mid \gamma \subseteq \sigma, \gamma \text{ finite}\}$. A topological space (M, τ), for which each element $x \in M$ can be identified by the set of its neighborhoods, is called a T_0-space. A T_0-space with countable base is called a *second countable* T_0-space [Eng89]. In the example of Fig. 3.3, $\{x\} = \bigcap \{A \in \sigma \mid x \in A\}$, but this is not true in general (Exercise 3.2.5).

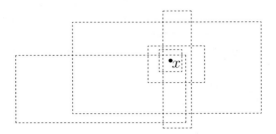

Fig. 3.3. Some properties $A \in \sigma$ of $x \in \mathbb{R}^2$, where $\sigma \subseteq 2^{\mathbb{R}^2}$ is the set of all open rectangular boxes with rational vertices

With each effective topological space we associate a representation:

Definition 3.2.2 (standard representation).
Let $\mathbf{S} = (M, \sigma, \nu)$ be an effective topological space. Define the standard representation $\delta_{\mathbf{S}} :\subseteq \Sigma^\omega \to M$ of $\mathbf{S}$ by

$$(p \in \mathrm{dom}(\delta_{\mathbf{S}}) \ and \ \iota(w) \lhd p) \Longrightarrow w \in \mathrm{dom}(\nu) \qquad (3.6)$$

$$\delta_{\mathbf{S}}(p) = x : \Longleftrightarrow \ \{A \in \sigma \mid x \in A\} = \{\nu(w) \mid \iota(w) \lhd p\} \qquad (3.7)$$

for all $w \in \Sigma^*$, $x \in M$ and $p \in \Sigma^\omega$ (Definition 2.1.7 and Sect. 1.4).

A $\delta_{\mathbf{S}}$-name p of an element $x \in M$ is a list of ν-names of *all* atomic properties $A \in \sigma$ of x. More precisely, p has subwords $\iota(w)$ (Definition 2.1.7) such that $x \in \nu(w)$. Informally, we will say that p is a list of all properties $A \in \sigma$ such that $x \in A$. Since x can be identified by its atomic properties, the representation $\delta_{\mathbf{S}}$ is well-defined. In general, any $x \in M$ has many $\delta_{\mathbf{S}}$-names, since the ν-names can be listed in any order. The set $\{A \in \sigma \mid x \in A\}$ can be finite or even empty. In such a case a $\delta_{\mathbf{S}}$-name $p \in \Sigma^\omega$ contains many repetitions or dummy regions. A $\delta_{\mathbf{S}}$-name p of an element $x \in M$ cannot contain subwords $\iota(w)$ such that $w \notin \mathrm{dom}(\nu)$. However, there can be words $w \in \mathrm{dom}(\nu)$ such that $x \in \nu(w)$ and $\iota(w)$ is not a subword of p ("incomplete names"). The representation $\delta_{\mathbf{S}}$ can be restricted to the set of complete names.

Lemma 3.2.3 (complete names). For an effective topological space $\mathbf{S} = (M, \sigma, \nu)$ define a representation $\delta_{\mathbf{S}}' :\subseteq \Sigma^\omega \to M$ by

$$\delta_{\mathbf{S}}'(p) = x : \Longleftrightarrow \ \{w \mid x \in \nu(w)\} = \{w \mid \iota(w) \lhd p\} = \mathrm{En}_\Sigma(p)$$

for all $x \in M$ and $p \in \Sigma^\omega$. Then $\delta_{\mathbf{S}}' \equiv_t \delta_{\mathbf{S}}$, and $\delta_{\mathbf{S}}' \equiv \delta_{\mathbf{S}}$, if $\mathbf{S}$ is computable.

Proof: See Exercise 3.2.16.   □

Therefore, for a computable topological space $\mathbf{S}$ the restriction of the standard representation $\delta_{\mathbf{S}}$ to the complete names is equivalent to $\delta_{\mathbf{S}}$. Most effective topological spaces of practical interest are computable.

If $\mathbf{S} = (M, \sigma, \nu)$ and $\mathbf{S}' = (M, \sigma, \nu')$ are computable topological spaces such that $\nu \equiv \nu'$, then $\delta'_{\mathbf{S}} \equiv \delta_{\mathbf{S}}$. Notice that for every notation ν satisfying (3.5) there is an equivalent one with recursive domain satisfying (3.5) (Exercises 3.2.14 and 15).

Example 3.2.4 (some computable topological spaces).

1. Let μ be a notation of a set M. Define an effective topological space $\mathbf{S} = (M, \sigma, \nu)$ by $\nu(w) := \{\mu(w)\}$. Then $\delta_{\mathbf{S}} \equiv_t \mu$, and $\tau_{\mathbf{S}} = \tau_\mu$ is the discrete topology on M (Theorem 3.1.9). If the equivalence problem $\{(u, v) \mid u, v \in \mathrm{dom}(\mu),\ \mu(u) = \mu(v)\}$ is r.e. , then $\mathbf{S}$ is a computable topological space and $\delta_{\mathbf{S}} \equiv \mu$.

2. Define a computable topological space $\mathbf{S} = (\mathbb{R}, \sigma, \nu)$ by
   $\sigma := \{(a; b) \subseteq \mathbb{R} \mid a, b \in \mathbb{Q} \text{ and } a < b\}$,
   $\mathrm{dom}(\nu) := \{\langle u, v \rangle \in \Sigma^* \mid u, v \in \mathrm{dom}(\nu_\mathbb{Q}) \text{ and } \nu_\mathbb{Q}(u) < \nu_\mathbb{Q}(v)\}$ and
   $\nu(\langle u, v \rangle) := (\nu_\mathbb{Q}(u); \nu_\mathbb{Q}(v)) \subseteq \mathbb{R}$. Then the topology $\tau_{\mathbf{S}}$ on $\mathbb{R}$ has the set of open intervals with rational endpoints as a subbase. Since σ is closed under intersection, it is a base, and obviously $\tau_{\mathbf{S}}$ is the set of all open subsets of the real numbers with "open" in the usual meaning. A $\delta_{\mathbf{S}}$-name p of a real number x is a list of all open intervals I with rational endpoints (more precisely, a list of ν-names of all open intervals I with rational endpoints) such that $x \in I$. Remember that in Sect. 1.3 a name of a real number was a sequence $(I_0, I_1, I_2, \ldots)$ of closed intervals with rational endpoints such that $I_{n+1} \subseteq I_n$ and $\{x\} = \bigcap_{n \in \mathbb{N}} I_n$. We can deduce from Lemma 4.1.6 that these names can be translated to $\delta_{\mathbf{S}}$-names and vice versa by machines, hence both naming systems induce the same computability theory on $\mathbb{R}$.

3. Define a computable topological space $\mathbf{S} = (\Sigma^\omega, \sigma, \nu)$ by $\nu(w) := w\Sigma^\omega$. The representation $\delta_{\mathbf{S}}$ is equivalent to the identity $\mathrm{id}_{\Sigma^\omega} : \Sigma^\omega \to \Sigma^\omega$ and $\tau_{\mathbf{S}}$ is the Cantor topology on Σ^ω.

4. Define a computable topological space $\mathbf{S} = (\Sigma^\omega, \sigma, \nu)$ by $\mathrm{dom}(\nu) := \{1^i a \mid i \in \mathbb{N}, a \in \Sigma\}$ and $\nu(1^i a) := \{p \in \Sigma^\omega \mid p(i) = a\}$. The representation $\delta_{\mathbf{S}}$ is equivalent to the identity $\mathrm{id}_{\Sigma^\omega} : \Sigma^\omega \to \Sigma^\omega$ and $\tau_{\mathbf{S}}$ is the Cantor topology on Σ^ω.

5. Define a computable topological space $\mathbf{S}_+ = (2^\mathbb{N}, \sigma, \nu)$ by $\mathrm{dom}(\nu) := \{0^n \mid n \in \mathbb{N}\}$ and $\nu(0^n) := \{A \subseteq \mathbb{N} \mid n \in A\}$. Then $\delta_{\mathbf{S}} \equiv \mathrm{En}$ and $\tau_{\mathbf{S}}$ is the Scott topology on $2^\mathbb{N}$ (Definition 3.1.2, Example 3.1.4.11).

6. Define a computable topological space $\mathbf{S}_0 = (2^\mathbb{N}, \sigma, \nu)$ by $\mathrm{dom}(\nu) := \{010^n, 0010^n \mid n \in \mathbb{N}\}$, $\nu(010^n) := \{A \subseteq \mathbb{N} \mid n \in A\}$ and $\nu(0010^n) := \{A \subseteq \mathbb{N} \mid n \notin A\}$. Then $\delta_{\mathbf{S}} \equiv \mathrm{Cf}$ (Definition 3.1.2) and $\tau_{\mathbf{S}}$ is the Cantor topology on $2^\mathbb{N}$, if we identify A with cf_A.

7. Define a computable topological space $\mathbf{S} = (\mathbb{N}^\omega, \sigma, \nu)$ by $\mathrm{dom}(\nu) := \{1^i 0^n \mid i, n \in \mathbb{N}\}$ and $\nu(1^i 0^n) := \{p \in \mathbb{N}^\omega \mid p(i) = n\}$. Then $\delta_\mathbf{S} \equiv \delta_\mathbb{B}$ (Definition 3.1.2).

The proofs are left as Exercise 3.2.9.   □

In Sect. 4.1 we will discuss various representations of the real numbers. Observe that an En-name of a set A lists all valid properties "$n \in A$" and that a Cf-name lists all valid properties "$n \in A$" as well as all valid properties "$m \notin A$".

For an effective topological space $\mathbf{S} = (M, \sigma, \nu)$, the set σ of atomic properties specifies a concept of *approximation* on the set M: an element $x \in M$ can be approximated by a sequence of its atomic properties. Formally, this concept of approximation is given by the topology $\tau_\mathbf{S}$. (Remember that mathematical topology is the general theory for studying approximation.) The notation ν of σ specifies computability on the set σ (it expresses in which way atomic properties can be handled by machines) and the representation $\delta_\mathbf{S}$ induces a computability concept on the set M. The two ingredients approximation and ("discrete") computability seem also to be necessary for defining computability on uncountable sets.

For a given set M (which is not too big) there are many computable topological spaces $\mathbf{S} = (M, \sigma, \nu)$, that is, many kinds of approximation and many kinds of computability. In applications the set M is the domain of an already given (algebraic or topological) structure $\mathbf{M}$, and the computability concept should be compatible with this structure. Therefore, for most of the fundamental sets used in analysis (real numbers, compact subsets of real numbers, Lebesgue integrable functions, ...) there are only very few interesting computability concepts most of which can be defined by computable topological spaces.

Consider a user who wants to compute a function $f : M \to M'$ where M and M' are not countable. Since the set M has no notation, the machine cannot read elements or names of elements entirely in a finite amount of time but only "finite portions of information" (properties) of the ideal input x. However, the user cannot choose the supply σ of properties arbitrarily, but instead has to consider those properties which are available in the specific situation (for example results of measurements, of a preceeding computation or of throwing the dice). Similarly, the user will not choose the notation of the set σ arbitrarily but will consider a "canonical" one or the one which is already given. Although in principle the computable topological space for the output of the machine can be chosen arbitrarily, only for very few cases the function f will turn out to be computable.

For an effective topological space $\mathbf{S} = (M, \sigma, \nu)$ the topology $\tau_\mathbf{S}$ and the representation $\delta_\mathbf{S}$ are very closely related.

Lemma 3.2.5 (properties of $\delta_{\mathbf{S}}$). Let $\mathbf{S} = (M, \sigma, \nu)$ be an effective topological space.

1. The representation $\delta_{\mathbf{S}}$ is $(\tau_C, \tau_{\mathbf{S}})$-continuous
   (that is, $X \in \tau_{\mathbf{S}} \Longrightarrow \delta_{\mathbf{S}}^{-1}[X]$ open in $\mathrm{dom}(\delta_{\mathbf{S}})$).
2. $\delta_{\mathbf{S}}$ is $(\tau_C, \tau_{\mathbf{S}})$-open (that is, $V \in \tau_C \Longrightarrow \delta_{\mathbf{S}}[V] \in \tau_{\mathbf{S}}$).
3. $\tau_{\mathbf{S}}$ is the final topology of $\delta_{\mathbf{S}}$.
4. $\zeta \leq_t \delta_{\mathbf{S}}$ for all $(\tau_C, \tau_{\mathbf{S}})$-continuous functions $\zeta :\subseteq \Sigma^\omega \to M$.
5. Let (M', τ') be a topological space and let $H :\subseteq M \to M'$ be a function such that $H \circ \delta_{\mathbf{S}} :\subseteq \Sigma^\omega \to M'$ is (τ_C, τ')-continuous. Then H is $(\tau_{\mathbf{S}}, \tau')$-continuous.

Proof: 1. Consider $X \in \tau_{\mathbf{S}}$. $\delta_{\mathbf{S}}^{-1}[X]$ is the set
$\{p \in \mathrm{dom}(\delta_{\mathbf{S}}) \mid \iota(w) \lhd p \text{ for some word } w \text{ with } \nu(w) \subseteq X\}$,
which is open in $\mathrm{dom}(\delta_{\mathbf{S}})$.

2. It suffices to consider $V = u\Sigma^\omega$. For $u \in \Sigma^*11$,

$$\delta_{\mathbf{S}}[u\Sigma^\omega] = \bigcap \{\nu(x) \mid \iota(x) \lhd u\} \in \tau_{\mathbf{S}} \ .$$

If $u \notin \Sigma^*11$ then there is some $v \in \Sigma^*11$ such that for all $x \in \Sigma^*$, $\iota(x) \lhd u$, iff $\iota(x) \lhd uv$. (Notice that w must have odd length, if $\iota(z) = 11w11$.) Then $uv \in \Sigma^*11$ and $\delta_{\mathbf{S}}[u\Sigma^\omega] = \delta_{\mathbf{S}}[uv\Sigma^\omega] \in \tau_{\mathbf{S}}$.

3. This follows from Properties 1 and 2.

4. Let ζ be $(\tau_C, \tau_{\mathbf{S}})$-continuous. Then for any $p \in \mathrm{dom}(\zeta)$ and any $X \in \sigma$ there is some number n with $\zeta[p_{<n}\Sigma^\omega] \subseteq X$, iff $\zeta(p) \in X$. Let $(w_0, w_1, \ldots)$ be a list of $\mathrm{dom}(\nu)$. Define $f : \Sigma^\omega \to \Sigma^\omega$ by $f(p) := h(0)h(1)\ldots$, where

$$h\langle i, n\rangle := \begin{cases} \iota(w_i) & \text{if } \zeta[p_{<n}\Sigma^\omega] \subseteq \nu(w_i) \\ 11 & \text{otherwise} \end{cases}$$

for all $i, n \in \mathbb{N}$. Then f is continuous and for $p \in \mathrm{dom}(\zeta)$, $f(p)$ is a list of all w_i with $\zeta(p) \in \nu(w_i)$. Therefore, f translates ζ to $\delta_{\mathbf{S}}$.

5. Consider $T \in \tau'$. Then $(H \circ \delta_{\mathbf{S}})^{-1}[T]$ is open in $\mathrm{dom}(H \circ \delta_{\mathbf{S}})$, hence $\delta_{\mathbf{S}}^{-1}[H^{-1}[T]] = V \cap \mathrm{dom}(H \circ \delta_{\mathbf{S}})$ for some open $V \subseteq \Sigma^\omega$. We obtain $H^{-1}[T] = \delta_{\mathbf{S}}[V \cap \mathrm{dom}(H \circ \delta_{\mathbf{S}})] = \delta_{\mathbf{S}}[V \cap \delta_{\mathbf{S}}^{-1}[\mathrm{dom}(H)]] = \delta_{\mathbf{S}}[V] \cap \mathrm{dom}(H)$. Since $\delta_{\mathbf{S}}[V]$ is open by Property 2, $H^{-1}[T]$ is open in $\mathrm{dom}(H)$. Therefore, H is continuous. $\qquad\square$

By the following lemma every second countable T_0-space can be extended to an effective topological space, and the continuous equivalence class of the standard representation $\delta_{\mathbf{S}}$ of an effective topological space $\mathbf{S}$ depends only on the topology $\tau_{\mathbf{S}}$.

Lemma 3.2.6.

1. For every second countable T_0-space (M, τ) there is an effective topological space $\mathbf{S} = (M, \sigma, \nu)$ with $\tau = \tau_{\mathbf{S}}$.
2. If $\mathbf{S} = (M, \sigma, \nu)$ and $\mathbf{S}' = (M, \sigma', \nu')$ are effective topological spaces such that $\tau_{\mathbf{S}} = \tau_{\mathbf{S}'}$, then $\delta_{\mathbf{S}} \equiv_t \delta_{\mathbf{S}'}$.

Proof: 1. Let σ be a countable subbase of τ and let ν be a notation of σ. Then $\tau = \tau_{\mathbf{S}}$.

2. By Lemma 3.2.5.1, $\delta_{\mathbf{S}}$ is $(\tau_C, \tau_{\mathbf{S}})$-continuous and so $(\tau_C, \tau_{\mathbf{S}'})$-continuous. By Lemma 3.2.5.4 (for $\mathbf{S}'$), $\delta_{\mathbf{S}} \leq_t \delta_{\mathbf{S}'}$. Correspondingly, $\delta_{\mathbf{S}'} \leq_t \delta_{\mathbf{S}}$ and so $\delta_{\mathbf{S}} \equiv_t \delta_{\mathbf{S}'}$. □

Properties 1 and 3 to 5 from Lemma 3.2.5 hold correspondingly for every naming system γ replacing $\delta_{\mathbf{S}}$ which is continuously equivalent to $\delta_{\mathbf{S}}$. Therefore, we are interested in all naming systems which are continuously equivalent to a standard representation $\delta_{\mathbf{S}}$.

Definition 3.2.7 (admissible naming systems).

1. Let (M, τ) be a second countable T_0-space. A naming system $\gamma :\subseteq Y \to M$ is called admissible w.r.t. τ, iff $\gamma \equiv_t \delta_{\mathbf{S}}$ for some effective topological space $\mathbf{S} = (M, \sigma, \nu)$ with $\tau = \tau_{\mathbf{S}}$.
2. A naming system $\gamma :\subseteq Y \to M$ is called admissible, iff it is admissible w.r.t. τ for some second countable T_0-space (M, τ).

The following theorem summarizes some important properties.

Theorem 3.2.8 (admissible naming systems). Let (M, τ) be a second countable T_0-space. Then:
1. The naming systems of M admissible w.r.t. τ form a non-empty continuous equivalence class.
2. If a naming system γ is admissible w.r.t. τ, then τ is the final topology of γ,
3. If γ is admissible w.r.t. τ and to τ', then $\tau = \tau'$.
4. A notation ν of M is admissible w.r.t. τ, iff τ is the discrete topology; in particular, every notation is admissible.

Proof: 1. Suppose, γ and γ' are admissible w.r.t. τ. Then there are effective topological spaces $\mathbf{S}$ and $\mathbf{S}'$ such that $\gamma \equiv_t \delta_{\mathbf{S}}$, $\gamma' \equiv_t \delta_{\mathbf{S}'}$ and $\tau = \tau_{\mathbf{S}} = \tau_{\mathbf{S}'}$. By Lemma 3.2.6.2, $\delta_{\mathbf{S}} \equiv_t \delta_{\mathbf{S}'}$, and so $\gamma \equiv_t \gamma'$. By Lemma 3.2.6.1, M has a representation admissible w.r.t. τ.

2. $\gamma \equiv_t \delta_{\mathbf{S}}$ for some effective topological space $\mathbf{S} = (M, \sigma, \nu)$ with $\tau = \tau_{\mathbf{S}}$. By Lemma 3.2.5.3, $\tau_{\mathbf{S}}$ is the final topology of $\delta_{\mathbf{S}}$. By Theorem 3.1.8, τ is the final topology of γ.

3. This follows immediately from Property 2.

4. If ν is admissible w.r.t. τ, then by Property 2, τ is the final topology of ν, that is, the discrete topology on M by Example 3.1.4.4. On the other hand, let τ be the discrete topology. Define an effective topological space $\mathbf{S} = (M, \sigma, \mu)$ by $\mu(w) := \{\nu(w)\}$. By Example 3.2.4.1, $\nu \equiv_t \delta_{\mathbf{S}}$, and so ν and $\delta_{\mathbf{S}}$ have the same final topology, that is, the discrete topology. Therefore, ν is admissible w.r.t. the final topology of $\delta_{\mathbf{S}}$, the discrete topology. □

By Properties 3 and 2, an admissible naming system γ is admissible w.r.t. a single topology τ which is the final topology of γ. Therefore, there is a one-to-one correspondence between the second countable T_0-spaces and the continuous equivalence classes of admissible naming systems. The converse of Property 2 is false in general (Theorem 4.1.13).

Sometimes we will say "γ is admissible with final topology τ" instead of "γ is admissible w.r.t. τ".

By definition every continuous equivalence class of admissible naming systems contains a standard representation $\delta_{\mathbf{S}}$. By Property 4 it contains all notations of M, if $\tau_{\mathbf{S}}$ is the discrete topology on M, and no notation, otherwise. Further properties follow immediately from Definition 3.2.7 and Lemma 3.2.5. In particular, from Property 3.2.5.4 we can derive a very interesting characterizations of the admissible representations.

Theorem 3.2.9 (characterization of admissible representations).
Let $\delta :\subseteq \Sigma^\omega \to M$ be a representation of a second countable T_0-space (M, τ). Then the following properties are equivalent:
1. δ is admissible w.r.t. τ.
2. a) δ is (τ_C, τ)-continuous and
   b) $\zeta \leq_t \delta$ for every (τ_C, τ)-continuous representation ζ of M.
3. For all representations ζ of M,

$$\zeta \text{ is } (\tau_C, \tau) \text{ -continuous} \iff \zeta \leq_t \delta \,.$$

Proof: 1.$\Longrightarrow$ 2. There is some effective topological space $\mathbf{S}$ such that $\delta \equiv_t \delta_{\mathbf{S}}$. Then δ is (τ_C, τ)-continuous, since $\delta \leq_t \delta_{\mathbf{S}}$ and $\delta_{\mathbf{S}}$ is continuous by Lemma 3.2.5.1. Let ζ be a (τ_C, τ)-continuous representation of M. By Lemma 3.2.5.4, $\zeta \leq_t \delta_{\mathbf{S}}$, and so $\zeta \leq_t \delta$.

2.$\Longrightarrow$ 1. By Lemma 3.2.6.1, there is some effective topological space $\mathbf{S}$ such that $\tau = \tau_{\mathbf{S}}$. By Lemma 3.2.5.1, $\delta_{\mathbf{S}}$ is (τ_C, τ)-continuous, and so $\delta_{\mathbf{S}} \leq_t \delta$ by b). Since by a), δ is (τ_C, τ)-continuous, $\delta \leq_t \delta_{\mathbf{S}}$ by Lemma 3.2.5.4. Therefore, $\delta \equiv_t \delta_{\mathbf{S}}$, and so δ is admissible w.r.t. τ.

2. $\Longleftrightarrow$ 3. See Exercise 3.2.7. $\square$

A representation δ is admissible w.r.t. τ, iff it is $\leq_t$-complete in the class of all (τ_C, τ)-continuous representations. It is the "poorest" continuous representation of M. Its continuous equivalence class is defined by continuity and a maximality property. Fig. 3.4 illustrates the theorem.

If ν is a notation of a subbase of the topology τ, then $\zeta :\subseteq \Sigma^\omega \to M$ is continuous, iff $G := \{(w, p) \in \Sigma^* \times \Sigma^\omega \mid \zeta(p) \in \nu(w)\}$ is open in $\mathrm{dom}(\nu) \times \mathrm{dom}(\zeta)$ (Exercise 3.2.8). Replacing "open" by "r.e. open" we obtain a computational version of Theorem 3.2.9 on admissible representations. It characterizes the equivalence class of $\delta_{\mathbf{S}}$ and hence the computability theory induced by it by a computability property and a maximality property.

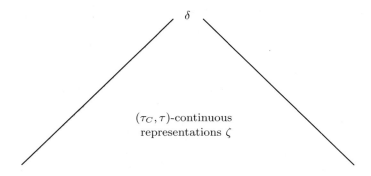

δ

(τ_C, τ)-continuous
representations ζ

Fig. 3.4. Characterization of admissible representations δ

Theorem 3.2.10 (computable characterization). Let $\mathbf{S} = (M, \sigma, \nu)$
be a computable topological space and let $\delta_{\mathbf{S}}$ be the associated standard
representation of M. Then

$$\{(x, A) \in M \times \sigma \mid x \in A\} \text{ is } (\zeta, \nu)\text{-r.e.} \iff \zeta \le \delta_{\mathbf{S}}$$

for all representations $\zeta :\subseteq \Sigma^\omega \to M$.

Proof: See Exercise 3.2.12.   □

Roughly speaking, the representation $\delta_{\mathbf{S}}$ is $\le$-complete in the set of all
representations ζ of M for which the property $\zeta(p) \in \nu(w)$ is r.e. open. Fig.
3.4 illustrates Theorem 3.2.10 accordingly.

Let $\mathbf{S}_i = (M_i, \sigma_i, \nu_i)$ be effective topological spaces with standard representations δ_i $(i = 1, 2)$ and let $g :\subseteq \Sigma^\omega \to \Sigma^\omega$ be a continuous (δ_1, δ_2)-realization of a function $f :\subseteq M_1 \to M_2$. Then g maps every δ_1-name p of any $x \in \mathrm{dom}(f)$ to a δ_2-name of $f(x)$. Roughly speaking, the function g maps every list $(A_0, A_1, \ldots)$ of all properties $(\in \sigma_1)$ of x to a list $(B_0, B_1, \ldots)$ of all properties $(\in \sigma_2)$ of $f(x)$. This induces continuity of f (cf. Fig. 1.1 and Theorem 1.3.4). By our "main theorem" for admissible naming systems continuity induced by admissible representations is the same as topological continuity.

Theorem 3.2.11 (MAIN THEOREM).   For $i = 0, \ldots, k$ let
$\delta_i :\subseteq \Sigma^\omega \to M_i$ be an admissible naming system w.r.t. a topology τ_i. Then
for any function $f :\subseteq M_1 \times \ldots \times M_k \to M_0$,

$$f \text{ is } (\tau_1, \ldots, \tau_k, \tau_0)\text{-continuous} \iff f \text{ is } (\delta_1, \ldots, \delta_k, \delta_0)\text{-continuous}.$$

Proof: First consider the case $k = 1$. By Definition 3.2.7 for $i = 0, 1$ there is
a standard representation $\delta_{\mathbf{S}_i}$ equivalent to δ_i. By Corollary 3.1.9 it suffices
to prove the theorem for $\delta_{\mathbf{S}_i}$ $(i = 0, 1)$ instead of δ_i $(i = 0, 1)$.

Let f be (τ_1, τ_0)-continuous. Then $f \circ \delta_{\mathbf{S}_1}$ is (τ_C, τ_0)-continuous. By Lemma 3.2.5.4, $f \circ \delta_{\mathbf{S}_1} \leq_t \delta_{\mathbf{S}_0}$, that is, there is a continuous function g with $f \circ \delta_{\mathbf{S}_1}(p) = \delta_{\mathbf{S}_0} \circ g(p)$ for all $p \in \mathrm{dom}(f \circ \delta_{\mathbf{S}_1})$. Therefore, f is $(\delta_{\mathbf{S}_1}, \delta_{\mathbf{S}_0})$-continuous.

Let f be $(\delta_{\mathbf{S}_1}, \delta_{\mathbf{S}_0})$-continuous. Then there is a continuous function g with $f \circ \delta_{\mathbf{S}_1}(p) = \delta_{\mathbf{S}_0} \circ g(p)$ for all $p \in \mathrm{dom}(f \circ \delta_1)$. Therefore, $f \circ \delta_{\mathbf{S}_1}$ is continuous. By Theorem 3.2.5.5, f is (τ_1, τ_0)-continuous.

Consider now $k \geq 2$. Define a representation δ of $M := M_1 \times \ldots \times M_k$ by $\delta \langle p_1. \ldots, p_k \rangle := (\delta_1(p_1), \ldots, \delta_k(p))$. For any representation ζ of M by Theorem 3.2.9 we have:

$$\zeta \text{ is continuous} \iff \mathrm{pr}_i \circ \zeta :\subseteq \Sigma^\omega \to M_i \text{ is continuous for all } i$$

$$\iff \mathrm{pr}_i \circ \zeta \leq_t \delta_i \text{ for all } i \iff \zeta \leq_t \delta .$$

Again by Theorem 3.2.9, δ is admissible and, therefore, f is continuous, iff it is (δ, δ_0)-continuous (Case $k = 1$). Finally, f is (δ, δ_0)-continuous, iff it is $(\delta_1, \ldots, \delta_k, \delta_0)$-continuous. $\qquad\square$

The theorem may fail for non-admissible representations. As an example consider the decimal representation $\rho_{\mathrm{b},10}$ of the real numbers . Its final topology is $\tau_{\mathbb{R}}$ (the real line topology, see Theorem 4.1.13), but the $(\tau_{\mathbb{R}}, \tau_{\mathbb{R}})$-continuous real function $x \mapsto 3 \cdot x$ is *not* $(\rho_{\mathrm{b},10}, \rho_{\mathrm{b},10})$-continuous (Example 2.1.4.7).

Corollary 3.2.12 (computable implies continuous). For $i = 0, \ldots, k$ let δ_i admissible w.r.t. τ_i. Then every $(\delta_1, \ldots, \delta_k, \delta_0)$-computable function is $(\delta_1, \ldots, \delta_k, \delta_0)$-continuous and hence $(\tau_1, \ldots, \tau_k, \tau_0)$-continuous.

This follows immediately from the above theorem, since every computable function on Σ^ω is continuous. Theorem 1.3.4 is a special case of the corollary. Although continuous but non-computable functions can be constructed, in almost all applications from analysis non-computable functions are not even continuous. In such a case the finiteness property (every finite portion of output information is determined already by a finite portion of input information), which is much more elementary than computability, is violated. Usually, proofs of discontinuity are not very difficult (Examples 2.1.4.6, 2.1.4.7, 3.1.4.9, 3.1.4.10, 3.1.4.13, Exercise 2.1.10).

A topological space (M, τ) is called *connected*, iff $(U \in \tau$ and $M \setminus U \in \tau)$ implies $(U = \emptyset$ or $U = M)$. The following simple corollary has many important applications.

Corollary 3.2.13. Let $\delta :\subseteq \Sigma^\omega \to M$ be an admissible representation with connected final topology and let $\nu :\subseteq \Sigma^* \to N$ be a notation. Then every (δ, ν)-continuous or (δ, ν)-computable total function $f : M \to N$ is constant.

Proof: The notation ν is admissible. Its final topology τ_ν is the discrete topology on N, that is, the set $\{x\}$ is open for every $x \in N$. By Theorem 3.2.11, the function f is (τ_δ, τ_ν)-continuous. Assume $x \in \text{range}(f)$. Since $\{x\} \in \tau_\nu$ and $N \setminus \{x\} \in \tau_\nu$, $f^{-1}[\{x\}]$ and $M \setminus f^{-1}[\{x\}] = f^{-1}[N \setminus \{x\}]$ are open, that is, $\in \tau_\delta$. Since τ_δ is connected and $f^{-1}[\{x\}] \neq \emptyset$, $f^{-1}[\{x\}] = M$, and so f is a constant function. $\square$

A subset of the set M can be considered as a *property* on M. As a consequence of the corollary, for an admissible representation of a set M with connected final topology, no non-trivial property $A \subseteq M$ can be decided (the characteristic function $\text{cf}_A : M \to \mathbb{N}$ cannot be $(\delta, \nu_\mathbb{N})$-computable). This statement resembles Rice's theorem from recursion theory for the standard numbering ϕ of the partial recursive functions [Rog67, Odi89]. Later we will use admissible representations of the real line $(\mathbb{R}, \tau_\mathbb{R})$, which is connected.

There is a very interesting topological characterization of computable functions on computable topologival spaces (P. Hertling, private communication).

Theorem 3.2.14 (topological computability). Let $\mathbf{S}_1 = (M_1, \sigma_1, \nu_1)$ and $\mathbf{S}_0 = (M_0, \sigma_0, \nu_0)$ be computable topological spaces. Define a notation β of a base of $\tau_{\mathbf{S}_1}$ by

$$\beta(w) := \bigcap \{\nu_1(u) \mid \iota(u) \lhd w \text{ and } u \in \text{dom}(\nu_1)\} .$$

Then a function $f :\subseteq M_1 \to M_0$ is $(\delta_{\mathbf{S}_1}, \delta_{\mathbf{S}_0})$-computable, iff there is an r.e. set $A \subseteq \Sigma^* \times \Sigma^*$ with

$$f^{-1}[\nu_0(v)] = \bigcup_{(v,w) \in A} \beta(w) \cap \text{dom}(f) .$$

This theorem effectivizes the topological definition of a continuous function f: the pre-image of every open set is open in $\text{dom}(f)$. We leave the proof as Exercise 3.2.13.

By Definition 3.2.2 a $\delta_{\mathbf{S}}$-name of an element x is a list of *all* atomic properties $A \in \sigma$ with $x \in A$. In almost all important examples there are equivalent representations δ, where δ-names list merely "sufficiently many" atomic properties. Later we will use such simpler representations.

If in Definition 3.2.1, Condition (3.4) does not hold, points cannot be identified by their atomic properties, formally the topology $\tau_{\mathbf{S}}$ is not T_0. In this case the representation $\delta_{\mathbf{S}}$ can be considered as a representation of equivalence classes on M, where

$$x \equiv y \iff \{A \in \sigma \mid x \in A\} = \{A \in \sigma \mid y \in A\} .$$

Earlier definitions of standard representations started with a (partial or total) numbering U of a *base* of the T_0-topology of M [KW85, Wei87] and the standard representation δ_U of M was defined by one of the following alternatives: $\delta_U(p) = x$, iff

- $i \in \mathbb{N}$ is enumerated by p, iff $x \in U(i)$,
- $\{U(i) \mid i$ is enumerated by $p\} = \{U(i) \mid x \in U(i)\}$,
- $\{U(i) \mid i$ is enumerated by $p\}$ is a subbase of the set of neighborhoods of x.

In each case the representation δ_U is continuously equivalent to $\delta_{\mathbf{S}}$ from Definition 3.2.2, hence admissible. In most applications for natural choices of ν and U, δ_U and $\delta_{\mathbf{S}}$ are even (computationally) equivalent.

Our definition of admissibility does not apply to spaces with uncountable base. M. Schröder [Sch00] has extended this concept to more general spaces. He uses Theorem 3.2.9.2 as a definition and then, in particular, can prove the main theorem 3.2.11 with *sequentially continuous* replacing continuous.

Exercises 3.2.

1. Let U and V be the notations of the recursive subsets of Σ^* defined at the beginning of this section. Show $U < V$.
2. Let S be the set of all notations ν of the set $\mathbb{N}$ of natural numbers, for which the successor function $x \mapsto x+1$ is (ν, ν)-computable. Then $\nu_{\mathbb{N}} \in S$ and $\nu_{\mathbb{N}} \leq \nu$ for all $\nu \in S$. This means that $\nu_{\mathbb{N}}$ is the smallest (or richest) notation of $\mathbb{N}$, for which the successor function is computable.
3. Let S be the set of all notations ν of the set $\mathbb{Z}$ of integers such that subtraction $(x, y) \mapsto x - y$ is (ν, ν, ν)-computable. Then $\nu_{\mathbb{Z}} \in S$ and $\nu_{\mathbb{Z}} \leq \nu$ for all $\nu \in S$. This means that $\nu_{\mathbb{Z}}$ is the smallest (or richest) notation of $\mathbb{Z}$, for which subtraction is computable.
4. Let S be the set of all notations ν of the set $\mathbb{Q}$ of rational numbers such that subtraction and division $(x, y) \to x/y$ are (ν, ν, ν)-computable. Then $\nu_{\mathbb{Q}} \in S$ and $\nu_{\mathbb{Q}} \leq \nu$ for all $\nu \in S$. This means that $\nu_{\mathbb{Q}}$ is the smallest (or richest) notation of $\mathbb{Q}$, for which subtraction and division are computable.
5. Consider the computable topological space from Example 3.2.4.5. Show that $\{\{0\}\} \neq \bigcap \{V \in \sigma \mid \{0\} \in V\}$.
6. Define computable topological spaces $\mathbf{S}_i = (\mathbb{N}, \sigma_i, \nu_i)$ $(i = 1, 2)$ by $\sigma_1(w) := \{n \in \mathbb{N} \mid n \leq \nu_{\mathbb{N}}(w)\}$ and $\sigma_2(w) := \{n \in \mathbb{N} \mid n \geq \nu_{\mathbb{N}}(w)\}$. Show: $\nu_{\mathbb{N}} < \delta_{\mathbf{S}_1}$, $\nu_{\mathbb{N}} < \delta_{\mathbf{S}_2}$, $\delta_{\mathbf{S}_1} \not\leq_t \delta_{\mathbf{S}_2}$, $\delta_{\mathbf{S}_2} \not\leq_t \delta_{\mathbf{S}_1}$.
7. $\Diamond$ Show that the second and third property in Theorem 3.2.9 are equivalent.
8. Let ν be a notation of a subbase of a topology τ on M. Show that a function $\zeta :\subseteq \Sigma^\omega \to M$ is continuous, iff $G := \{(w, p) \in \Sigma^* \times \Sigma^\omega \mid \zeta(p) \in \nu(w)\}$ is open in $\mathrm{dom}(\nu) \times \mathrm{dom}(\zeta)$.
9. Prove the statements from Example 3.2.4.

◇10. Show that Properties 1 and 3 to 5 from Lemma 3.2.5 hold correspondingly for every naming system γ replacing $\delta_{\mathbf{S}}$ which is continuously equivalent to $\delta_{\mathbf{S}}$.

11. Let $\delta_{\mathbf{S}}$ be the standard representation for some computable topological space $\mathbf{S}$ and let $\delta \equiv \delta_{\mathbf{S}}$. Show that δ has a restriction δ' which is equivalent to δ and maps open sets to open sets (cf. Lemma 3.2.5.2).

12. Prove Theorem 3.2.10.

13. Prove Theorem 3.2.14.

14. Let $\mathbf{S} = (M, \sigma, \nu)$ and $\mathbf{S}' = (M, \sigma, \nu')$ be computable topological spaces with $\nu \equiv \nu'$. Show that $\delta_{\mathbf{S}} \equiv \delta_{\mathbf{S}'}$.

15. Let ν be a notation such that $\{(u, v) \mid \nu(u = \nu(v) \neq \text{div}\}$ is r.e. Show that there is some notation ν' with recursive domain which is equivalent to ν.

16. Prove Lemma 3.2.3. The representations $\delta_{\mathbf{S}}$ and $\delta'_{\mathbf{S}}$ may be non-equivalent, if the equivalence problem of ν is not r.e.

◆17. In this book we are using infinite sequences $p \in \Sigma^\omega$ of symbols from a finite alphabet Σ as names of infinite objects. Such sequences can be considered as inscriptions of Turing tapes which have immediate physical realizations. Only slightly more abstract names are infinite sequences $p \in \mathbb{B} := \{q \mid q : \mathbb{N} \to \mathbb{N}\}$ which are used, for example in [Wei85, KW85, Wei87]. Computability on Baire space $\mathbb{B}$ can be defined via the standard embedding $\alpha : \mathbb{B} \to \Sigma^\omega$, $\alpha(i_0, i_1, \ldots) := 0^{i_0} 1 0^{i_1} 1 \ldots$ (Exercise 2.2.9).
Another possibility are subsets $A \subseteq \mathbb{N}$ as names.
Define a topological space $(2^{\mathbb{N}}, \tau)$ by the basic sets $O_E := \{A \subseteq \mathbb{N} \mid E \subseteq A\}$ with finite $E \subseteq \mathbb{N}$. A function $f : 2^{\mathbb{N}} \to 2^{\mathbb{N}}$ is called computable, iff it is continuous and $\{(i, j) \mid D_j \subseteq f(D_i)\}$ is r.e. (where D is a standard numbering of the finite subsets of $\mathbb{N}$). These computable functions are called *enumeration operators* [Rog67, Odi89]. Let En be the representation from Definition 3.1.2. Show:
a) f is continuous, iff f is (En, En)-continuous,
b) f is computable, iff f is (En, En)-computable.
A *set-representation* is a surjective function $\psi :\subseteq 2^{\mathbb{N}} \to M$. Definition 3.1.3.3 can be transferred to set-representations. Let ψ, ψ' be set-representations. Show that for injective set-representations $\psi :\subseteq 2^{\mathbb{N}} \to M$ and $\psi' :\subseteq 2^{\mathbb{N}} \to M'$, a function $f :\subseteq M \to M'$ is
a) (ψ, ψ')-continuous, iff it is $(\psi \circ \text{En}, \psi' \circ \text{En})$-continuous,
b) (ψ, ψ')-computable, iff it is $(\psi \circ \text{En}, \psi' \circ \text{En})$-computable.
However, computability of multi-valued functions cannot be defined directly as in Definition 3.1.3.4 via enumeration operators. For a computable topological space $\mathbf{S} = (M, \sigma, \nu)$, the (injective) standard set-representation $\psi_{\mathbf{S}} : 2^{\mathbb{N}} \to M$ is defined by $\psi_{\mathbf{S}}^{-1}(x) := \{i \in \mathbb{N} \mid x \in \nu \circ \nu_\Sigma(i)\}$. It is a homeomorphism from $\text{dom}(\psi_{\mathbf{S}})$ to M (cf. Lemma 3.2.5.1/2). Show: $\psi_{\mathbf{S}} \circ \text{En} \equiv \delta_{\mathbf{S}}$. For generalizations, see [Bla97, Bla99].

3.3 Constructions of New Naming Systems

In this section we present canonical methods for constructing naming systems from given ones. The most important constructs are the representations of products and of spaces of continuous functions. The other parts may be skipped in a first reading.

For a naming system $\delta :\subseteq \Sigma^\omega \to M$ and a subset $N \subseteq M$ consider the restriction $\delta|^N$ (Sect. 1.4). Since the identities $i : N \to M$ and $j :\subseteq M \to N$ are $(\delta|^N, \delta)$-computable and $(\delta, \delta|^N)$-computable, respectively, we can use δ instead of $\delta|^N$ for mappings from or to N. However, the $\delta|^N$-r.e. subsets of N are not δ-r.e. in general.

Lemma 3.3.1 (restriction). Let $\delta :\subseteq \Sigma^a \to M$ be a naming system and let $N \subseteq M$.
1. If $X \subseteq N$ is δ-r.e, then it is $\delta|^N$-r.e.
2. Every $\delta|^N$-r.e. set $X \in N$ is δ-r.e. , iff N is δ-r.e.

The properties hold accordingly with "open" instead of "r.e." .

Proof: 1. If X is δ-r.e, then $\delta^{-1}[X] = U \cap \text{dom}(\delta)$ for some r.e. open set $U \subseteq \Sigma^a$. Then $(\delta|^N)^{-1}[X] = \delta^{-1}[X] \cap \text{dom}(\delta|^N) = U \cap \text{dom}(\delta|^N)$.

2. If every $\delta|^N$-r.e. set $X \subseteq N$ is δ-r.e. , then N is δ-r.e. since it is $\delta|^N$-r.e. On the other hand, if N is δ-r.e. and $X \subseteq N$ is $\delta|^N$-r.e. , then $\delta^{-1}[X] = (\delta|^N)^{-1}[X] = V \cap \delta^{-1}[N] = V \cap U \cap \text{dom}(\delta)$ for some r.e. open sets U, V.
For "open" instead of "r.e." the proof is similar.   □

The restriction of an admissible representation is admissible.

Lemma 3.3.2 (admissible restriction). If δ is an admissible representation of M and $N \subseteq M$, then $\delta|^N$ is an admissible representation of N.

The proof is left as Exercise 3.3.1. We introduce naming systems of Cartesian products by means of tupling functions (Definition 2.1.7).

Definition 3.3.3 (products of naming systems). *For all $i \in \mathbb{N}$ let $\delta_i :\subseteq \Sigma^a \to M_i$ be a naming system.*
1. *For $k \geq 1$ define the naming system $[\delta_1, \ldots, \delta_k]$ of $M_1 \times \ldots \times M_k$ by*

$$[\delta_1, \ldots, \delta_k]\langle p_1, \ldots, p_k \rangle := (\delta_1(p_1), \ldots, \delta_k(p_k)) .$$

2. *If for all i, δ_i is a notation or for all i, δ_i is a representation, define a representation $[\delta_0, \delta_1, \ldots]$ of $M_0 \times M_1 \times \ldots$ by*

$$[\delta_0, \delta_1, \ldots]\langle p_0, p_1, \ldots \rangle = (\delta_0(p_0), \delta_1(p_1), \ldots) .$$

We will use the abbreviations $[\delta]^k := [\delta, \ldots, \delta]$ (k times) and $[\delta]^\omega := [\delta, \delta, \ldots]$.

Therefore, $p = \langle p_1, p_2 \rangle$ is a $[\delta_1, \delta_2]$-name of (x_1, x_2), iff p_1 is a δ_1-name of x_1 and p_2 is a δ_2-name of x_2. For a family $(M_i, \tau_i)_{i \in I}$ of topological spaces, the product topology $\bigotimes_{i \in I} \tau_i$ on $\times_{i \in I} M_i$ is defined by the subbase $\{V_{kU} \mid k \in I, U \in \tau_k\}$ where $V_{kU} = \times_{i \in I} U_i$ with $U_k = U$ and $U_i = M_i$ for all $i \neq k$. It is the coarsest (that is, smallest as a set) topology τ such that for every $k \in I$ the projection $\mathrm{pr}_k : \times_{i \in I} M_i \to M_k$ is (τ, τ_k)-continuous. We characterize the equivalence class of a product naming system correspondingly.

Lemma 3.3.4 (characterization of product).   For all $i \in \mathbb{N}$ let $\delta_i :\subseteq \Sigma^a \to M_i$ be a naming system.
1. For all naming systems δ of $M_1 \times M_2$ we have:
   The projection $\mathrm{pr}_i : M_1 \times M_2 \to M_i$ is (δ, δ_i)-computable for every $i \in \{1, 2\}$, iff $\delta \leq [\delta_1, \delta_2]$.
2. For all naming systems δ of $M_0 \times M_1 \times \ldots$ we have:
   There is a computable function $f :\subseteq \Sigma^* \times \Sigma^\omega \to \Sigma^\omega$ such that for each $i \in \mathbb{N}$ the function $f(0^i, \cdot)$ is a (δ, δ_i)-realization of the projection $\mathrm{pr}_i : M_0 \times M_1 \times \ldots \to M_i$, iff $\delta \leq [\delta_0, \delta_1, \ldots]$.

Proof: 1. Assume that f_i is a computable (δ, δ_i)-realization of pr_i for $i \in \{1, 2\}$. Then f with $f(p) := \langle f_1(p), f_2(p) \rangle$ is a computable translation from δ to $[\delta_1, \delta_2]$. On the other hand, for each $i \in \{1, 2\}$, pr_i is $([\delta_1, \delta_2], \delta_i)$-computable. If $\delta \leq [\delta_1, \delta_2]$, then each pr_i is also (δ, δ_i)-computable.
2. The proof for infinite products is similar. $\square$

Roughly speaking, the product naming system is, up to equivalence, the greatest (or "poorest") naming system δ of the Cartesian product for which the i-th projection becomes (δ, δ_i)-computable for all i. Compare this characterization with Theorem 2.3.14.4, Corollary 2.3.17 and Theorems 3.2.9 and 3.2.10. The product is associative on equivalence classes: $[[\delta_1, \delta_2], \delta_3] \equiv [\delta_1, [\delta_2, \delta_3]]$. The product of admissible representations is admissible where the final topology is the product topology.

Lemma 3.3.5 (admissible product). For $i \in \mathbb{N}$ let $\mathbf{S}_i = (M_i, \sigma_i, \nu_i)$ be computable topological spaces.
1. Define a computable topological space $\mathbf{S}_{12} = (M_1 \times M_2, \sigma, \nu)$ by
   $\nu(01w) := \nu_1(w) \times M_2$ and $\nu(001w) := M_1 \times \nu_2(w)$.
   Then $\delta_{\mathbf{S}_{12}} \equiv [\delta_{\mathbf{S}_1}, \delta_{\mathbf{S}_2}]$ and $\tau_{\mathbf{S}_{12}} = \tau_{\mathbf{S}_1} \otimes \tau_{\mathbf{S}_2}$.
2. Assume that $\{(i, w) \mid \nu_i(w) = M_i\}$ is r.e.
   Define a computable topological space $\mathbf{S}_\infty = (M_0 \times M_1 \times \ldots, \sigma, \nu)$ by
   $\nu(0^i 1w) := U_0 \times U_1 \times \ldots$ with $U_i := \nu_i(w)$ and $U_j := M_j$ otherwise.
   Then $\delta_{\mathbf{S}_\infty} \equiv [\delta_{\mathbf{S}_0}, \delta_{\mathbf{S}_1}, \ldots]$ and $\tau_{\mathbf{S}_\infty} = \tau_{\mathbf{S}_0} \otimes \tau_{\mathbf{S}_1} \otimes \ldots$.

Proof: 1. Since $\{w \mid \nu_i(w) = M_i\}$ is r.e. for $i = 1, 2$, $\mathbf{S}_{12}$ is a computable topological space and range$(\nu) = \sigma$ is a subbase of the product topology $\tau_{\mathbf{S}_1} \otimes \tau_{\mathbf{S}_2}$. For every $x \in M_1 \times M_2$ every $\delta_{\mathbf{S}_{12}}$-name as well as every $[\delta_{\mathbf{S}_1}, \delta_{\mathbf{S}_2}]$-name lists all atomic properties of every component of x. Programs translating the two representations to each other can be specified easily.

2. The second part can be proved similarly.   □

For naming systems the product of the final topologies is not equal to the final topology of the product in general but at least one inclusion holds (Exercise 3.3.7). The product on naming systems generalizes pairing on Σ^* and Σ^ω.

Lemma 3.3.6.

1. A tuple $(x_1, \ldots, x_k)$ is $(\delta_1, \ldots, \delta_k)$-computable,
   iff it is $[\delta_1, \ldots, \delta_k]$-computable.
2. A set is $(\delta_1, \ldots, \delta_k)$-decidable (-r.e. , -clopen, -open),
   iff it is $[\delta_1, \ldots, \delta_k]$-decidable (-r.e. , -clopen, -open).
3. A function or multi-valued function is $(\delta_1, \ldots, \delta_k, \delta_0)$-computable
   (-continuous), iff it is $([\delta_1, \ldots, \delta_k], \delta_0)$-computable (-continuous).
4. A function f is $(\delta_0, [\delta_1, \ldots, \delta_k])$-computable (-continuous),
   iff $\mathrm{pr}_i \circ f$ is (δ_0, δ_i)-computable (-continuous) for $i = 1, \ldots . k$.

The proof is left as Exercise 3.3.3. The infinite product can be used to define computability of ω-ary functions: f is $(\delta_0, \delta_1, \ldots, \delta)$-computable, iff it is $([\delta_0, \delta_1, \ldots], \delta)$-computable.

Definition 3.3.7 (conjunction of naming systems). *For all $i \in \mathbb{N}$ let* $\delta_i :\subseteq \Sigma^a \to M_i$ *be a naming system.*

1. *For $k \geq 1$ define a naming system $\delta_1 \wedge \ldots \wedge \delta_k$ of the set $M_1 \cap \ldots \cap M_k$*
   by

$$\delta_1 \wedge \ldots \wedge \delta_k \langle p_1, \ldots, p_k \rangle = x : \Longleftrightarrow \delta_1(p_1) = \ldots = \delta_k(p_k) = x .$$

2. *If for all i, δ_i is a notation or for all i, δ_i is a representation, define a*
   representation $(\delta_0 \wedge \delta_1 \wedge \ldots)$ of $M_0 \cap M_1 \cap \ldots$ by:

$$(\delta_0 \wedge \delta_1 \wedge \ldots)\langle p_0, p_1, \ldots \rangle = x, \text{ iff } \delta_0(p_0) = \delta_1(p_1) = \ldots = x .$$

Therefore, $p = \langle p_1, p_2 \rangle$ is a $(\delta_1 \wedge \delta_2)$-name of x, iff p_1 is a δ_1-name and p_2 is a δ_2-name of x. For a family $(M_i, \tau_i)_{i \in I}$ of topological spaces, the conjunction topology $\bigwedge_{i \in I} \tau_i$ on $\bigcap_{i \in I} M_i$ is defined by the subbase $\{U \cap \bigcap_{i \in I} M_i \mid (\exists k \in I) U \in \tau_k\}$. It is the coarsest (that is, smallest as a set) topology τ such that for every $k \in I$ the canonical injection $\mathrm{id}_k : \bigcap_{i \in I} M_i \to M_k$ is (τ, τ_k)-continuous. We characterize the equivalence class of a conjunction naming system correspondingly.

Lemma 3.3.8 (characterization of conjunction). For all $i \in \mathbb{N}$ let $\delta_i :\subseteq \Sigma^a \to M_i$ be a naming system.
1. For all naming systems δ we have:
   $\delta \leq \delta_1$ and $\delta \leq \delta_2$, iff $\delta \leq \delta_1 \wedge \delta_2$.
2. For all naming systems δ we have:
   There is a computable function $f :\subseteq \Sigma^* \times \Sigma^\omega \to \Sigma^\omega$ such that for each $i \in \mathbb{N}$ the function $f(0^i, \cdot)$ translates δ to δ_i, iff $\delta \leq (\delta_0 \wedge \delta_1 \wedge \ldots)$.

Proof: 1. For $i = 1, 2$ let f_i be a computable translation from δ to δ_i. Then $\delta(p) = (\delta_1 \wedge \delta_2)\langle f_1(p), f_2(p)\rangle$. Therefore, the computable function f, defined by $f(p) := \langle f_1(p), f_2(p)\rangle$, translates δ to $\delta_1 \wedge \delta_2$. On the other hand, if f is a computable translation from δ to $\delta_1 \wedge \delta_2$, then $\pi_i^2 \circ f$ with $\pi_i^2 \langle p_1, p_2 \rangle = p_i$ is a computable translation from δ to δ_i.

2. The proof is similar to that of Statement 1. $\square$

Obviously, $\delta_1 \wedge \delta_2$ is the, up to equivalence unique, greatest lower bound of δ_1 and δ_2. Conjunction is commutative and associative on equivalence classes: $\delta_1 \wedge \delta_2 \equiv \delta_2 \wedge \delta_1$ and $(\delta_1 \wedge \delta_2) \wedge \delta_3 \equiv \delta_1 \wedge (\delta_2 \wedge \delta_3)$.

Lemma 3.3.9 (admissible conjunction). For $i \in \mathbb{N}$ let $\mathbf{S}_i = (M, \sigma_i, \nu_i)$ be computable topological spaces for a single set M.
1. Define $\sigma := \sigma_1 \cup \sigma_2$, $\nu(01w) := \nu_1(w)$ and $\nu(001w) := \nu_2(w)$. If $\{(u, v) \mid \nu(u) = \nu(v)\}$ is r.e. , then $\mathbf{S} := (M, \sigma, \nu)$ is a computable topological space with $\delta_{\mathbf{S}} \equiv \delta_{\mathbf{S}_1} \wedge \delta_{\mathbf{S}_2}$ and $\tau_{\mathbf{S}} = \tau_{\mathbf{S}_1} \wedge \tau_{\mathbf{S}_2}$.
2. Define $\sigma := \sigma_0 \cup \sigma_1 \cup \ldots$ and $\nu(0^i 1w) := \nu_i(w)$ for all i, w. If $\{(u, v) \mid \nu(u) = \nu(v)\}$ is r.e. , then $\mathbf{S} := (M, \sigma, \nu)$ is a computable topological space with $\delta_{\mathbf{S}} \equiv (\delta_{\mathbf{S}_0} \wedge \delta_{\mathbf{S}_1} \wedge \ldots)$ and $\tau_{\mathbf{S}} = (\tau_{\mathbf{S}_0} \wedge \tau_{\mathbf{S}_1} \wedge \ldots)$.

Proof: 1. Since σ_1 identifies points in M, σ identifies points in M. Since the equivalence problem of ν is r.e. , $\mathbf{S}$ is a computable topological space.
We show $\delta'_{\mathbf{S}} \leq \delta_{\mathbf{S}_1}$, $\delta'_{\mathbf{S}}$ from Lemma 3.2.3:
There is a Type-2 machine L such that for any $p \in \Sigma^\omega$, $f_L(p)$ is a list of all words $\iota(w)$ such that $\iota(01w) \lhd p$. Then f_L translates $\delta'_{\mathbf{S}}$ to $\delta_{\mathbf{S}_1}$.
For a similar reason $\delta'_{\mathbf{S}} \leq \delta_{\mathbf{S}_2}$. By Lemmas 3.2.3 and 3.3.8 we obtain $\delta_{\mathbf{S}} \leq \delta_{\mathbf{S}_1} \wedge \delta_{\mathbf{S}_2}$.
On the other hand, let L be a Type-2 machine, which on input $\langle p_1, p_2 \rangle$ writes all words $\iota(01u)$ for which $\iota(u) \lhd p_1$ and all words $\iota(001v)$ for which $\iota(v) \lhd p_2$. This way, the machine L translates every $\delta_{\mathbf{S}_1} \wedge \delta_{\mathbf{S}_2}$-name of some $x \in M$ to a $\delta_{\mathbf{S}}$-name of x. Therefore, $\delta_{\mathbf{S}_1} \wedge \delta_{\mathbf{S}_2} \leq \delta_{\mathbf{S}}$.
Since $\sigma := \sigma_1 \cup \sigma_2$ is a subbase of $\tau_{\mathbf{S}}$, $\tau_{\mathbf{S}} = \tau_{\mathbf{S}_1} \wedge \tau_{\mathbf{S}_2}$.

2. The proof is similar to that of Statement 1. $\square$

If the equivalence problem for ν is not r.e. , then still $\tau_{\mathbf{S}} = \tau_{\mathbf{S}_1} \wedge \tau_{\mathbf{S}_2}$ (Exercise 3.3.7). For the conjunction of arbitrary naming systems at least one inclusion holds (Exercise 3.3.7).

Example 3.3.10. (See Exercise 3.1.15) Consider the enumeration representation En and the characteristic function representation Cf of the set $2^{\mathbb{N}}$ (Definition 3.1.2). Define a representation En^c of $2^{\mathbb{N}}$ by $\text{En}^c(p) := \mathbb{N} \setminus \text{En}(p)$. A straightforward proof shows $\text{Cf} \equiv \text{En} \wedge \text{En}^c$: There are Type-2 machines which compute an enumeration of A and an enumeration of $\mathbb{N} \setminus A$ from the characteristic function of any set A. On the other hand, there is a Type-2 machine which computes the characteristic function of A from an enumeration of A and an enumeration of $\mathbb{N} \setminus A$. We obtain as an immediate corollary that a subset of $A \subseteq \mathbb{N}$ is recursive, iff A and $\mathbb{N} \setminus A$ are r.e.   □

Definition 3.3.11 (disjunction of naming systems).
Let $\delta_i :\subseteq \Sigma^* \to M_i$ be a notation for all i or let $\delta_i :\subseteq \Sigma^\omega \to M_i$ be a representation for all i.
1. For $k \geq 1$ define a naming system $\delta_1 \vee \ldots \vee \delta_k$ of the set $M_1 \cup \ldots \cup M_k$ by
$$\delta_1 \vee \ldots \vee \delta_k(0^n 1p) := \delta_n(p) \quad (1 \leq n \leq k) \,.$$

2. Define a naming system $(\delta_0 \vee \delta_1 \vee \ldots)$ of $M_0 \cup M_1 \cup \ldots$ by:
$$(\delta_0 \vee \delta_1 \vee \ldots)(0^n 1p) := \delta_n(p) \,.$$

For a family $(M_i, \tau_i)_{i \in I}$ of topological spaces define the disjunction topology $\bigvee_{i \in I} \tau_i$ on $\bigcup_{i \in I} M_i$ by $X \in \bigvee_{i \in I} \tau_i :\Longleftrightarrow (\forall i \in I)X \cap M_i \in \tau_i$. It is the finest (that is, biggest as a set) topology τ such that for every $k \in I$ the canonical injection $\text{id}_k : M_k \to \bigcup_{i \in I} M_i$ is (τ_k, τ)-continuous. We characterize the equivalence class of a disjunction naming system correspondingly.

Lemma 3.3.12 (characterization of disjunction). Let $\delta_i :\subseteq \Sigma^* \to M_i$ be a notation for all i or let $\delta_i :\subseteq \Sigma^\omega \to M_i$ be a representation for all i.
1. For all naming systems δ we have:
   $\delta_1 \leq \delta$ and $\delta_2 \leq \delta$, iff $\delta_1 \vee \delta_2 \leq \delta$.
2. For all naming systems δ we have:
   There is a computable function $f :\subseteq \Sigma^* \times \Sigma^a \to \Sigma^b$ such that for each $i \in \mathbb{N}$ the function $f(0^i, \cdot)$ translates δ_i to δ, iff $(\delta_0 \vee \delta_1 \vee \ldots) \leq \delta$.

The proof is easy (Exercise 3.3.6). Obviously, $\delta_1 \vee \delta_2$ is the, up to equivalence unique, least upper bound of δ_1 and δ_2. Disjunction is commutative and associative on equivalence classes: $\delta_1 \vee \delta_2 \equiv \delta_2 \vee \delta_1$ and $(\delta_1 \vee \delta_2) \vee \delta_3 \equiv \delta_1 \vee (\delta_2 \vee \delta_3)$. In general, $\tau_{\delta_1} \vee \tau_{\delta_2}$ is not second countable. However, if δ_1 and δ_2 are admissible and $M_1 \cap M_2 = \emptyset$, then $\delta_1 \vee \delta_2$ is admissible (Exercise 3.3.9).

In the following we use any $\eta^{\omega\omega}$-name of a (δ_1, δ_2)-realization g of a function f as a name of f (Theorem 2.3.8, Definition 3.1.3). Since g realizes also every restriction of f, we fix $N = \text{dom}(f)$ in advance.

Definition 3.3.13 (space of continuous functions). *For any two naming systems $\gamma_1 :\subseteq \Sigma^a \to M_1$ and $\gamma_2 :\subseteq \Sigma^b \to M_2$ $(a, b \in \{*, \omega\})$ and any set $N \subseteq M_1$ let*

1. *$[\gamma_1 \to \gamma_2]_N$ be the representation of the set $C(\gamma_1, \gamma_2, N)$ of the (γ_1, γ_2)-continuous functions $f :\subseteq M_1 \to M_2$ with $N = \mathrm{dom}(f)$, defined by*

$$[\gamma_1 \to \gamma_2]_N(p) = f : \iff \left(f \circ \gamma_1(q) = \gamma_2 \circ \eta_p^{ab}(q) \text{ whenever } \gamma_1(q) \in N \right),$$

2. *$[\gamma_1 \to \gamma_2]_N^c$ be the notation of the set $C^c(\gamma_1, \gamma_2, N)$ of the (γ_1, γ_2)-computable functions $g :\subseteq M_1 \to M_2$ with $N = \mathrm{dom}(g)$, defined by*

$$[\gamma_1 \to \gamma_2]_N^c(w) = g : \iff \left(g \circ \gamma_1(q) = \gamma_2 \circ \xi_w^{ab}(q) \text{ whenever } \gamma_1(q) \in N \right)$$

for all $p \in \Sigma^\omega$, $q \in \Sigma^a$ and $w \in \Sigma^*$.

By the following characterization the representations $[\gamma_1 \to \gamma_2]_N$ and $[\gamma_1 \to \gamma_2]_N^c$ are tailor-made for making the evaluation function computable.

Lemma 3.3.14 (characterization). Define the function
apply $:\subseteq C(\gamma_1, \gamma_2, N) \times M_1 \to M_2$ by $\mathrm{apply}(f, x) := f(x)$. Then, for any representation δ of any subset of $C(\gamma_1, \gamma_2, N)$ and notation ν of any subset of $C^c(\gamma_1, \gamma_2, N)$,

$$\text{apply is } (\delta, \gamma_1, \gamma_2)\text{-computable, iff } \delta \leq [\gamma_1 \to \gamma_2]_N ,$$

$$\text{apply is } (\delta, \gamma_1, \gamma_2)\text{-continuous, iff } \delta \leq_t [\gamma_1 \to \gamma_2]_N ,$$

$$\text{apply is } (\nu, \gamma_1, \gamma_2)\text{-computable, iff } \nu \leq [\gamma_1 \to \gamma_2]_N^c .$$

Therefore, $[\gamma_1 \to \gamma_2]_N$ is the "poorest" representation which makes evaluation computable etc. .

Proof: Assume that apply is $(\delta, \gamma_1, \gamma_2)$-computable. Then there is a computable function $h :\subseteq \Sigma^\omega \times \Sigma^a \to \Sigma^b$ with $\delta(p) \circ \gamma_1(q) = \mathrm{apply}(\delta(p), \gamma_1(q)) = \gamma_2 \circ h(p, q)$. By the smn-theorem, there is a computable function $r : \Sigma^\omega \to \Sigma^\omega$ with $h(p, q) = \eta_{r(p)}^{ab}(q)$, hence $\delta(p) \circ \gamma_1(q) = \gamma_2 \circ \eta_{r(p)}^{ab}(q)$. We obtain $\delta(p) = [\gamma_1 \to \gamma_2]_N(r(p))$ by Definition 3.3.13, hence $\delta \leq [\gamma_1 \to \gamma_2]_N$.

On the other hand, $\mathrm{apply}([\gamma_1 \to \gamma_2]_N(p), \gamma_1(q)) = [\gamma_1 \to \gamma_2]_N(p) \circ \gamma_1(q) = \gamma_2 \circ \eta_p^{ab}(q)$ by Definition 3.3.13. By the utm-theorem, $(p, q) \mapsto \eta_p^{ab}(q)$ is computable. Therefore, the function apply is $([\gamma_1 \to \gamma_2]_N, \gamma_1, \gamma_2)$-computable. If $\delta \leq [\gamma_1 \to \gamma_2]_N$, then apply is also $(\delta, \gamma_1, \gamma_2)$-computable.

The other two statements can be proved similarly.   □

Lemma 3.3.14 can be considered as the naming system version of the smn-theorem and the utm-theorem.

Theorem 3.3.15 (type conversion). For $i = 1, 2, 3$ let δ_i be a representation of M_i. For any total function $f : M_1 \times M_2 \to M_3$ define a function $T(f) : M_1 \to M_3^{M_2}$ by

$$T(f)(x)(y) := f(x, y) .$$

Then
1. $T \circ [[\delta_1, \delta_2] \to \delta_3] \equiv [\delta_1 \to [\delta_2 \to \delta_3]]$,
2. f is $([\delta_1, \delta_2], \delta_3)$-computable $\iff$ $T(f)$ is $(\delta_1, [\delta_2 \to \delta_3])$-computable

where we use the abbreviations $[\gamma_1 \to \gamma_2] := [\gamma_1 \to \gamma_2]_{\text{range}(\gamma_1)}$.

Proof: 1. We have

$$\big(T \circ [[\delta_1, \delta_2] \to \delta_3]\big)(p)\big(\delta_1(q)\big)\delta_2(r) = [[\delta_1, \delta_2] \to \delta_3](p)\big(\delta_1(q), \delta_2(r)\big)$$
$$= [[\delta_1, \delta_2] \to \delta_3](p)[\delta_1, \delta_2]\langle q, r\rangle$$
$$= \delta_3 \eta_p^{\omega\omega}\langle q, r\rangle$$

and

$$[\delta_1 \to [\delta_2 \to \delta_3]](s)\big(\delta_1(q)\big)\delta_2(r) = \big([\delta_2 \to \delta_3]\eta_s^{\omega\omega}(q)\big)\delta_2(r)$$
$$= \delta_3 \eta_{\eta_s^{\omega\omega}(q)}^{\omega\omega}(r) .$$

By the utm-theorem and the smn-theorem for $\eta^{\omega\omega}$ (Definition 2.3.3, Theorem 2.3.13), there are computable functions a, b, c, d, e with

$$\eta_p^{\omega\omega}\langle q, r\rangle = a(\langle p, q\rangle, r) = \eta_{b\langle p, q\rangle}^{\omega\omega}(r) = \eta_{\eta_{c(p)}^{\omega\omega}(q)}^{\omega\omega}(r) ,$$

$$\eta_{\eta_p^{\omega\omega}(q)}^{\omega\omega}(r) = d(p, \langle q, r\rangle) = \eta_{e(p)}^{\omega\omega}\langle q, r\rangle .$$

Then c translates $T \circ [[\delta_1, \delta_2] \to \delta_3]$ to $[\delta_1 \to [\delta_2 \to \delta_3]]$ and e translates $[\delta_1 \to [\delta_2 \to \delta_3]]$ to $T \circ [[\delta_1, \delta_2] \to \delta_3]$.
2. This follows from Property 1. $\qquad\square$

The set of infinite sequences on M can be considered as an infinite product (Definition 3.3.3) or as the set of functions $f : \mathbb{N} \to M$ (Definition 3.3.13). The canonical representations are equivalent.

Lemma 3.3.16 (sequences). For any representation $\delta :\subseteq \Sigma^\omega \to M$,

$$[\nu_{\mathbb{N}} \to \delta]_{\mathbb{N}} \equiv [\delta]^\omega .$$

The proof is left as Exercise 3.3.14. In general, not very much can be said about the final topology of $[\gamma_1 \to \gamma_2]_N$ in terms of the final topologies of γ_1 and γ_2, even for admissible representations. However, we will get an interesting characterization for real functions and standard representations of real numbers (Lemma 6.1.7).

The restriction to a class of functions with a single domain $N \subseteq M$ is inconvenient in many applications. In general a continuous function $g \in \mathrm{F}^{ab}$ is a (γ, γ_0)-realization of many functions $f :\subseteq M \to M_0$. We can associate with each function $g \in \mathrm{F}^{ab}$

- the function $f :\subseteq M \to M_0$ strongly (γ, γ_0)-realized by g (if it exists) (Exercise 3.1.5),
- the function $f :\subseteq M \to M_0$ with maximal domain (γ, γ_0)-realized by g (Exercise 3.1.6),
- the set of all functions $f :\subseteq M \to M_0$ (γ, γ_0)-realized by g (Exercise 3.1.10.

Via the representation η^{ab} we obtain three representations of (γ, γ_0)-continuous functions.

Remark 3.3.17 (representations of continuous functions).
1. Consider Definition 3.1.3, where $Y = \Sigma^a$ and $Y_0 = \Sigma^b$ ($a, b \in \{*, \omega\}$). Let $[\gamma \to \gamma_0]_s$ denote the standard representation of the set of all *strongly* $(\gamma \to \gamma_0)$-continuous functions $f :\subseteq M \to M_0$ (Exercise 3.1.5) defined by $[\gamma \to \gamma_0]_s(p) = f$, iff η_p^{ab} is a (γ, γ_0)-realization of f and $\eta_p^{ab}(q) = \mathrm{div}$, if $q \in \mathrm{dom}(\gamma)$ and $\gamma(q) \notin \mathrm{dom}(f)$. Strong representations have been considered, for example in [Wei87, Wei93].
2. Consider Definition 3.1.3, where $Y = \Sigma^a$ and $Y_0 = \Sigma^b$ ($a, b \in \{*, \omega\}$). Define a representation $[\gamma \to \gamma_0]_m$ of functions $f :\subseteq M \to M_0$ by $[\gamma \to \gamma_0]_m(p) = f$, iff f is the function with maximal domain (γ, γ_0)-realized by η_p^{ab} (Exercise 3.1.6).
3. Consider Definition 3.1.3, where $Y = \Sigma^a$ and $Y_0 = \Sigma^b$ ($a, b \in \{*, \omega\}$). Define a representation $[\gamma \to \gamma_0]_{set}$ of sets of functions $f :\subseteq M \to M_0$ by $[\gamma \to \gamma_0]_{set}(p)$ is the set of all functions $f :\subseteq M \to M_0$ (γ, γ_0)-realized by η_p^{ab}. Thus, $[\gamma \to \gamma_0]_{set}(p)$ is the set of all restrictions of $[\gamma \to \gamma_0]_m(p)$ (see 2 above). Define reducibility for set-valued naming systems by

$$\delta \leq \delta' \iff (\forall p)\delta(p) \subseteq \delta' \circ f(p) \text{ for some computable function } f$$

(Exercise 3.1.7). In particular,

$$[\gamma \to \gamma_0]_{set} \leq [\gamma' \to \gamma_0']_{set} \quad \text{if} \quad \gamma' \leq \gamma \quad \text{and} \quad \gamma_0 \leq \gamma_0' .$$

Representations can be adjusted to make given functions computable ([Wei87], Chap. 2.7).

Remark 3.3.18 (adjustment of representations). For $i = 1, 2$ let $\delta_i :\subseteq \Sigma^\omega \to M_i$ be representations and let $f : M_1 \to M_2$. If f is not (δ_1, δ_2)-computable, there are two possibilities to make it computable: either to strengthen δ_1 or to weaken δ_2.
1. Define a representation δ_f of M_1 by

$$\delta_f(p) = x, \text{ iff } p = \langle p_1, p_2 \rangle \text{ with } \delta_1(p_1) = x \text{ and } \delta_2(p_2) = f(x) .$$

Then for any representation γ of M_1 with $\gamma \le \delta_1$:

$$f \text{ is } (\gamma, \delta_2)\text{-computable} \iff \gamma \le \delta_f .$$

Therefore, δ_f is the weakest representation $\gamma \le \delta_1$ for which f is (γ, δ_2)-computable. There are straightforward extensions to finitely many functions and to an infinite sequence of functions. As we will see, a generalization to the case $f : M_1 \times M_1 \to M_2$ is not possible in general (Theorem 4.1.16).

2. Define a representation δ^f of M_2 by

$$\delta^f := f \circ \delta_1 \vee \delta_2 .$$

Then for any representation γ of M_2 with $\delta_2 \le \gamma$:

$$f \text{ is } (\delta_1, \gamma)\text{-computable} \iff \delta^f \le \gamma .$$

Therefore, δ^f is the strongest representation γ with $\delta_2 \le \gamma$ for which f is (δ_1, γ)-computable. The definition of δ^f can be generalized to functions with two or more arguments. For two or more functions use disjunctions (Lemma 3.3.12).

The properties of δ_f and δ^f hold accordingly for "$\le_t$" and "-continuous" instead of "$\le$" and "-computable", respectively.

Exercises 3.3.

1. Prove Lemma 3.3.2. The equivalence problem of ν_N is not r.e. in general. Find a counter example.
   (Hint: $\nu(0w) := \{\nu_\mathbb{N}(w)\}$, $\nu(1w) := \{\nu_\mathbb{N}(w), \nu_\mathbb{N}(w) + 1\}$, $\mathbb{N} \setminus N$ not r.e.)
2. Consider Lemma 3.3.5. Define $\mathbf{S}'$ by substituting ν' for ν in the definition of $\mathbf{S}_{12}$, where $\nu'\langle u, v \rangle := \nu_1(u) \times \nu_2(v)$. Show $\delta_{\mathbf{S}'} \equiv \delta_{\mathbf{S}_{12}}$.
3. Prove Lemma 3.3.6. Can Property 4 be generalized to multi-valued functions?
4. Prove Lemma 3.3.8.2.
5. Define computable topological spaces $\mathbf{S}_i = (\mathbb{N}, \sigma_i, \nu_i)$ $(i = 1, 2)$ by $\sigma_1(w) := \{n \in \mathbb{N} \mid n \le \nu_\mathbb{N}(w)\}$ and $\sigma_2(w) := \{n \in \mathbb{N} \mid n \ge \nu_\mathbb{N}(w)\}$. Show: $\nu_\mathbb{N} \equiv \delta_{\mathbf{S}_1} \wedge \delta_{\mathbf{S}_2}$.
6. Prove Lemma 3.3.12.
7. Show that for any two representations δ_1 and δ_2 with final topologies τ_{δ_1} and τ_{δ_2},
   a) $\tau_{\delta_1} \otimes \tau_{\delta_2} \subseteq \tau_{[\delta_1, \delta_2]}$,
   b) $\tau_{\delta_1} \wedge \tau_{\delta_2} \subseteq \tau_{\delta_1 \wedge \delta_2}$,
   c) $\tau_{\delta_1} \vee \tau_{\delta_2} = \tau_{\delta_1 \vee \delta_2}$,
   d) $\tau_{\delta_1} \wedge \tau_{\delta_2} = \tau_{\delta_1 \wedge \delta_2}$, if δ_1 and δ_2 are admissible.

(The product of the final topologies is a subset of the final topology of the product etc. , see Lemma 3.4.5 and Corollary 3.4.16 in [Wei87].)

8. Show that in Lemma 3.3.9.1 the equivalence problem of ν is r.e. , if $\sigma_1 \cap \sigma_2 = \emptyset$.

9. Show that $\delta_1 \vee \delta_2$ is admissible, if δ_1 and δ_2 are admissible and $M_1 \cap M_2 = \emptyset$.

10. Complete the proof of Lemma 3.3.14.

11. For $i = 1, 2, 3$ let δ_i be a representation of M_i. Then the composition

$$H : (g, f) \mapsto g \circ f$$

for total functions $f : M_1 \to M_2$ and $g : M_2 \to M_3$ is

$$([\delta_2 \to \delta_3], [\delta_1 \to \delta_2], [\delta_1 \to \delta_3]) \text{ -computable}$$

(where $[\gamma_1 \to \gamma_2] := [\gamma_1 \to \gamma_2]_{\text{range}(\gamma_1)}$).

12. Prove $\delta_{\mathbb{B}} \equiv [\nu_{\mathbb{N}} \to \nu_{\mathbb{N}}]_{\mathbb{N}}$.

13. Show that the operations "product", "conjunction" and "disjunction" are monotone w.r.t. "$\leq$" and "$\leq_t$" and, furthermore, that

$$(\delta_0 \leq \delta_1 \text{ and } \delta_2 \leq \delta_3) \implies [\delta_1 \to \delta_2]_N \leq [\delta_0 \to \delta_3]_N \ ,$$

$$(\delta_0 \leq_t \delta_1 \text{ and } \delta_2 \leq_t \delta_3) \implies [\delta_1 \to \delta_2]_N \leq_t [\delta_0 \to \delta_3]_N \ .$$

14. Let $\delta :\subseteq \Sigma^\omega \to M$ be a representation. Show $[\nu_{\mathbb{N}} \to \delta]_{\mathbb{N}} \equiv [\delta]^\omega$.

4. Computability on the Real Numbers

Real numbers are the basic objects in analysis. For most non-mathematicians a real number is an infinite decimal fraction, for example $\pi = 3.14159\ldots$. Mathematicians prefer to define the real numbers *axiomatically* as follows: $(\mathbb{R}, +, \cdot, 0, 1, <)$ is, up to isomorphism, the only Archimedean ordered field satisfying the axiom of continuity [Die60]. The set of real numbers can also be *constructed* in various ways, for example by means of Dedekind cuts or by completion of the (metric space of) rational numbers. We will neglect all foundational problems and assume that the real numbers form a well-defined set $\mathbb{R}$ with all the properties which are proved in analysis. We will denote the real line topology, that is, the set of all open subsets of $\mathbb{R}$, by $\tau_{\mathbb{R}}$.

In Sect. 4.1 we introduce several representations of the real numbers, three of which (and the equivalent ones) will survive as useful. We introduce a representation ρ^n of $\mathbb{R}^n$ by generalizing the definition of the main representation ρ of the set $\mathbb{R}$ of real numbers. In Sect. 4.2 we discuss the computable real numbers. Sect. 4.3 is devoted to computable real functions. We show that many well known functions are computable, and we show that partial summation of sequences is computable and that limit operator on sequences of real numbers is computable, if a modulus of convergence is given. We also prove a computability theorem for power series.

Convention 4.0.1. We still assume that Σ is a fixed finite alphabet containing all the symbols we will need.

4.1 Various Representations of the Real Numbers

According to the principles of TTE we introduce computability on $\mathbb{R}$ by naming systems. Since the set $\mathbb{R}$ is not countable, it has no notation $\nu :\subseteq \Sigma^* \to \mathbb{R}$ (onto) but only representations. Most of its numerous representations have no applications. In this section we introduce three representations $\rho, \rho_<$ and $\rho_>$ of the set of real numbers (and some equivalent ones) which induce the most important computability concepts. We will also discuss some other representations which, for various reasons, are only of little interest in computable analysis.

Since the set $\mathbb{Q}$ of rational numbers is dense in $\mathbb{R}$, every real number has arbitrarily tight lower and arbitrarily tight upper rational bounds. Every real number x can be identified by the set

$$\{(a;b) \mid a,b \in \mathbb{Q},\ a < x < b\}$$

of all open intervals with rational endpoints containing x, by the set

$$\{a \in \mathbb{Q} \mid a < x\}$$

of all rational numbers smaller than x or by the set

$$\{a \in \mathbb{Q} \mid a > x\}$$

of all rational numbers greater than x. According to the concept of standard admissible representations

(Definitions 3.2.1, 3.2.2) a name of x will be a list of all open intervals with rational endpoints containing x, a list of all rational lower bounds of x or a list of all rational upper bounds of x, respectively.

Convention 4.1.1. In the following we will abbreviate $\nu_{\mathbb{Q}}(w)$ by $\overline{w}$ where $\nu_{\mathbb{Q}}$ is our standard notation of the rational numbers (Definition 3.1.2).

First, we introduce a standard notation I^n of all rational n-dimensional cubes with edges parallel to the coordinate axes and rational vertices.

Definition 4.1.2 (notation of rational cubes). *Assume $n \geq 1$.*
  1. For $(a_1, \ldots, a_n) \in \mathbb{R}^n$ define the (maximum) norm

$$||(a_1, \ldots, a_n)|| := \max |a_1|, \ldots, |a_n|$$

  and for $x, y \in \mathbb{R}^n$ define the (maximum) distance by

$$d(x,y) := ||x - y|| .$$

  2. Let $\mathrm{Cb}^{(n)} := \{B(a,r) \mid a \in \mathbb{Q}^n, r \in \mathbb{Q}, r > 0\}$ be the set of open rational balls (or cubes), where $B(a,r) := \{x \in \mathbb{R}^n \mid d(x,a) < r\}$.
  3. Define a notation I^n of the set $\mathrm{Cb}^{(n)}$ by

$$I^n(\iota(v_1) \ldots \iota(v_n)\iota(w)) := B((\overline{v_1}, \ldots, \overline{v_n}), \overline{w}) .$$

  4. By $\overline{I}^n(w)$ we denote the closure of the cube $I^n(w)$.

In particular, $\mathrm{Cb}^{(1)}$ is the set of all open intervals with rational endpoints and $I^1(\iota(v)\iota(w))$ is the open interval $(\overline{v} - \overline{w}; \overline{v} + \overline{w})$, $\mathrm{Cb}^{(2)}$ is the set of all open squares with rational vertices (edges parallel to the coordinate axes) and $I^2(\iota(v_1)\iota(v_2)\iota(w))$ is the open square $(\overline{v_1} - \overline{w}; \overline{v_1} + \overline{w}) \times (\overline{v_2} - \overline{w}; \overline{v_2} + \overline{w})$, $\mathrm{Cb}^{(3)}$ is the set of all open cubes with rational vertices (edges parallel to the coordinate axes), etc. .

We introduce representations $\rho, \rho_<$ and $\rho_>$ as our standard representations of computable topological spaces as follows.

Definition 4.1.3 (the representations $\rho, \rho_<$ and $\rho_>$). *Define computable topological spaces*

- $\mathbf{S}_= := (\mathbb{R}, \mathrm{Cb}^{(1)}, \mathrm{I}^1)$,
- $\mathbf{S}_< := (\mathbb{R}, \sigma_<, \nu_<)$, $\nu_<(w) := (\overline{w}; \infty)$,
- $\mathbf{S}_> := (\mathbb{R}, \sigma_>, \nu_>)$, $\nu_>(w) := (-\infty; \overline{w})$,

and let $\rho := \delta_{\mathbf{S}_=}$, $\rho_< := \delta_{\mathbf{S}_<}$ and $\rho_> := \delta_{\mathbf{S}_>}$.

(The sets $\sigma_<$ and $\sigma_>$ are defined implicitly.) Notice that $\mathbf{S}_=$, $\mathbf{S}_<$ and $\mathbf{S}_>$ are computable topological spaces (Definition 3.2.1), since the properties $\nu_{\mathbb{Q}}(u) = \nu_{\mathbb{Q}}(v)$ and $\mathrm{I}^1(u) = \mathrm{I}^1(v)$ are decidable in (u, v). By Definition 3.2.2,

$$\rho(p) = x \quad \Longleftrightarrow \quad \{J \in \mathrm{Cb}^{(1)} \mid x \in J\} = \{\mathrm{I}^1(w) \mid \iota(w) \lhd p\} ,$$

or roughly speaking, iff p is a list of all $J \in \mathrm{Cb}^{(1)}$ such that $x \in J$.

Similarly, $\rho_<(p) = x$, iff p is a list of all rational numbers a such that $a < x$, and $\rho_>(p) = x$, iff p is a list of all rational numbers a such that $a > x$. Fig. 4.1 shows some open intervals $J \in \mathrm{Cb}^{(1)}$ with $x \in J$ and some rational numbers a with $a < y$.

Fig. 4.1. Some open intervals $J \in \mathrm{Cb}^{(1)}$ with $x \in J$ and some rational numbers a with $a < y$

The final topologies of the above representations can be characterized easily (Definition 3.1.3.2, Lemma 3.2.5.3):

Lemma 4.1.4 (final topologies of $\rho, \rho_<$ and $\rho_>$).
1. The final topology of ρ is the real line topology $\tau_{\mathbb{R}}$.
2. The final topology of $\rho_<$ is $\tau_{\rho_<} := \{(x; \infty) \mid x \in \mathbb{R}\}$.
3. The final topology of $\rho_>$ is $\tau_{\rho_>} := \{(-\infty; x) \mid x \in \mathbb{R}\}$.

Proof: $\mathrm{Cb}^{(1)}$ generates $\tau_{\mathbb{R}}$, $\sigma_<$ generates $\tau_{\rho_<}$ and $\sigma_>$ generates $\tau_{\rho_>}$.   □

Another important representation of the real numbers is the *Cauchy representation*. Since every real number is the limit of a Cauchy sequence of rational numbers, such sequences can be used as names of real numbers. However, the "naive Cauchy representation" which considers *all* converging sequences of rational numbers as names is not very useful (Example 4.1.14.1).

For the Cauchy representation we consider merely the "rapidly converging" sequences of rational numbers.

Definition 4.1.5 (Cauchy representation).   *Define the Cauchy representation $\rho_C :\subseteq \Sigma^\omega \to \mathbb{R}$ by*

$$\rho_C(p) = x : \Longleftrightarrow \begin{cases} \text{there are words } \ w_0, w_1 \ldots \in \text{dom}(\nu_{\mathbb{Q}}) \\ \text{such that} \ \ p = \iota(w_0)\iota(w_1)\iota(w_2)\ldots, \\ |\overline{w}_i - \overline{w}_k| \leq 2^{-i} \ \ \text{for} \ \ i < k \ \ \text{and} \ \ x = \lim_{i \to \infty} \overline{w}_i \ . \end{cases}$$

The representation ρ, the Cauchy representation ρ_C and many variants of them are equivalent:

Lemma 4.1.6 (representations equivalent to ρ). The following representations of the real numbers are equivalent to the standard representation $\rho :\subseteq \Sigma^\omega \to \mathbb{R}$:
1. the Cauchy representation ρ_C,
2. the representations ρ_C', ρ_C'' and ρ_C''' obtained by substituting

$$|\overline{w}_i - \overline{w}_k| < 2^{-i}, \quad |\overline{w}_i - x| < 2^{-i} \quad \text{or} \quad |\overline{w}_i - x| \leq 2^{-i} \ ,$$

   respectively, for $|\overline{w}_i - \overline{w}_k| \leq 2^{-i}$ in the definition of ρ_C,
3.

$$\rho^a(p) = x : \Longleftrightarrow \begin{cases} \text{there are words} \ \ u_0, u_1 \ldots \in \text{dom}(\text{I}^1) \\ \text{such that} \ \ p = \iota(u_0)\iota(u_1)\ldots, \\ (\forall k) \left(\overline{\text{I}}^1(u_{k+1}) \subseteq \text{I}^1(u_k) \ \text{and length}(\text{I}^1(u_k)) < 2^{-k} \right) \\ \text{and} \ \ \{x\} = \text{I}^1(u_0) \cap \text{I}^1(u_1) \cap \ldots , \end{cases}$$

4.

$$\rho^b(p) = x \ : \Longleftrightarrow \ \ \{x\} = \bigcap \left\{ \overline{\text{I}}^1(v) \mid \iota(v) \lhd p \right\} \ .$$

The representations ρ_C', ρ_C'' and ρ_C''' are inessential modifications of the Cauchy representation, and ρ^a is a representation by strongly nested, rapidly converging sequences of open intervals. A ρ^b-name of x is a list of closed rational intervals for which x is the only common point. While a ρ-name of x is a list of *all* open rational intervals containing x, these characterizations show that it suffices to list merely "sufficiently many" of them.

Proof:

$\rho^b \leq \rho^a$: Define representations δ_1 and δ_2 of $\mathbb{R}$ by $\delta_1(p) = x$, iff

$$p = \iota(u_0)\iota(u_1)\ldots, \quad \overline{\text{I}}^1(u_{k+1}) \subseteq \overline{\text{I}}^1(u_k) \ \text{and} \ \{x\} = \overline{\text{I}}^1(u_0) \cap \overline{\text{I}}^1(u_1) \cap \ldots , $$

and $\delta_2(p) = x$, iff

$$p = \iota(u_0)\iota(u_1)\dots, \quad \overline{\mathrm{I}}^1(u_{k+1}) \subseteq \mathrm{I}^1(u_k) \text{ and } \{x\} = \mathrm{I}^1(u_0) \cap \mathrm{I}^1(u_1) \cap \dots .$$

There is a computable function $h :\subseteq \Sigma^\omega \to \Sigma^\omega$ such that

$$h(p) = \iota(v_0)\iota(v_1)\dots \text{ where } \overline{\mathrm{I}}^1(v_i) = \bigcap \left\{ \overline{\mathrm{I}}^1(v) \mid \iota(v) \vartriangleleft p_{<i+i_p} \right\},$$

where i_p is the smallest number k such that $\iota(v) \vartriangleleft p_{<k}$ for some $v \in \mathrm{dom}(\mathrm{I}^1)$. Then obviously, the function h translates ρ^b to δ_1.
There is a computable function $g : \mathbb{N} \times \Sigma^* \to \Sigma^*$ such that

$$\mathrm{I}^1(w) = B(a,r) \implies \mathrm{I}^1 \circ g(i,w) = B(a, (1 + 2^{-i}) \cdot r) .$$

There is a computable function $f :\subseteq \Sigma^\omega \to \Sigma^\omega$ such that

$$f(\iota(w_0)\iota(w_1)\iota(w_2)\dots) = \iota \circ g(0, w_0)\iota \circ g(1, w_1)\iota \circ g(2, w_2)\dots .$$

Then f translates δ_1 to δ_2. It remains to select from any δ_2-name a rapidly converging subsequence of intervals. There is a Type-2 machine M which on input $p = \iota(u_0)\iota(u_1)\dots \in \mathrm{dom}(\delta_2)$ computes a sequence $q = \iota(u_{m_0})\iota(u_{m_1})\dots$ where m_k is the smallest number $i > m_{k-1}$ such that $\mathrm{length}(\mathrm{I}^1(u_i)) < 2^{-k}$. Then f_M, the function computed by the machine M, translates δ_2 to ρ^a. Therefore, $\rho^b \leq \rho^a$.

$\rho^a \leq \rho'_C$ and $\rho^a \leq \rho''_C$: There is a computable function $f :\subseteq \Sigma^\omega \to \Sigma^\omega$ mapping any $p = \iota(u_0)\iota(u_1)\iota(u_1)\dots \in \mathrm{dom}(\rho^a)$ to $q = \iota(v_0)\iota(v_1)\iota(v_1)\dots$ such that $\nu_\mathbb{Q}(v_i) = \inf(\mathrm{I}^1(u_i))$ for all i. If $\rho^a(p) = x$, then $\overline{v}_0 < \overline{v}_1 < \overline{v}_2 < \dots < x$ and $|\overline{v}_i - x| < 2^{-i}$ for all i. Therefore, f translates ρ^a to ρ'_C and ρ''_C.

$\rho'_C \leq \rho_C$ and $\rho''_C \leq \rho'''_C$: The identity in Σ^ω translates ρ'_C into ρ_C and ρ''_C into ρ'''_C.

$\rho_C \leq \rho'''_C$: Suppose $|\overline{w}_i - \overline{w}_k| \leq 2^{-i}$ for $i < k$ and $x = \lim_{i \to \infty} \overline{w}_i$. Since

$$|\overline{w}_i - x| \leq |\overline{w}_i - \overline{w}_k| + |\overline{w}_k - x| \leq 2^{-i} + |\overline{w}_k - x|$$

for all $i < k$, $|\overline{w}_i - x| \leq 2^{-i}$, and so the identity translates ρ_C to ρ'''_C.

$\rho'''_C \leq \rho$: If $\rho'''_C(\iota(w_0)\iota(w_1)\iota(w_2)\dots) = x$, then for any $v \in \mathrm{dom}(\mathrm{I}^1)$:

$$x \in \mathrm{I}^1(v) \iff (\exists i)[\overline{w}_i - 2^{-i}; \overline{w}_i + 2^{-i}] \subseteq \mathrm{I}^1(v) .$$

Let ν_Σ be the standard numbering of Σ^* (Sect. 1.4). There is a Type-2 machine M mapping every $p = \iota(w_0)\iota(w_1)\iota(w_2)\dots \in \mathrm{dom}(\rho'''_C)$ to a sequence $q = \iota(v_0)\iota(v_1)\iota(v_2)\dots$ such that

$$v_{\langle i,k \rangle} = \begin{cases} \nu_\Sigma(k) & \text{if } \nu_\Sigma(k) \in \mathrm{dom}(\mathrm{I}^1) \\ & \text{and } [\overline{w}_i - 2^{-i}; \overline{w}_i + 2^{-i}] \subseteq \mathrm{I}^1(\nu_\Sigma(k)) \\ u & \text{otherwise} \end{cases}$$

where u is a word with $\mathrm{I}^1(u) = (\overline{w}_0 - 2; \overline{w}_0 + 2)$. If $\rho'''_C(p) = x$, then q is a list of all words v such that $x \in \mathrm{I}^1(v)$. Therefore, the function f_M translates ρ'''_C to ρ.

$\rho \le \rho^b$: The identity on Σ^ω translates ρ to ρ^b.
Therefore, the given representations are equivalent.   □

The representation δ informally introduced in Sect. 1.3.2 has the property $\rho^a \le \delta \le \rho^b$ (Lemma 4.1.6) and so is equivalent to ρ.

By definition, a ρ-name of x is a list of all intervals $(a; b)$ with rational endpoints such that $x \in (a; b)$. Every arbitrarily tight lower and every arbitrarily tight upper rational bound of x can be obtained from a finite prefix of p. This is the characteristic common property of all representations equivalent to ρ:

Lemma 4.1.7 (characterization of ρ). For every representation $\delta :\subseteq \Sigma^\omega \to \mathbb{R}$,

$$\delta \le \rho \iff \begin{cases} \{(x, a) \in \mathbb{R} \times \mathbb{Q} \mid a < x\} & \text{is } (\delta, \nu_\mathbb{Q})\text{-r.e. and} \\ \{(x, a) \in \mathbb{R} \times \mathbb{Q} \mid x < a\} & \text{is } (\delta, \nu_\mathbb{Q})\text{-r.e. .} \end{cases}$$

This is essentially a special case of Theorem 3.2.10. For a proof see Exercise 4.1.3. Therefore, ρ is up to equivalence the "poorest" representation δ of the real numbers such that the properties "$a < x$" and "$x < a$" are r.e.

Also the representations $\rho_<$ and $\rho_>$ can be simplified.

Lemma 4.1.8 (representations equivalent to $\rho_<$). The following representations of the real numbers are equivalent to the representation $\rho_< :\subseteq \Sigma^\omega \to \mathbb{R}$:

1. $\rho_<^a(p) = x :\iff \begin{cases} \text{there are } u_0, u_1 \ldots \in \text{dom}(\nu_\mathbb{Q}) \\ \text{such that } p = \iota(u_0)\iota(u_1)\ldots, \\ \overline{u_0} < \overline{u_1} < \ldots < x \text{ and } x = \lim_{i \to \infty} \overline{u_i} , \end{cases}$

2. $\rho_<^b(p) = x :\iff x = \sup\{\overline{v} \mid \iota(v) \vartriangleleft p\} .$

The proof is left as Exercise 4.1.5. However, if we force rapid convergence, we obtain a representation equivalent to ρ (Exercise 4.1.6). Lemma 4.1.7 holds correspondingly for $\rho_<$ replacing ρ. Therefore, $\rho_<$ is the "poorest" representation δ of the real numbers such that the property "$a < x$" is r.e. The above considerations hold correspondingly for $\rho_>$ replacing $\rho_<$.

The following lemma is obvious already from our informal characterizations of ρ, $\rho_<$ and $\rho_>$.

Lemma 4.1.9.
1. $\rho \equiv \rho_< \wedge \rho_>$ (that is, ρ is the greatest lower bound of $\rho_<$ and $\rho_>$), in particular, $\rho \le \rho_<$ and $\rho \le \rho_>$.
2. $\rho_< \not\le_t \rho$, $\rho_> \not\le_t \rho$, $\rho_< \not\le_t \rho_>$ and $\rho_> \not\le_t \rho_<$.

Proof: 1. By Definition 3.3.7, $(\rho_< \wedge \rho_>)\langle p, q\rangle = x \iff \rho_<(p) = \rho_>(q) = x$. There is a Type-2 machine M which on input $p \in \mathrm{dom}(\rho)$ produces a list of all $\iota(w)$ for which there is some word v such that $\iota(v) \lhd p$ and $\overline{w}$ is the left endpoint of the interval $\mathrm{I}^1(v)$. Then the function f_M translates ρ to $\rho_<$. For a similar reason, $\rho \le \rho_>$. By Lemma 3.3.8, $\rho \le \rho_< \wedge \rho_>$.

For the other direction it suffices to prove $\rho_<^a \wedge \rho_>^a \le \rho^b$. There is a Type-2 machine M which on input $\langle p, q\rangle \in \mathrm{dom}(\rho_<^a \wedge \rho_>^a)$, where $p = \iota(u_0)\iota(u_1)\ldots$ and $q = \iota(v_0)\iota(v_1)\ldots$, produces a sequence $\iota(w_0)\iota(w_1)\ldots$, such that $\mathrm{I}^1(w_i) = (\overline{u}_i; \overline{v}_i)$. The function f_M translates $\rho_<^a \wedge \rho_>^a$ to ρ^b.

2. This follows from the simple observation that a lower bound cannot be obtained from a finite set of upper bounds and vice versa. More formally we can use the fact $\gamma' \le_t \gamma \implies \tau_\gamma \subseteq \tau_{\gamma'}$ from Theorem 3.1.8: $\rho_< \not\le_t \rho$ since $\tau_\rho \not\subseteq \tau_{\rho_<}$ etc.. $\qquad \square$

To sum up, one can roughly say that for a real number x, a $\rho_<$-name consists of all rational lower bounds of x, a $\rho_>$-name consists of all rational upper bounds of x, and a ρ-name consists of all rational lower bounds and all rational upper bounds of x. Since by Corollary 3.2.12 for admissible

representations only continuous functions can be computable, Lemma 4.1.4 tells us which of the three computability concepts is adequate in a given topological setting.

Computability of elements and functions induced by the representations ρ, $\rho_<$ and $\rho_>$ will be discussed in the following sections. As an instructive example we discuss the problem of finding a rational upper bound for a real number.

Example 4.1.10. Consider the multi-valued function

$$F : \mathbb{R} \rightrightarrows \mathbb{Q}, \quad \mathrm{R}_F := \{(x, a) \in \mathbb{R} \times \mathbb{Q} \mid x < a\} .$$

We prove the following effectiveness properties (Definition 3.1.3):
1. F is not $(\rho_<, \nu_\mathbb{Q})$-continuous.
2. F is $(\rho_>, \nu_\mathbb{Q})$-computable (and therefore, $(\rho, \nu_\mathbb{Q})$-computable).
3. F has no $(\rho, \nu_\mathbb{Q})$-continuous choice function (and therefore, no $(\rho_>, \nu_\mathbb{Q})$-continuous choice function).

Remember, that $\nu_\mathbb{Q}$ is admissible with discrete final topology $\tau_{\nu_\mathbb{Q}}$ (Example 3.2.4.1) .

1. Suppose F is $(\rho_<, \nu_\mathbb{Q})$-continuous. Then F has a continuous $(\rho_<^b, \nu_\mathbb{Q})$-realization $g :\subseteq \Sigma^\omega \to \Sigma^*$ (Lemma 4.1.8). Consider $\rho_<^b(p) = x$. By assumption, $g(p) = w$ with $x < \nu_\mathbb{Q}(w) = \overline{w}$ for some $w \in \Sigma^*$. Since g is continuous, there is some number n with $g[p_{<n}\Sigma^\omega] = \{w\}$. For some $k \ge n$, 11 is the suffix of $p_{<k}$. Choose $u \in \mathrm{dom}(\nu_\mathbb{Q})$ with $\overline{w} < \overline{u}$ and define $q := p_{<k}\iota(u)\iota(u)\ldots$. Then $\rho_<^b(q) = \overline{u}$ and $\nu_\mathbb{Q} \circ g(q) = \overline{w}$, since $q \in p_{<n}\Sigma^\omega \cap \mathrm{dom}(g)$. However, by assumption on g we must have $\overline{u} = \rho_<^b(q) < \nu_\mathbb{Q} \circ g(q) = \overline{w}$. Contradiction!

2. Let M be a Type-2 machine which for any input $p = \iota(w_0)\iota(w_1)\ldots$ prints the word w_0 and halts. Then $\rho_>(p) < \overline{w}_0 = \nu_\mathbb{Q} \circ f_M(p)$ for all

$p = \iota(w_0)\iota(w_1)\ldots \in \mathrm{dom}(\rho_>)$. Therefore, f_M is a $(\rho_>, \nu_{\mathbb{Q}})$-realization of the relation R.

3. Since the final topology $\tau_{\mathbb{R}}$ of the admissible representation ρ is connected, every $(\rho, \nu_{\mathbb{Q}})$-continuous function is constant by Corollary 3.2.13. But a choice function of F cannot be constant. $\qquad\square$

The definitions of ρ, ρ_C, $\rho_<$ and $\rho_>$ (and their variants) can be modified in various other ways without affecting the induced continuity or computability, respectively. So far we have used the set $\mathbb{Q}$ of the rational numbers as a "standard" dense countable subset of the real numbers and $\nu_{\mathbb{Q}}$ as its standard notation. Can we replace $\nu_{\mathbb{Q}}$ by some other notation ν_Q of a dense subset Q of $\mathbb{R}$ such that the resulting representations induce the same continuity or computability concepts on $\mathbb{R}$? The following "robustness" theorem gives an answer.

Theorem 4.1.11 (robustness). For any representation δ of the real numbers introduced in Definitions 4.1.3, 4.1.5, 4.1.6 and 4.1.8 consider δ as a function of $\nu_{\mathbb{Q}}$, that is, $\delta = D(\nu_{\mathbb{Q}})$. Then for every notation ν_Q of a dense subset $Q \subseteq \mathbb{R}$,
1. $\delta \equiv_t D(\nu_Q)$,
2. $\delta \equiv D(\nu_Q)$, if $\nu_{\mathbb{Q}}$ and ν_Q are r.e.-related
   (that is, if $\{(v, w, i) \mid |\nu_{\mathbb{Q}}(v) - \nu_Q(w)| < 2^{-i}\}$ is r.e.).

The proof is left as Exercise 4.1.8. Examples for notations r.e.-related to $\nu_{\mathbb{Q}}$ are any notation of $\mathbb{Q}$ equivalent to $\nu_{\mathbb{Q}}$ and any standard notation of the binary rational numbers $\mathbb{Q}_2 := \{z/2^n \mid z \in \mathbb{Z}, n \in \mathbb{N}\}$.

Computability concepts introduced via robust definitions are not sensitive to "inessential" modifications. It can be expected that they occur in many applications. On the other hand, computability concepts introduced via non-robust definitions are not very relevant.

The *Turing machine* is another famous example of a robust definition. Numerous variants of the original definition are used in the literature, all of which define the same notion of computable functions. Usually the representation by infinite decimal fractions is considered to be the most natural representation of the real numbers.

Definition 4.1.12 (finite and infinite base-n fractions). *For $n \geq 2$ define the notation $\nu_{\mathrm{b},n}$ of the finite base-n fractions and the representation $\rho_{\mathrm{b},n} :\subseteq \Sigma^\omega \to \mathbb{R}$ of the real numbers by infinite $-n$ fractions as follows:*

$$\mathrm{dom}(\nu_{\mathrm{b},n}) := \{\lambda, -\}(\Gamma^* \setminus 0\Gamma^*)\bullet\Gamma^*,$$

$$\mathrm{dom}(\rho_{\mathrm{b},n}) := \{\lambda, -\}(\Gamma^* \setminus 0\Gamma^*)\bullet\Gamma^\omega,$$

$$\nu_{\mathrm{b},n}(sa_k \ldots a_0 \bullet a_{-1} a_{-2} \ldots a_{-m}) := \overline{s} \cdot \sum_{k \geq i \geq -m} a_i \cdot n^i,$$

$$\rho_{\mathrm{b},n}(sa_k \ldots a_0 \bullet a_{-1} a_{-2} \ldots) := \overline{s} \cdot \sum_{i \leq k} a_i \cdot n^i,$$

where $a_i \in \Gamma_n := \{0, 1, \ldots, n-1\}$ for all $i \le k$, $\overline{s} := 1$, if $s = \lambda$, and $\overline{s} := -1$, if $s = -$. (We assume tacitly $\Gamma_n \subseteq \Sigma$.)

(Remember that λ is the empty word.) The following theorem summarizes some interesting properties of the representations by infinite base-n fractions.

Theorem 4.1.13 (infinite base-n fractions). For any $m, n \ge 2$ (assuming $\Gamma_m, \Gamma_n \subseteq \Sigma$)
  1. $\rho_{\mathrm{b},n}|^{\mathrm{Ir}} \equiv \rho|^{\mathrm{Ir}}$ (where $\mathrm{Ir} := \mathbb{R} \setminus \mathbb{Q}$ is the set of irrational real numbers),
  2. x is $\rho_{\mathrm{b},n}$-computable, iff x is ρ-computable.
  3. $\rho_{\mathrm{b},n} \le \rho$ and $\rho \not\le_t \rho_{\mathrm{b},n}$,
  4. $\rho_{\mathrm{b},m} \le \rho_{\mathrm{b},n}$, if $\mathrm{pd}(n) \subseteq \mathrm{pd}(m)$, $\rho_{\mathrm{b},m} \not\le_t \rho_{\mathrm{b},n}$ otherwise
     (where $\mathrm{pd}(n)$ denotes the set of prime divisors of n),
  5. $\rho_{\mathrm{b},n}$ has the final topology $\tau_\mathbb{R}$.
  6. $\rho_{\mathrm{b},n}$ is not admissible,
  7. $f :\subseteq \mathbb{R} \to \mathbb{R}$ is $(\rho_{\mathrm{b},n}, \rho)$-computable (-continuous),
     iff it is (ρ, ρ)-computable (-continuous).

Proof: 1. and 2. See Exercise 4.1.10

3. There is a Type-2 machine M which maps any sequence
$p = sa_k \ldots a_0 \bullet a_{-1} a_{-2} \ldots \in \mathrm{dom}(\rho_{\mathrm{b},n})$ to a sequence $\iota(u_0)\iota(u_1)\ldots$ with
$\overline{u}_j = \nu_{\mathrm{b},n}(sa_k \ldots a_0 \bullet a_{-1} a_{-2} \ldots a_{-j})$. Then M translates $\rho_{\mathrm{b},n}$ to ρ_C.
The relation $\rho \not\le_t \rho_{\mathrm{b},n}$ will be concluded from Property 4.

4. Suppose, $\mathrm{pd}(n) \subseteq \mathrm{pd}(m)$. Then for each $k \in \mathbb{N}$ numbers $l_k, b_k \in \mathbb{N}$ can be determined such that $m^{l_k} = b_k \cdot n^k$. For computing a $\rho_{\mathrm{b},n}$-name of $x = \rho_{\mathrm{b},m}(p)$, it suffices to compute integers $c_0, c_1, \ldots$ such that $c_k \le n^k \cdot x \le c_k + 1$ for all k. There is a Type-2 machine M, which on input (p, u) ($x := \rho_{\mathrm{b},m}(p)$, $k := \nu_\mathbb{N}(u)$) determines (names of) numbers $l_k, b_k \in \mathbb{N}$ such that $m^{l_k} = b_k \cdot n^k$ and then (a name of) a number c_k such that

$$c_k \cdot b_k \le m^{l_k} \cdot x \le (c_k + 1) \cdot b_k .$$

(For this purpose, M must read only the first l_k digits of p after the dot.) Since $m^{l_k} = b_k \cdot n^k$, we obtain immediately $c_k \le n^k \cdot x \le c_k + 1$, as required. Therefore, $\rho_{\mathrm{b},m}$ can be translated to $\rho_{\mathrm{b},n}$ by a Type-2 machine.

Now consider that the prime number $r \in \mathbb{N}$ divides n but not m. There is some $c \in \mathbb{N}$, $1 \le c < n$, with $n = c \cdot r$. The number $1/r$ has a $\rho_{\mathrm{b},m}$-name $p := \bullet a_{-1} a_{-2} \ldots$ which has neither the period 0 nor the period $(m-1)$. Suppose some continuous function $f :\subseteq \Sigma^\omega \to \Sigma^\omega$ translates $\rho_{\mathrm{b},m}$ to $\rho_{\mathrm{b},n}$. Then $f(p) = \bullet c 00 \ldots$ or $f(p) = \bullet 0(c-1)(n-1)(n-1) \ldots$. Consider the first case $f(p) = \bullet c 00 \ldots$. By continuity of f there is some l such that $f[\bullet a_{-1} a_{-2} \ldots a_{-l} \Sigma^\omega] \subseteq \bullet c \Sigma^\omega$. Choose $p' := \bullet a_{-1} a_{-2} \ldots a_{-l} 00 \ldots$. Then $f(p') \in \bullet c \Sigma^\omega$. We have $\rho_{\mathrm{b},m}(p') < 1/r$ (since p does not have the period 0) and $\rho_{\mathrm{b},n}(f(p')) \in \rho_{\mathrm{b},n}[\bullet c \Sigma^\omega]$, hence $\rho_{\mathrm{b},n}(f(p')) \ge 1/r$ (contradiction). Therefore, $\rho_{\mathrm{b},m} \not\le_t \rho_{\mathrm{b},n}$.
The second case can be treated similarly.

3. (continued) Suppose $\rho \leq_t \rho_{b,n}$. Choose some prime number m which does not divide n. Then $\rho_{b,m} \leq_t \rho \leq_t \rho_{b,n}$ by Property 1, but $\rho_{b,m} \not\leq_t \rho_{b,n}$ as proved above (contradiction). This shows $\rho \not\leq_t \rho_{b,n}$.

5. If $X \subseteq \mathbb{R}$ is open, then it is ρ-open by Lemma 4.1.4. Since $\rho_{b,n} \leq \rho$ by Property 3, X is $\rho_{b,n}$-open by Theorem 3.1.8. Let $X \subseteq \mathbb{R}$ be $\rho_{b,n}$-open. Then $\rho_{b,n}^{-1}[X] = A\Sigma^\omega \cap \text{dom}(\rho_{b,n})$ for some $A \subseteq \Sigma^*$. Consider $x \in X$, $x > 0$.

Case 1: $x = a/n^j$ for some integer a and some natural number j.

Then x has two $\rho_{b,n}$-names $p := $ "$a_k \ldots a_0 \bullet a_{-1} \ldots$" and $q := $ "$b_k \ldots b_0 \bullet b_{-1} \ldots$", such that there is some some $m \leq k$ with $a_m > 0$, $a_i = 0$ for $i < m$, $b_m = a_m - 1$ and $b_i = n - 1$ for $i < m$. Since $\rho_{b,n}^{-1}[X]$ is open in $\text{dom}(\rho_{b,n})$, there is a number $l > m$ such that $\rho_{b,n}[$"$a_k \ldots a_0 \bullet a_{-1} \ldots a_{-l}$" $\Sigma^\omega] \subseteq X$ and $\rho_{b,n}[$"$b_k \ldots b_0 \bullet b_{-1} \ldots b_{-l}$" $\Sigma^\omega] \subseteq X$. We obtain $[x; x + n^{-l}) \subseteq X$ and $(x - n^{-l}; x] \subseteq X$, i,e, x has the open neighborhood $(x - n^{-l}; x + n^{-l}) \subseteq X$.

Case 2: Not Case 1.

Then x has a $\rho_{b,n}$-name $p := $ "$a_k \ldots a_0 \bullet a_{-1} \ldots$" which has neither the period 0 nor the period $(n-1)$. Since $\rho_{b,n}^{-1}[X]$ is open in $\text{dom}(\rho_{b,n})$, there is a number l such that $\rho_{b,n}[$"$a_k \ldots a_0 \bullet a_{-1} \ldots a_{-l}$" $\Sigma^\omega] \subseteq X$. Then $x \in (y; z) \subseteq X$, where $y := \rho_{b,n}($"$a_k \ldots a_0 \bullet a_{-1} \ldots a_{-l}00 \ldots$"$)$ and $z := \rho_{b,n}($"$a_k \ldots a_0 \bullet a_{-1} \ldots a_{-l}(n-1)(n-1) \ldots$"$)$, hence x has an open neighborhood in X.

For the case $x < 0$ the proof is similar. If $x = 0$ consider the two names "$\bullet 00 \ldots$" and "$-\bullet 00 \ldots$".

6. By Lemma 4.1.4, the representation ρ is admissible with final topology $\tau_\mathbb{R}$. By Theorem 3.2.8.1 any two admissible representations with final topology $\tau_\mathbb{R}$ are continuously equivalent. Since the representation $\rho_{b,n}$ has the final topology $\tau_\mathbb{R}$ but is not continuously equivalent to ρ, it cannot be admissible.

7. See Exercise 4.1.12. $\square$

Restricted to the irrational numbers, $\rho_{b,n}$ and ρ are equivalent. Therefore, a real number x is $\rho_{b,n}$-computable, iff it is ρ-computable (notice that the rational numbers are $\rho_{b,n}$-computable and ρ-computable).

Since $\rho_{b,n}$ is not admissible, a $\rho_{b,n}$-name cannot be interpreted as a (sufficiently rich) list of atomic properties from a subbase of its final topology $\tau_\mathbb{R}$. Although $\rho_{b,n}$-names are richer than ρ-names by Property 3, this additional information is useless for computing real functions by Property 7. From Property 7 we conclude also that every $(\rho_{b,n}, \rho_{b,m})$-computable function is (ρ, ρ)-computable. However, already the simple (ρ, ρ)-computable real function $x \mapsto 3 \cdot x$ is not $(\rho_{b,10}, \rho_{b,10})$-continuous (Example 2.1.4.7).

We discuss some further representations of the real numbers which, however, have only very few applications.

Example 4.1.14 (further representations of $\mathbb{R}$).

1. *Naive Cauchy representation:* Define the naive Cauchy representation ρ_{Cn} of the real numbers by

$$\rho_{\text{Cn}}(p) = x : \iff p = \iota(w_0)\iota(w_1)\ldots \text{ and } \lim_{i \to \infty} \overline{w_i} = x .$$

Since ρ_C, $\rho_<$ and $\rho_>$ are restrictions of ρ_{Cn}, we have $\rho_C, \rho_<, \rho_> \leq \rho_{Cn}$. ρ_{Cn} has the final topology $\{\emptyset, \mathbb{R}\}$ (no property of $x = \rho_{Cn}(p)$ can be concluded from a finite prefix w of p). From this fact and Theorem 3.1.8 we conclude $\rho_{Cn} \not\leq_t \rho_C$, $\rho_{Cn} \not\leq_t \rho_<$ and $\rho_{Cn} \not\leq_t \rho_>$.

2. *Cut representations:* Define computable topological spaces
   - $\mathbf{S}_\leq = (\mathbb{R}, \sigma_\leq, \nu_\leq)$ by $\nu_\leq(w) := [\nu_\mathbb{Q}(w); \infty)$ and
   - $\mathbf{S}_\geq = (\mathbb{R}, \sigma_\geq, \nu_\geq)$ by $\nu_\geq(w) := (-\infty; \nu_\mathbb{Q}(w)]$

   (cf. Definition 4.1.3). Define the *left cut* representation and the *right cut* representation by $\rho_\leq := \delta_{\mathbf{S}_\leq}$ and $\rho_\geq := \delta_{\mathbf{S}_\geq}$, respectively. If $\rho_\leq(p) = x$ $(\rho_\geq(p) = x)$, then p is a list of all $a \in \mathbb{Q}$ with $a \leq x$ $(a \geq x)$. The final topology of $\rho_\leq$ is $\tau_{\rho_<} \cup \sigma_\leq$, that is, also intervals ("atomic properties") $[a; \infty)$ with $a \in \mathbb{Q}$ are called "open". We have $\rho_\leq \leq \rho_<$, but neither $\rho_<$, $\rho_>$, ρ_C nor $\rho_\geq$ are t-reducible to $\rho_\leq$.
   As an example assume $\rho_C \leq_t \rho_\leq$. Then $\tau_{\rho_\leq} \subseteq \tau_\rho$ by Theorem 3.1.8. But $[0; \infty) \in \tau_{\rho_\leq} \setminus \tau_\rho$. Restricted to the irrational numbers $\rho_<$ and $\rho_\leq$ are equivalent: $\rho_<|^{Ir} \equiv \rho_\leq|^{Ir}$. (The properties hold accordingly for $\rho_\geq$.) The definitions of $\rho_\leq$ and $\rho_\geq$ are not topologically robust, that is, replacement of $\mathbb{Q}$ in the definitions by another dense subset Q yields representations which are not continuously equivalent to $\rho_\leq$ and $\rho_\geq$, respectively.

3. *Continued fraction representation:* For any real number $x \geq 0$ define its continued fraction $\mathrm{fr}(x) := [a_0, a_1, \ldots]$ $(a_i \in \mathbb{N})$ inductively by:

$$x_0 := x,$$

$$a_n := \lfloor x_n \rfloor, \quad x_{n+1} := \begin{cases} \frac{1}{x_n - a_n} & \text{if } a_n \neq x_n \\ 0 & \text{otherwise.} \end{cases}$$

Then, informally

$$x = a_0 + \cfrac{1}{a_1 + \cfrac{1}{a_2 + \ldots}},$$

where the fraction is finite ($a_n = 0$ for all $n \geq n_0$), iff the number x is rational. Define the continued fraction representation ρ_{cf} of the real numbers by $\rho_{cf}(p) = x$, iff ($x \geq 0$ and $p = 1^{a_0}01^{a_1}0\ldots$ where $\mathrm{fr}(x) := [a_0, a_1, \ldots]$) or ($x < 0$ and $p = \text{-}1^{a_0}01^{a_1}0\ldots$ where $\mathrm{fr}(-x) := [a_0, a_1, \ldots]$). One can show

$$\rho_{cf} \equiv \rho_\leq \wedge \rho_\geq,$$

in particular, $\rho_{cf} \leq \rho_\leq$ and $\rho_{cf} \leq \rho_\geq$. Furthermore, $\rho_{cf} \leq \rho_{b,n}$ for all $n \geq 2$. Neither $\rho_\leq$, $\rho_\geq$ nor $\rho_{b,n}$ for $n \geq 2$ are t-reducible to ρ_{cf}. Restricted to the irrational numbers, ρ and ρ_{cf} are equivalent: $\rho|^{Ir} \equiv \rho_{cf}|^{Ir}$. □

The definitions of the cut representations $\rho_\leq$ and $\rho_\geq$ are not even topologically robust. Neither the representations by infinite base-n fractions and the naive Cauchy representation nor the cut representations $\rho_\leq$ and $\rho_\geq$ and their greatest lower bound ρ_{cf} are of much interest in computable analysis. Fig. 4.2 shows the reducibility order of the representations of the real numbers we have introduced so far. Many other representations of the real numbers are introduced and compared in [Hau73, Dei84].

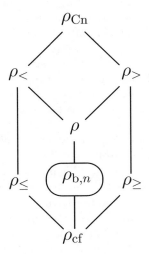

$$\rho_{Cn}$$
$$\rho_< \qquad \rho_>$$
$$\rho$$
$$\rho_\le \quad \boxed{\rho_{b,n}} \quad \rho_\ge$$
$$\rho_{cf}$$

Fig. 4.2. Reducibility order of some representations of $\mathbb{R}$

Our main representation $\rho :\subseteq \Sigma^\omega \to \mathbb{R}$ of the real numbers is not a total function and it is not injective. The same holds for all representations equivalent to it which we have discussed. It would be convenient to have an injective representation δ of the real numbers which is equivalent to ρ. Unfortunately this is not possible.

Theorem 4.1.15.
1. There is no total representation $\delta : \Sigma^\omega \to \mathbb{R}$ with $\delta \equiv \rho$.
2. There is no injective representation $\delta :\subseteq \Sigma^\omega \to \mathbb{R}$ with $\delta \equiv \rho$.
The two statements hold accordingly for $\rho_<$ and $\rho_>$ instead of ρ.

Proof: 1. Let $\delta : \Sigma^\omega \to \mathbb{R}$ be a total function with $\delta \equiv \rho$. δ is continuous, since ρ is continuous. By Lemma 2.2.5 the metric space (Σ^ω, d) with Cantor topology τ_C is compact. A continuous function maps compact sets to compact sets, therefore, range(δ) $= \delta[\Sigma^\omega]$ is compact. Since $\mathbb{R}$ is not compact (for example the open cover $\{(z; z+2) \mid z \in \mathbb{Z}\}$ has no finite subcover), $\mathbb{R} \neq$ range(δ).

2. Let $\delta :\subseteq \Sigma^\omega \to \mathbb{R}$ be an injective function with $\delta \equiv \rho$. By Lemma 3.2.5, $\tau_\mathbb{R}$ is the final topology of δ, that is, X is open, iff $\delta^{-1}[X]$ is open in dom(δ). Since δ is injective, there is some $w \in \Sigma^*$ such that $0 \in \delta[w\Sigma^\omega]$ and $1 \notin \delta[w\Sigma^\omega]$. Since $w\Sigma^\omega$ and $\Sigma^\omega \setminus w\Sigma^\omega$ are open, $A := \delta[w\Sigma^\omega]$ and

$B := \delta[\Sigma^\omega \setminus w\Sigma^\omega]$ are open subsets of $\mathbb{R}$ such that $0 \in A$, $1 \in B$, $A \cup B = \mathbb{R}$ and $A \cap B = \emptyset$. This is impossible.

The proofs for $\rho_<$ and $\rho_>$ are left for Exercise 4.1.17. $\qquad\square$

In the "real-RAM" model of computation which is used, for example, in computational geometry [PS85] and studied in detail by Blum et al. [BCSS98] one uses the test $x < y$ for real numbers x, y as a basic operation. We show that no representation makes this test decidable. The test $x < y$ is "absolutely non-decidable", at least in the framework of TTE.

Theorem 4.1.16 ($x \leq y$ is absolutely undecidable). For every representation $\delta :\subseteq \Sigma^\omega \to \mathbb{R}$ the relations "$x = y$" and "$x \leq y$" are not (δ, δ)-open and the relation "$x < y$" is not (δ, δ)-clopen.

Proof: Assume that the relation "$x = y$" is (δ, δ)-open. By Definitions 2.4.1 and 3.1.3.2 there is a continuous function $f :\subseteq \Sigma^\omega \times \Sigma^\omega \to \Sigma^*$ such that $f(p, q) = 0$, if $\delta(p) = \delta(q)$, and $f(p, q) = \text{div}$ otherwise for all $p, q \in \text{dom}(\delta)$. Consider z and p with $\delta(p) = z$. We obtain $f(p, p) = 0$. Since f is continuous, $f[w\Sigma^\omega \times w\Sigma^\omega] = \{0\}$ for some prefix $w \in \Sigma^*$ of p. We obtain $x = y$ for any $x, y \in \delta[w\Sigma^\omega]$, hence $\{z\} = \delta[w\Sigma^\omega]$. Therefore, for every $z \in \mathbb{R}$ there is some $w \in \Sigma^*$ with $\{z\} = \delta[w\Sigma^\omega]$. But this is impossible, since Σ^* is countable and $\mathbb{R}$ is uncountable. If "$x \leq y$" is (δ, δ)-open, then also "$x \geq y$" is (δ, δ)-open, hence "$x = y$" is (δ, δ)-open by Theorem 2.4.5. If "$x < y$" is (δ, δ)-clopen, then "$x \geq y$" is (δ, δ)-open. $\qquad\square$

Among the representations of the real numbers discussed in this section, $\rho, \rho_<$ and $\rho_>$ (and equivalent ones) are the most natural ones. For representations of $\mathbb{R}^n$ ($n \geq 2$) we generalize Definition 4.1.3 straightforwardly.

Definition 4.1.17 (standard representation ρ^n of $\mathbb{R}^n$). *For $n \geq 1$ let ρ^n be the standard representation of $\mathbb{R}^n$ derived from the computable topological space $\mathbf{S}^n := (\mathbb{R}^n, \text{Cb}^{(n)}, \text{I}^n)$ (Definition 4.1.2).*

A sequence $p \in \Sigma^\omega$ is a ρ^n-name of $x \in \mathbb{R}^n$, iff it is a list of all n-dimensional open rational cubes $J \in \text{Cb}^{(n)}$ such that $x \in J$. Fig. 4.3 shows some rational squares $J \in \text{Cb}^{(2)}$ and a point $x \in \mathbb{R}^2$ such that $x \in J$.

The definitions of the other representations equivalent to ρ given above can be generalized straightforwardly to n dimensions by substituting the n-dimensional norm $|| \cdot ||$ for the absolute value $| \cdot |$ and the notation I^n for I^1.

Lemma 4.1.18 (representations equivalent to ρ^n). For $n \geq 2$ generalize the definitions of $\rho, \rho_C, \rho'_C, \rho''_C, \rho'''_C, \rho^a$ and ρ^b from Definitions 4.1.3, 4.1.5 and Lemma 4.1.6, respectively, from $\mathbb{R}$ to $\mathbb{R}^n$ by substituting the B-dimensional norm $|| \cdot ||$ for the absolute value $| \cdot |$ and the notation I^n for I^1. Then all the resulting representations are equivalent to ρ^n.

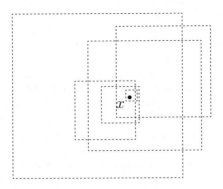

Fig. 4.3. Some open rational squares $J \in Cb^{(2)}$ such that $x \in J$

Proof: The proof of Lemma 4.1.6 can be generalized straightforwardly.   □

Instead of maximum norm, distance and balls, sometimes we will use Euclidean norm (absolute value), distance and balls:

- $|(a_1, \ldots, a_n)| = \sqrt{a_1^2 + \ldots + a_n^2}$,
- $d^e(x, y) := |x - y| = \sqrt{(x_1 - y_1)^2 + \ldots + (x_n - y_n)^2}$,
- $B^e(a, r) := \{x \in \mathbb{R}^n \mid |x - a| < r\}$.

The two metrics are related by

$$d(x, y) \leq d^e(x, y) \leq \sqrt{n}\, d(x, y)$$

and generate the same topology on the set $\mathbb{R}^n$.

If we replace the maximum metric by the Euclidean metric in the above definitions, we obtain representations of $\mathbb{R}^n$ which are equivalent to ρ^n. We leave the proofs to the reader. According to Definition 3.3.3, the product $[\rho]^n = [\rho, \ldots, \rho]$ is defined by

$$[\rho]^n \langle p_1, \ldots, p_n \rangle = (\rho(p_1), \ldots, \rho(p_n)) \, .$$

The following lemma summarizes some useful properties.

Lemma 4.1.19.
1. $[\rho]^n \equiv \rho^n$, ρ^n is admissible with final topology $\tau_{\mathbb{R}^n}$.
2. A tuple $(x_1, \ldots, x_n)$ is $(\rho, \ldots, \rho)$-computable, iff it is ρ^n-computable.
3. A set $X \subseteq \mathbb{R}^n$ is $(\rho, \ldots, \rho)$-decidable (-r.e. , -clopen, -open), iff it is ρ^n-decidable (-r.e. , -clopen, -open).
4. A function or multi-valued function $f :\subseteq \mathbb{R}^n \rightrightarrows M$ is $(\rho, \ldots, \rho, \delta)$-computable (-continuous), iff it is (ρ^n, δ)-computable (-continuous).
5. A function $f :\subseteq M \rightarrow \mathbb{R}^k$ is (δ, ρ^k)-computable (-continuous), iff $pr_i \circ f$ is (δ, ρ)-computable (-continuous) for $i = 1, \ldots, k$.

Proof: 1. Every ρ^n-name of $(x_1, \ldots, x_n)$ essentially consists of arbitrarily tight upper and arbitrarily tight lower bounds for every component x_i. The same is true for every $[\rho]^n$-name of $(x_1, \ldots, x_n)$. Translations from ρ^n to $[\rho]^n$ and from $[\rho]^n$ to ρ^n can be constructed straightforwardly.

2.-5. These properties follow from Property 1 and Lemma 3.3.6.   □

Convention 4.1.20. From now on we will consider as standard the notations $\nu_{\mathbb{N}}, \nu_{\mathbb{Z}}, \nu_{\mathbb{Q}}$ and id_{Σ^*} of the natural numbers, the integers, rational numbers and the set Σ^*, respectively, and the representations $\mathrm{id}_{\Sigma^\omega} : \Sigma^\omega \to \Sigma^\omega$, ρ and ρ^n of Σ^ω, $\mathbb{R}$ and $\mathbb{R}^n$, respectively. Occasionally, we will omit prefixes $\nu_{\mathbb{N}}$-, $\nu_{\mathbb{Z}}$-, $\nu_{\mathbb{Q}}$-, id_{Σ^*}-, $\mathrm{id}_{\Sigma^\omega}$- ρ- and ρ^n- and say *computable* instead of ρ-computable, *recursively enumerable (r.e.)* instead of $(\nu_{\mathbb{N}}, \rho)$-r.e., *computable* instead of $(\rho, \nu_{\mathbb{Q}}, \rho)$-computable etc.. Often we will use representations which we have proved to be equivalent to ρ or ρ^n.

The representations ρ, $\rho_<$ and $\rho_>$ can be extended to representations of $\overline{\mathbb{R}}$, the closure of $\mathbb{R}$ under supremum and infimum. We now modify Definition 4.1.3.

Definition 4.1.21 (representations of $\overline{\mathbb{R}}$). *For $\overline{\mathbb{R}} := \mathbb{R} \cup \{-\infty, \infty\}$ define computable topological spaces*

- $\mathbf{S}_< := (\overline{\mathbb{R}}, \sigma_<, \nu_<)$, $\quad \nu_<(w) := (\overline{w}; \infty]$,
- $\mathbf{S}_> := (\overline{\mathbb{R}}, \sigma_>, \nu_>)$, $\quad \nu_>(w) := [-\infty; \overline{w})$,

and let $\overline{\rho}_< := \delta_{\mathbf{S}_<}$, $\overline{\rho}_> := \delta_{\mathbf{S}_>}$ and $\overline{\rho} := \overline{\rho}_< \wedge \overline{\rho}_>$.

Exercises 4.1.

◇ 1. Show that the set of real numbers is not countable.

2. Show $\rho|^{\mathbb{N}} \equiv \nu_{\mathbb{N}}$, $\rho|^{\mathbb{Z}} \equiv \nu_{\mathbb{Z}}$, $\nu_{\mathbb{Q}} \leq \rho$ and $\rho|^{\mathbb{Q}} \not\leq_t \nu_{\mathbb{Q}}$ (Sect. 1.4).

3. Prove Lemma 4.1.7.

4. Show that a representation δ of the real numbers is reducible to $\rho_<$, iff the set $\{(x, a) \in \mathbb{R} \times \mathbb{Q} \mid a < x\}$ is $(\delta, \nu_{\mathbb{Q}})$-r.e. (cf. Lemma 4.1.7).

5. Prove Lemma 4.1.8. (See the proof of Lemma 4.1.6.)

6. Define a representation $\rho_{C<} : \Sigma^\omega \to \mathbb{R}$ by: $\rho_{C<}(p) = x$, iff $\rho_C(p) = x$ and $\overline{w}_0 < \overline{w}_1 < \ldots < x$ (w_i from Definition 4.1.5). Show $\rho_{C<} \equiv \rho$.

7. Show that $\rho \equiv \rho_G$ where

$$\rho_G(p) = x : \Longleftrightarrow \begin{cases} \text{there are words } w_0, w_1 \ldots \in \mathrm{dom}(\nu_{\mathbb{Z}}) \\ \text{such that } p = \iota(w_0)\iota(w_1)\iota(w_2)\ldots \\ \text{and } \left| x - \frac{\nu_{\mathbb{Z}}(w_i)}{i+1} \right| < \frac{1}{i+1} \text{ for all } i. \end{cases}$$

8. Prove that the definitions of ρ, $\rho_<$ and $\rho_>$ are topologically and computationally robust (Theorem 4.1.11).

9◊ a) Show that multiplication by 3 is not continuous with respect to $\rho_{b,2}$.

◊ b) Show that neither addition nor multiplication are continuous with respect to $\rho_{b,n}$ for $n \geq 2$.

10. Prove Theorem 4.1.13.1 and 4.1.13.2.

♦11. The representations by infinite base-n fractions can be studied as members of a larger class of representations [Wei92a]:

Let $\nu : \mathbb{N} \to Q$ be a numbering of a dense subset $Q \subseteq \mathbb{R}$. Define a representation ϑ_ν of the real numbers by

$$\vartheta_\nu(p) = x :\iff (\forall i) \begin{cases} \nu(i) < x \implies p(i) = 0 \\ \nu(i) > x \implies p(i) = 1 . \end{cases}$$

Show:

a) If μ is a standard numbering of the finite base-n fractions, for example $\mu\langle i, j, k \rangle := (i - j)/n^k$, then $\rho_{b,n} \equiv \vartheta_\mu$.

b) For any two numberings $\mu : \mathbb{N} \to P$ and $\nu : \mathbb{N} \to Q$ of dense subsets of $\mathbb{R}$ we have:

- $\vartheta_\nu \leq_t \rho$, $\rho \not\leq_t \vartheta_\nu$,
- $\vartheta_\nu \leq \rho$, if ν and $\nu_\mathbb{Q}$ are r.e.-related,
- $Q \subseteq P \iff \vartheta_\mu \leq_t \vartheta_\nu$,
- $\nu \leq \mu \implies \vartheta_\mu \leq \vartheta_\nu$.

c) Define $\nu_0 : \mathbb{N} \to \mathbb{Q}$ by $\nu_0\langle i, j, k \rangle := (i - j)/(1 + k)$. Then $\vartheta_{\nu_0} \equiv \rho_{b,2} \wedge \rho_{b,3} \wedge \ldots$, that is, ϑ_{ν_0} is the greatest lower bound of all $\rho_{b,n}$ (Definition 3.3.7).

d) Let ν and $\nu_\mathbb{Q}$ be r.e.-related. Then $f :\subseteq \mathbb{R} \to \mathbb{R}$ is (ϑ_ν, ρ)-computable, iff it is (ρ, ρ)-computable.

e) $f :\subseteq \mathbb{R} \to \mathbb{R}$ is $(\rho_{b,n}, \rho)$-computable, iff it is (ρ, ρ)-computable.

♦12. Prove Theorem 4.1.13.7 without using Exercise 4.1.11 [Her99b].

13. Let ρ_{Cn} be the naive Cauchy representation. Prove:

a) $\rho_{Cn} \not\leq_t \rho_<$,

b) ρ_{Cn} has the final topology $\{\emptyset, \mathbb{R}\}$,

c) there is a (ρ_{Cn}, ρ_{Cn})-computable function, which is not (ρ, ρ)-computable (hint: consider a constant function with value x_A (Example 1.3.2) where A is an r.e. non-recursive set),

♦ d) A real function is continuous, iff it is (ρ_{Cn}, ρ_{Cn})-continuous [BH00].

14. Prove the properties of the cut representations $\rho_<$ and $\rho_>$ stated in Example 4.1.14.2.

15. For an arbitrary notation $\mu :\subseteq \Sigma^* \to Q$ of a dense subset $Q \subseteq \mathbb{R}$ define the effective topological space $\mathbf{S}_\mu = (\mathbb{R}, \sigma_\mu, \nu_\mu)$ by $\nu_\mu(w) := [\mu(w); \infty)$ and $\delta_\mu := \delta_{\mathbf{S}_\mu}$ (cf. Example 4.1.14.2). For notations $\mu :\subseteq \Sigma^* \to Q$ and $\mu' :\subseteq \Sigma^* \to Q'$ show: $\delta_\mu \leq_t \delta_{\mu'}$, iff $Q' \subseteq Q$.

16. Prove the properties of the continued fraction representation stated in Example 4.1.14.3.

17. Show that there is neither a total nor an injective representation of the real numbers which is equivalent to $\rho_<$.

18. Complete the proof of Lemma 4.1.18.

◇19. Let δ and δ' be representations of an uncountable set M. Show that the set $\{(x,y) \mid x,y \in M, x = y\}$ is not (δ, δ')-open.

20. Show that the definition of ρ^n is computationally robust, that is, replacement of $\nu_{\mathbb{Q}}$ in Definition 4.1.17 by a notation ν_Q of a dense subset of $\mathbb{R}$ which is r.e.-related to it (Theorem 4.1.11) yields an equivalent representation.

◇21. Show that $\rho_<$ and $\rho_>$ are the restrictions of $\overline{\rho}_<$ and $\overline{\rho}_>$, respectively, to $\mathbb{R}$ and that ρ is equivalent to the restriction of $\overline{\rho}$ to $\mathbb{R}$. (See Definition 4.1.21.)

◇22. Show that the function $f : \mathbb{R} \to \mathbb{R}$, $f(x) := 1/x^2$, is $(\overline{\rho}, \overline{\rho}_<)$-computable.

4.2 Computable Real Numbers

We will denote the set of computable (that is, ρ-computable) real numbers by $\mathbb{R}_c$. Informally, a real number x is computable, iff arbitrarily tight lower and arbitrarily tight upper rational bounds of x can be computed. This is formalized by each of the following characterizations.

Lemma 4.2.1. For any $x \in \mathbb{R}$ the following properties are equivalent:
1. x is ρ-computable.
2. [Tur36] x is $\rho_{b,n}$-computable, that is, the number x has a computable infinite base-n fraction ($n \in \mathbb{N}, n \geq 2$).
3. There is a computable function $g : \mathbb{N} \to \Sigma^*$ such that

$$|x - \nu_{\mathbb{Q}} g(n)| \leq 2^{-n} \quad \text{for all } n \in \mathbb{N} .$$

4. [Grz55] There is a computable function $f : \mathbb{N} \to \mathbb{N}$ such that

$$\left| |x| - \frac{f(n)}{n+1} \right| < \frac{1}{n+1} \quad \text{for all } n \in \mathbb{N} .$$

5. [PER89] There are computable functions $s, a, b, e : \mathbb{N} \to \mathbb{N}$ with

$$\left| x - (-1)^{s(k)} \frac{a(k)}{b(k)} \right| \leq 2^{-N}, \text{ if } k \geq e(N), \quad \text{for all } k, N \in \mathbb{N} .$$

Proof: 1 $\Longleftrightarrow$ 2: This follows from Theorem 4.1.13.

1 $\Longrightarrow$ 3: Let $\iota(u_0)\iota(u_1)\dots$ be a computable ρ_C-name of x. Then $|x - \nu_{\mathbb{Q}}(u_n)| \leq 2^{-n}$ for all n. Define $g(n) := u_n$.

$3 \Longrightarrow 5$: There are computable functions s, a, b with
$\nu_{\mathbb{Q}}g(k) = (-1)^{s(k)}\frac{a(k)}{b(k)}$. Define $e(N) := N$.

$5 \Longrightarrow 4$: From Property 5 we obtain $\big| |x| - a_N/b_N \big| \leq 2^{-(N+2)}$
$< 1/(2(N+1))$, where $a_N := a \circ e(N+2)$ and $b_N := b \circ e(N+2)$. There is
a computable function $f : \mathbb{N} \to \mathbb{N}$ with $|a_N(N+1)/b_N - f(N)| \leq 1/2$. We
obtain

$$\left| |x| - \frac{f(N)}{N+1} \right| \leq \left| |x| - \frac{a_N}{b_N} \right| + \left| \frac{a_N}{b_N} - \frac{f(N)}{N+1} \right| < 2\frac{1}{2(N+1)} \leq \frac{1}{N+1} \; .$$

$4 \Longrightarrow 1$: Assume $x \geq 0$. There is a computable function $g : \mathbb{N} \to \Sigma^*$ such
that $\nu_{\mathbb{Q}}g(n) = f(2^{n+1})/(2^{n+1} + 1)$. We obtain

$$|x - \nu_{\mathbb{Q}}g(n)| = \left| |x| - \frac{f(2^{n+1})}{2^{n+1} + 1} \right| < \frac{1}{2^{n+1} + 1} < 2^{-n-1} \; .$$

Define $p \in \Sigma^\omega$ by $p := \iota(g(0))\iota(g(1))\dots$. Then p is computable,
$|\nu_{\mathbb{Q}}g(n) - \nu_{\mathbb{Q}}g(m)| \leq |\nu_{\mathbb{Q}}g(n) - x| + |x - \nu_{\mathbb{Q}}g(m)| \leq 2^{-n}$ for $m > n$ and
$\lim_{n\to\infty} \nu_{\mathbb{Q}}g(n) = x$, hence $\rho_C(p) = x$. If $x < 0$, define g such that
$\nu_{\mathbb{Q}}g(n) = -f(2^n)/(2^n + 1)$. $\qquad\square$

Every rational number a is computable (if $\nu_{\mathbb{Q}}(u) = a$, define $g(n) := u$
for all n in Lemma 4.2.1.3). In Example 1.3.1 we have shown that $\sqrt{2}$ and
$\log_3 5$ are computable real numbers. Many other examples will follow from
theorems below (Example 4.3.13.8). By Lemma 4.1.19, a vector $(x_1, \dots, x_n)$
of real numbers is ρ^n-computable, iff all its components x_i are computable.

Definition 4.2.2 (modulus of convergence). *A function* $e : \mathbb{N} \to \mathbb{N}$ *is
called a modulus of convergence of a sequence* $(x_i)_{i\in\mathbb{N}}$, *iff* $|x_i - x_k| \leq 2^{-n}$
for $i, k \geq e(n)$.

If e is a modulus of convergence then e' with $e'(n) := \max_{k\leq n} e(k)$ is
a modulus of convergence, which is computable, if e is computable. Therefore we may assume in most cases that the modulus of convergence is non-
decreasing. If e is a modulus of convergence then $|x - x_i| \leq 2^{-n}$ for $i \geq e(n)$,
where $x = \lim_{i\to\infty} x_i$. If e' is a function with $|x - x_i| \leq 2^{-n}$ for $i \geq e'(n)$, then
e with $e(n) := e'(n+1)$ is a modulus of convergence, which is computable, if
e' is computable.

Therefore, it follows from Lemma 4.2.1.5 that the limit of any computa-
ble (more precisely $(\nu_{\mathbb{N}}, \nu_{\mathbb{Q}})$-computable) sequence of rational numbers with
computable modulus of convergence is a computable real number. This ob-
servation can be generalized as follows:

Theorem 4.2.3. Let $(x_i)_{i\in\mathbb{N}}$ be a $(\nu_{\mathbb{N}}, \rho)$-computable sequence of real
numbers with computable modulus of convergence $e : \mathbb{N} \to \mathbb{N}$. Then its
limit $x = \lim_{i\to\infty} x_i$ is computable.

Proof: Since the sequence is $(\nu_\mathbb{N}, \rho_C)$-computable, for any $i, j \in \mathbb{N}$, a word u_{ij} can be computed such that $x_i = \rho_C(\iota(u_{i0})\iota(u_{i1})\ldots)$. Define $v_i := u_{e(i+2),i+2}$ and $q := (\iota(v_0)\iota(v_1)\ldots)$. For all $k < m$ we obtain

$$\begin{aligned}
|\bar{v}_k - x| &\leq |\bar{u}_{e(k+2),k+2} - x_{e(k+2)}| + |x_{e(k+2)} - x| \\
&\leq 2^{-k-2} + 2^{-k-2} \\
&\leq 2^{-k-1}
\end{aligned}$$

and $|\bar{v}_k - \bar{v}_m| \leq |\bar{v}_k - x| + |\bar{v}_m - x| \leq 2^{-k-1} + 2^{-m-1} \leq 2^{-k}$. Therefore, $x = \rho_C(q)$ and x is ρ_C-computable, since $q \in \Sigma^\omega$ is computable. $\qquad\square$

Example 4.2.4.
1. Define $x_i := \sum_{k=0}^{i} 1/k!$. Then $e = \lim_{i\to\infty} x_i$. Obviously the sequence $(x_i)_{i\in\mathbb{N}}$ is $(\nu_\mathbb{N}, \nu_\mathbb{Q})$-computable, hence $(\nu_\mathbb{N}, \rho)$-computable. Define $e(n) := n+1$. Then $|x_i - x_j| \leq 2^{-n}$ for $i, j \geq e(n)$. By Theorem 4.2.3, the number e is computable.
2. A famous result by Leibnitz states $\pi/4 = 1 - 1/3 + 1/5 - 1/7\ldots$. Define $x_i := \sum_{k=0}^{i}(-1)^k a_k$ where $a_k = 1/(2k+1)$. Then the sequence $(x_i)_{i\in\mathbb{N}}$ is computable. For $i \leq j$ and even $j - i$,

$$\begin{aligned}
0 &\leq (a_i - a_{i+1}) + (a_{i+2} - a_{i+3}) + \ldots + (a_{j-2} - a_{j-1}) + a_j \\
&= (-1)^i \sum_{k=i}^{j}(-1)^k a_k \\
&= a_i - (a_{i+1} - a_{i+2}) - \ldots - (a_{j-1} - a_j) \\
&\leq a_i
\end{aligned}$$

and similarly for odd $j-i$, $0 \leq (-1)^i \sum_{k=i}^{j}(-1)^k a_k \leq a_i - a_j \leq a_i$. In both cases, $|\sum_{k=i}^{j}(-1)^k a_k| \leq a_i$. Define $e(n) := 2^n$. Then for $e(n) \leq i < j$, $|x_j - x_i| = |\sum_{k=i+1}^{j}(-1)^k a_k| \leq a_{i+1} = 1/(2i+3) < 1/i \leq 2^{-n}$. By Theorem 4.2.3, the number $\pi/4$ is computable, hence π is computable.
3. By Example 1.3.2 for any set $A \subseteq \mathbb{N}$ of numbers, the sum

$$x_A := \sum_{i\in A} 2^{-i}$$

is a computable real number, iff A is a recursive set.
Let A be recursively enumerable but not recursive. Then there is a computable injective function $f : \mathbb{N} \to \mathbb{N}$ with $A = \text{range}(f)$. We obtain

$$x_A := \sum_{j\in\mathbb{N}} 2^{-f(j)} = \lim_{n\to\infty} \sum_{j=0}^{n} 2^{-f(j)}.$$

Therefore, $(a_n)_{n\in\mathbb{N}}$ with $a_n := \sum_{j=o}^{n} 2^{-f(j)}$ is a computable increasing sequence of rational numbers. Its limit x_A is $\rho_<$-computable. Since x_A is not computable, by Theorem 4.2.3 the sequence $(a_n)_{n \in \mathbb{N}}$ cannot have a computable modulus of convergence. Since the set A is not recursive, the enumerating function f cannot have a computable lower bound which is monotone and unbounded. Therefore, unforeseeable values $f(n)$ are very small and terms $2^{-f(n)}$ are large. □

The $\rho_<$-computable numbers are also called *left-computable* or *left-r.e.* and the $\rho_>$-computable numbers are also called *right-computable* or *right-r.e..* By Specker's example there is a left-computable real number which is not computable (Examples 1.3.2, 4.2.4.3).

Lemma 4.2.5. A real number x
1. is left-computable, iff $-x$ is right-computable,
2. is computable, iff it is left-computable and right-computable.

Proof: A direct proof is easy. Property 2 follows also from $\rho \equiv \rho_< \wedge \rho_>$ (Lemma 4.1.9). □

The computable real numbers are a countable set which, however, cannot be enumerated "effectively". We prove a "positive" version of this statement by diagonalization.

Theorem 4.2.6.
1. Let $(x_i)_{i \in \mathbb{N}}$ be a $(\nu_{\mathbb{N}}, \rho)$-computable sequence of real numbers. Then there is a computable real number x such that $x \neq x_i$ for all $i \in \mathbb{N}$.
2. There is no numbering or notation ν of the set $\mathbb{R}_c$ of the computable real numbers with r.e. domain such that $\nu \leq \rho$.

Proof: 1. We construct x by diagonalization. We may assume that the sequence is $(\nu_{\mathbb{N}}, \rho_C)$-computable. For any $i \in \mathbb{N}$ we can determine a sequence $q_i := \iota(u_{i0}) \iota(u_{i1}) \ldots$ with $x_i = \rho_C(q_i)$. Therefore, there is a computable function $g : \Sigma^* \to \Sigma^*$ with $g(0^i) = u_{i,2i+2}$. We obtain $|\nu_{\mathbb{Q}} \circ g(0^i) - x_i| \leq 2^{-2i-2}$. We compute a nested sequence $([\overline{u}_i; \overline{v}_i])_{i \in \mathbb{N}}$ of closed intervals with $\overline{v}_i - \overline{u}_i = 3^{-i}$ such that $x_i \notin [\overline{u}_i; \overline{v}_i]$ as follows:

$$(\overline{u}_0, \overline{v}_0) := (\nu_{\mathbb{Q}} \circ g(\lambda) + 1, \overline{u}_0 + 1)$$

$$(\overline{u}_{i+1}, \overline{v}_{i+1}) := \begin{cases} (\overline{u}_i, \overline{u}_{i+1} + \frac{1}{3} \cdot 3^{-i}) & \text{if } \nu_{\mathbb{Q}} \circ g(0^{i+1}) \geq \overline{u}_i + \frac{1}{2} \cdot 3^{-i} \\ (\overline{u}_i + \frac{2}{3} \cdot 3^{-i}, \overline{v}_i) & \text{otherwise.} \end{cases}$$

If $\nu_{\mathbb{Q}} \circ g(0^{i+1}) \geq \overline{u}_i + \frac{1}{2} \cdot 3^{-i}$, then $x_{i+1} \geq \overline{u}_i + \frac{1}{2} \cdot 3^{-i} - 2^{-2(i+1)-2} > \overline{u}_i + \frac{1}{3} \cdot 3^{-i}$, and if $\nu_{\mathbb{Q}} \circ g(0^{i+1}) \not\geq \overline{u}_i + \frac{1}{2} \cdot 3^{-i}$, then $x_{i+1} < \overline{u}_i + \frac{2}{3} \cdot 3^{-i}$. We obtain $x_{i+1} \notin [\overline{u}_{i+1}; \overline{v}_{i+1}]$. There is a computable sequence $i \mapsto w_i$ of words such that $\mathrm{I}^1(w_i) = [u_i; v_i]$. Therefore, $q := \iota(w_0) \iota(w_1) \ldots$ is computable and $x := \rho^b(q)$ (ρ^b from Lemma 4.1.6) differs from all numbers x_i. Since $\rho^b \equiv \rho$, x is computable.

2. Assume that there is such a numbering $\nu :\subseteq \mathbb{N} \to \mathbb{R}_c$. There is a computable function $f : \mathbb{N} \to \mathbb{N}$ with range$(f) = \text{dom}(\nu)$. Then νf is a total numbering of $\mathbb{R}_c$ with $\nu f \leq \nu \leq \rho$. Therefore, $(\nu f(i))_{i \in \mathbb{N}}$ is a $(\nu_{\mathbb{N}}, \rho)$-computable sequence of all computable real numbers. This is impossible by Property 1.

If ν is a notation, there is a computable function $f : \mathbb{N} \to \Sigma^*$ with range$(f) = \text{dom}(\nu)$. Continue as above. $\qquad\square$

We derive a notation of the computable real numbers canonically from the representation ρ using the standard notation $\xi^{*\omega}$ of the computable functions $f :\subseteq \Sigma^* \to \Sigma^\omega$ (Definition 2.3.4).

Definition 4.2.7. *Define the notation* $\nu_\rho :\subseteq \Sigma^* \to \mathbb{R}_c$ *of the computable real numbers by*

$$\nu_\rho(w) := \rho \circ \xi_w^{*\omega}(\lambda) .$$

For $w \in \text{dom}(\nu_\rho)$, $\nu_\rho(w) = \rho(p)$ where $p \in \Sigma^\omega$ is the sequence computed by the Turing machine with code w on input λ. By the utm-theorem for $\xi^{*\omega}$, $\nu_\rho \leq \rho$. By Theorem 4.2.6 the domain $\text{dom}(\nu_\rho) = \{w \in \Sigma^* \mid \xi_w^{*\omega}(\lambda) \in \text{dom}(\rho)\}$ is not r.e.

Every countable subset $X \subseteq \mathbb{R}$ can be covered by arbitrarily small open sets. The (countable) set $\mathbb{R}_c$ of computable real numbers can be covered by arbitrarily small "r.e. open" sets [Spe59] (cf. Sect. 5.1).

Theorem 4.2.8. For each $N \in \mathbb{N}$ there is a computable sequence $i \mapsto w_i$ of words $w_i \in \text{dom}(\text{I}^1)$, such that

$$\mathbb{R}_c \subseteq U_N := \bigcup_{i \in \mathbb{N}} \text{I}^1(w_i) \quad \text{and} \quad \sum_{i \in \mathbb{N}} \text{length}(\text{I}^1(w_i)) \leq 2^{-N} .$$

Proof: For each $i = \langle k, t \rangle$ define the word w_i as follows:
If the Turing machine with code $\nu_\Sigma(k)$ on input λ in t steps (but not in $(t-1)$ steps) writes an output $\iota(u_0)\iota(u_1)\ldots\iota(u_{k+N+4}) \in \Sigma^*$ with $|\bar{u}_j - \bar{u}_m| \leq 2^{-j}$ for $0 \leq j \leq m \leq k + N + 4$, then

$$\text{I}^1(w_i) = \left(\bar{u}_{k+N+4} - 2^{-k-N-3} \; ; \; \bar{u}_{k+N+4} + 2^{-k-N-3} \right) ,$$

otherwise

$$\text{I}^1(w_i) = \left(0 \; ; \; 2^{-i-N-2} \right) .$$

Suppose x is computable. Then $x = \rho_C(p)$ for some computable $p \in \Sigma^\omega$. There is some number k such that $\xi_{\nu_\Sigma(k)}^{*\omega}(\lambda) = p$. Then there is some t such that the machine with code $\nu_\Sigma(k)$ on input λ in t steps computes a prefix $\iota(u_0)\iota(u_1)\ldots\iota(u_{k+N+4}) \in \Sigma^*$ of p with $u_i \in \text{dom}(\nu_\mathbb{Q})$. For $i = \langle k, t \rangle$ we obtain $x \in I^1(w_i)$. Therefore, $\mathbb{R}_c \subseteq \bigcup_{i \in \mathbb{N}} I^1(w_i)$.

For each k there is at most one t such that the first condition holds, therefore,

$$\sum_{i \in \mathbb{N}} \text{length}\left(I^1(w_i)\right) \leq \sum_{i \in \mathbb{N}} 2^{-i-N-2} + \sum_{k \in \mathbb{N}} 2^{-k-N-2} = 2^{-N} .$$

Since the function $(w, t) \mapsto v$, where v is the word which the machine with code w on input λ produces in t steps, is computable, the sequence $i \mapsto w_i$ is computable (cf. Lemma 2.1.5). $\square$

Exercises 4.2.

$\Diamond$ 1. Show that a real number x is computable, iff there are computable functions $f, g, h : \mathbb{N} \to \mathbb{N}$ with $|x - (f(n) - g(n))/(1 + h(n))| \leq 2^{-n}$ for all $n \in \mathbb{N}$.

2. a) Define a $(\nu_{\mathbb{N}}, \rho_<)$-computable sequence $(x_i)_{i \in \mathbb{N}}$ which lists all $\rho_<$-computable real numbers.

   b) Construct by diagonalization a $\rho_>$-computable real number which is not $\rho_<$-computable.

3. Let $a : \mathbb{N} \to \mathbb{R}$ be a computable sequence of real numbers with infinite range. Then there is an *injective* computable sequence of real numbers $b : \mathbb{N} \to \mathbb{R}$ with $\text{range}(a) = \text{range}(b)$.

4. Let $(a_k)_{k \in \mathbb{N}}$ be a computable sequence of real numbers converging to 0.

   a) Show that the sequence $(a_k)_{k \in \mathbb{N}}$ has a computable modulus of convergence, if $a_k \leq a_{k+1}$ for all k.

   b) Show that there is an increasing computable function $f : \mathbb{N} \to \mathbb{N}$ with $|a_{f(n)}| \leq 2^{-n}$ for all n (hence $n \mapsto n$ is a modulus of convergence of the computable subsequence $n \mapsto a_{f(n)}$).

   c) Show that there is a computable sequence $(b_k)_{k \in \mathbb{N}}$ of real numbers converging to 0 which has no computable modulus of convergence. (Hint: define $b_k := 2^{-f(k)}$ where f is an injective enumeration of a non-recursive r.e. set.)

5. Show that the sequence $(a_n)_{n \in \mathbb{N}}$ in Example 4.2.4.3 is $(\nu_{\mathbb{N}}, \nu_{\mathbb{Q}})$-computable, $(\nu_{\mathbb{N}}, \rho)$-computable and $(\nu_{\mathbb{N}}, \rho_\leq)$-computable (Example 4.1.14.2).

6. Consider $n \in \mathbb{N}$, $n > 2$. Is $\sum_{i \in A} n^{-i}$ computable, if $A \subseteq \mathbb{N}$ is recursive (r.e.)?

7. Show that there is a sequence $(a_i)_{i \in \mathbb{N}}$ of rational numbers which is $(\nu_{\mathbb{N}}, \rho)$-computable but not $(\nu_{\mathbb{N}}, \nu_{\mathbb{Q}})$-computable. Hint: Let $a : \mathbb{N} \to \mathbb{N}$ be a computable injective function such that $\text{range}(a)$ is not recursive. Define $x_n := 0$, if $n \notin \text{range}(a)$, $x_n := 2^{-k}$, if $a(k) = n$.

8. Define a $(\nu_{\mathbb{Q}}, \rho)$-computable function $f : \mathbb{Q} \to \mathbb{R}$ such that $\text{range}(f) \subseteq \mathbb{Q}$ and the restriction $f|^{\mathbb{Q}} : \mathbb{Q} \to \mathbb{Q}$ is not $(\nu_{\mathbb{Q}}, \nu_{\mathbb{Q}})$-computable. [Hau87]

9. Show that there is a computable sequence $y : \mathbb{N} \to \mathbb{R}$ of real numbers, such that the sequence $\mathrm{sgn} \circ y$ is not computable (where $\mathrm{sgn}(x) :=$ 0, if $x \le 0$, 1 otherwise).

♦10. By Taylor's theorem, $\sqrt{1+t} = \sum_{k=0}^{\infty} \binom{\frac{1}{2}}{k} \cdot t^k$ for all $t \in \mathbb{R}$ with $|t| < 1$. The series converges also for $|t| = 1$. For $t = -1$ we obtain

$$0 = 1 - \frac{1}{2} - \frac{1}{2 \cdot 4} - \frac{1 \cdot 3}{2 \cdot 4 \cdot 6} - \frac{1 \cdot 3 \cdot 5}{2 \cdot 4 \cdot 6 \cdot 8} - \cdots .$$

Determine a modulus of convergence (find an expression in elementary functions and do not use summation $\sum$) for the sequence $i \mapsto x_i$ with $x_i := \sum_{k=0}^{i} \binom{\frac{1}{2}}{k} \cdot (-1)^k$. (Consult, for example, [BB85].)

◇11. Show that for every bounded $(\nu_{\mathbb{N}}, \rho_<)$-computable sequence $(x_i)_{i \in \mathbb{N}}$ of real numbers, $\sup_{i \in \mathbb{N}} x_i$ is $\rho_<$-computable.

12. There is a left-computable real number $x \in (0; 2)$ such that $x \ne \sum_{i \in A} 2^{-i}$ for every r.e. set $A \subseteq \mathbb{N}$. Hint: Let $B \subseteq \mathbb{N}$ be recursively enumerable but not recursive and define

$$y_B := \sum_{i \in B} 2^{-(2i+1)} + \sum_{i \notin B} 2^{-(2i+2)} .$$

13. Let $(x_i)_{i \in \mathbb{N}}$ be a computable sequence of real numbers such that $\sum_{i \in \mathbb{N}} |x_{i+1} - x_i|$ is finite. Show that x is the sum of a left-computable and a right-computable real number. There are real numbers of this type, which are neither left- nor right-computable. Left-computable and more general types of real numbers are investigated in [WZ98a].

14. Show that the multi-valued function $F :\subseteq \mathbb{R} \times \mathbb{R} \rightrightarrows \mathbb{Q}$, defined by $R_F := \{((x,y), a) \mid x < a < y\}$, is $(\rho_>, \rho_<, \nu_{\mathbb{Q}})$-computable.

15. The open set $U_N := \bigcup_{i \in \mathbb{N}} \mathrm{I}^1(w_i)$ from Theorem 4.2.8 containing all computable real numbers can be written as a disjoint union of open intervals. Let $K \subseteq U_N$ be the interval of this partition of U_N containing the (computable) number $0 \in \mathbb{R}$ and let $c := \sup(K)$ be its right-hand endpoint.
   a) Show c is not computable, that is, $c \notin \mathbb{R}_c$.
   b) Let c_n be the right-hand endpoint of the longest interval $J \subseteq \bigcup_{i=0}^{n} \mathrm{I}^1(w_i)$ with $0 \in J$. Show that $(c_n)_{n \in \mathbb{N}}$ is a non-decreasing computable sequence with $c = \sup_{n \in \mathbb{N}} c_n$. (Hence c is left-computable.)

   c) Show that $c_n \le a$ or $c_n \ge b$, if $n \ge i$ and $(a, b) := \mathrm{I}^1(w_i)$.
   d) Show that $f :\subseteq \mathbb{R} \rightrightarrows \mathbb{N}$, defined by

   $$R_f := \{(x, n) \in \mathbb{R}_c \times \mathbb{N} \mid (\forall k > n)|c_k - x| \ge 2^{-n}\} ,$$

   is $(\rho, \nu_{\mathbb{N}})$-computable.

16. Effectivize Theorem 4.2.6.1. Define a multi-valued function

$$D : \mathbb{R}^{\mathbb{N}} \rightrightarrows \mathbb{R} \quad \text{by} \quad \mathrm{R}_D := \{((x_0, x_1, \dots,), x) \mid (\forall i) \; x \neq x_i\} \,.$$

   a) Show that D is $([\rho]^\omega, \rho)$-computable.
   b) Show that D has no $([\rho]^\omega, \rho)$-continuous choice function.
17. By Theorem 4.1.13.2, $x \in \mathbb{R}$ is $\rho_{\mathrm{b},2}$-computable, iff it is $\rho_{\mathrm{b},10}$-computable. By Theorem 4.1.13.4, $\rho_{\mathrm{b},2} \not\leq_{\mathrm{t}} \rho_{\mathrm{b},10}$. Show that there is a $[\rho_{\mathrm{b},2}]^\omega$-computable sequence $(x_0, x_1, \dots)$ which is not $[\rho_{\mathrm{b},10}]^\omega$-computable [Mos57]. Hint: There are injective computable functions $a, b : \mathbb{N} \to \mathbb{N}$ such that $A := \mathrm{range}(a)$ and $B := \mathrm{range}(b)$ are recursively inseparable, that is, $A \cap B = \emptyset$ and for no recursive set C, $A \subseteq C$ and $B \subseteq \mathbb{N} \setminus C$ [Rog67, Odi89]. Let $\rho_{\mathrm{b},2}(0{\bullet}d_1 d_2 \dots) = 1/5$ and define

$$x_n := \begin{cases} \rho_{\mathrm{b},2}(0{\bullet}d_1 d_2 \dots) & \text{for } n \notin A \cup B \,, \\ \rho_{\mathrm{b},2}(0{\bullet}d_1 d_2 \dots d_k 00 \dots) & \text{for } a(k) = n \,, \\ \rho_{\mathrm{b},2}(0{\bullet}d_1 d_2 \dots d_k 11 \dots) & \text{for } b(k) = n \,. \end{cases}$$

4.3 Computable Real Functions

A real function $f :\subseteq \mathbb{R} \to \mathbb{R}$ is computable (more precisely (ρ, ρ)-computable), iff some Type-2 machine transforms any ρ-name p of any $x \in \mathrm{dom}(f)$ to a ρ-name of $f(x)$, where a ρ-name of y is a list of all rational open intervals $J \in \mathrm{Cb}^{(1)}$ such that $y \in J$. Since equivalent representations induce the same computability, ρ can be replaced by any representation equivalent to it, for example by the Cauchy representation ρ_C (Lemma 4.1.6). Computable functions with n real arguments are realized accordingly by machines with n input tapes. Since the representation ρ is admissible with final topology $\tau_{\mathbb{R}}$ (Lemma 4.1.4), we obtain as a special case of our Main Theorem 3.2.11:

Theorem 4.3.1 (continuity). Every computable real function is continuous.

More precisely, every (ρ^n, ρ)-computable real function is continuous w.r.t. the real line topology $\tau_{\mathbb{R}}$. Many easily definable functions like the step function or the Gauß staircase (Fig. 1.3) are not (ρ, ρ)-computable, since they are not continuous. Some people reject TTE and similar approaches to computable analysis since they think that a reasonable computability theory for analysis should make such functions computable. This objection can be removed, since TTE admits various natural computable topological spaces which make the step function, the Gauß staircase and similar functions computable (for example the Gauß staircase is $(\rho, \rho_>)$-computable).

The following theorem lists some computable real functions.

Theorem 4.3.2 (some computable real functions). The following real functions are computable:
1. $(x_1, \ldots, x_n) \mapsto c$ (where $c \in \mathbb{R}$ is a computable constant),
2. $(x_1, \ldots, x_n) \mapsto x_i$ $(1 \leq i \leq n)$,
3. $x \mapsto -x$,
4. $(x, y) \mapsto x + y$,
5. $(x, y) \mapsto x \cdot y$,
6. $x \mapsto 1/x$,
7. $(x, y) \mapsto \min(x, y)$, $(x, y) \mapsto \max(x, y)$,
8. $x \mapsto |x|$,
9. every polynomial function in n variables with computable coefficients,
10. $(i, x) \mapsto x^i$ for $i \in \mathbb{N}$ and $x \in \mathbb{R}$ $(0^0 := 1)$.

Proof: We will use the representations ρ_C and ρ_C''' which are equivalent to ρ by Lemma 4.1.6.

1. Let M be a Type-2 machine with n input tapes which on every input computes some computable ρ-name of c. Then f_M realizes the constant function with value c.

2. Let M be a Type-2 machine with n input tapes which copies the i-th input tape to the output tape. Then f_M realizes the i-th projection.

3. There is a computable word function $f : \Sigma^* \to \Sigma^*$ such that $-\nu_{\mathbb{Q}}(w) = \nu_{\mathbb{Q}} f(w)$ for all $w \in \mathrm{dom}(\nu_{\mathbb{Q}})$. There is a Type-2 machine M, which transforms any input $p := \iota(u_0)\iota(u_1)\ldots \in \mathrm{dom}(\rho_C)$ (where $u_i \in \mathrm{dom}(\nu_{\mathbb{Q}})$) to the sequence $q := \iota(f(u_0))\iota(f(u_1))\ldots$. Obviously, $\rho_C(q) = -x$, if $\rho_C(p) = x$. Therefore, f_M realizes negation on $\mathbb{R}$.

4. Since addition on $\mathbb{Q}$ is $(\nu_{\mathbb{Q}}, \nu_{\mathbb{Q}}, \nu_{\mathbb{Q}})$-computable, there is a computable function $f :\subseteq \Sigma^* \times \Sigma^* \to \Sigma^*$ such that $\overline{u} + \overline{v} = \nu_{\mathbb{Q}} f(u, v)$ for all $u, v \in \mathrm{dom}(\nu_{\mathbb{Q}})$. There is a Type-2 machine M, which transforms any input (p, q), $p := \iota(u_0)\iota(u_1)\ldots \in \mathrm{dom}(\rho_C''')$ and $q := \iota(v_0)\iota(v_1)\ldots \in \mathrm{dom}(\rho_C''')$ (where $u_i, v_i \in \mathrm{dom}(\nu_{\mathbb{Q}})$) to the sequence $r := \iota(y_0)\iota(y_1)\ldots$ with $y_i := f(u_{i+1}, v_{i+1})$. Since $\overline{y}_i = \overline{u}_{i+1} + \overline{v}_{i+1}$, we have

$$|\overline{y}_i - (\rho_C'''(p) + \rho_C'''(q))| \leq |\overline{u}_{i+1} - \rho_C'''(p)| + |\overline{v}_{i+1} - \rho_C'''(q)| \leq 2 \cdot 2^{-i-1} = 2^{-i} \ .$$

We obtain $r = f_M(p, q) \in \mathrm{dom}(\rho_C''')$ and $\rho_C'''(r) = x + y$. Therefore, f_M is a $(\rho_C''', \rho_C''', \rho_C''')$-realization of addition.

5. Since multiplication on $\mathbb{Q}$ is $(\nu_{\mathbb{Q}}, \nu_{\mathbb{Q}}, \nu_{\mathbb{Q}})$-computable, there is a computable function $f :\subseteq \Sigma^* \times \Sigma^* \to \Sigma^*$ such that $\overline{u} \cdot \overline{v} = \nu_{\mathbb{Q}} f(u, v)$ for all $u, v \in \mathrm{dom}(\nu_{\mathbb{Q}})$. There is a Type-2 machine M which transforms any input (p, q), $p := \iota(u_0)\iota(u_1)\ldots \in \mathrm{dom}(\rho_C''')$ and $q := \iota(v_0)\iota(v_1)\ldots \in \mathrm{dom}(\rho_C''')$, to the sequence $r := \iota(y_0)\iota(y_1)\ldots$ with $y_i := f(u_{m+i}, v_{m+i})$, where $m \in \mathbb{N}$ is the smallest number with $|\overline{u}_0| + 2 \leq 2^{m-1}$ and $|\overline{v}_0| + 2 \leq 2^{m-1}$. For all n we have

$$|\overline{u}_n| \leq |\overline{u}_n - \rho_C'''(p)| + |\rho_C'''(p) - \overline{u}_0| + |\overline{u}_0| \leq 2 + |\overline{u}_0| \leq 2^{m-1}$$

and accordingly $|\overline{v}_n| \leq 2^{m-1}$. With $x := \rho_C'''(p)$ and $y := \rho_C'''(q)$ we obtain

$$
\begin{aligned}
|\overline{y}_i - x \cdot y| &= |\overline{u}_{m+i} \cdot \overline{v}_{m+i} - x \cdot y| \\
&\leq |\overline{u}_{m+i} \cdot \overline{v}_{m+i} - \overline{u}_{m+i} \cdot y| + |\overline{u}_{m+i} \cdot y - x \cdot y| \\
&\leq |\overline{u}_{m+i} \cdot (\overline{v}_{m+i} - y)| + |(\overline{u}_{m+i} - x) \cdot y| \\
&\leq 2 \cdot 2^{m-1} \cdot 2^{-m-i} \\
&= 2^{-i} \ .
\end{aligned}
$$

We obtain $r = f_M(p,q) \in \text{dom}(\rho_C''')$ and $\rho_C'''(r) = x \cdot y$. Therefore, f_M is a $(\rho_C''', \rho_C''', \rho_C''')$-realization of multiplication.

6. There is a Type-2 machine M which works as follows on input $p := \iota(u_0)\iota(u_1)\ldots \in \text{dom}(\rho_C''')$: First M searches for the smallest $N \in \mathbb{N}$ with $|\overline{u}_N| > 3 \cdot 2^{-N}$. As soon as such a number N has been found, M starts to write the sequence $r := \iota(y_0)\iota(y_1)\ldots$ with $\overline{y}_i = 1/\overline{u}_{2N+i}$. Suppose that $x := \rho_C'''(p) \neq 0$. Then N exists and $|\overline{u}_k| > 2^{-N}$ for all $k \geq N$ and hence $|x| \geq 2^{-N}$. We obtain

$$
\left|\overline{y}_i - \frac{1}{x}\right| = \left|\frac{1}{\overline{u}_{2N+i}} - \frac{1}{x}\right| = \frac{|x - \overline{u}_{2N+i}|}{|\overline{u}_{2N+i}| \cdot |x|} \leq 2^{-2N-i} \cdot 2^N \cdot 2^N = 2^{-i} \ .
$$

Therefore, f_M is a (ρ_C''', ρ_C''')-realization of inversion. Notice that $f_M(p)$ does not exist, if $\rho_C'''(p) = 0$.

7. There is a Type-2 machine M which transforms any input (p,q), $p := \iota(u_0)\iota(u_1)\ldots \in \text{dom}(\rho_C''')$ and $q := \iota(v_0)\iota(v_1)\ldots \in \text{dom}(\rho_C''')$, to the sequence $r := \iota(y_0)\iota(y_1)\ldots$ with $\overline{y}_i = \min(\overline{u}_i, \overline{v}_i)$. If $x = \min(x, y, \overline{u}_i, \overline{v}_i)$, then

$$
|\overline{y}_i - \min(x,y)| = \min(\overline{u}_i, \overline{v}_i) - x \leq \overline{u}_i - x \leq 2^{-i} \ .
$$

If $\overline{u}_i = \min(x, y, \overline{u}_i, \overline{v}_i)$, then

$$
|\overline{y}_i - \min(x,y)| = \min(x,y) - \overline{u}_i \leq x - \overline{u}_i \leq 2^{-i} \ .
$$

For the cases $y = \min(x, y, \overline{u}_i, \overline{v}_i)$ and $\overline{v}_i = \min(x, y, \overline{u}_i, \overline{v}_i)$, $|\overline{y}_i - \min(x,y)| \leq 2^{-i}$ can be concluded similarly. Therefore, f_M is a $(\rho_C''', \rho_C''', \rho_C''')$-realization of min. Since $\max(x,y) = (x+y) - \min(x,y)$, max is computable by Properties 3 and 4 and the composition theorem 3.1.6.

8. $|x| = \max(x, -x)$, apply Properties 3 and 7.

9. We apply the composition theorem 3.1.6. Every monomial f of degree 0 (that is, $f(x_1, \ldots, x_n) := c$) with computable constant c is computable by Property 1. Suppose that all monomials of degree k with computable coefficients are computable. If f_{k+1} is a monomial of degree $k+1$ with computable coefficient, then $f_{k+1}(x_1, \ldots, x_n) = f_k(x_1, \ldots, x_n) \cdot \text{pr}_i(x_1, \ldots, x_n)$ for some monomial of degree k with computable coefficient and some i. By induction and Properties 2 and 5, f_{k+1} is computable.

Another easy induction shows that every polynomial function with computable coefficients is computable.

10. For $h(i, x) := x^i$ we have

$$h(0, x) = 1 \ ,$$

$$h(n + 1, x) = x \cdot h(n, x) \ .$$

If we define $f(x) := 1$ and $f'(n, y, x) := x \cdot y$, then f and f' are computable by Properties 1 and 5 above, and h is computable by Theorem 3.1.7.2 on primitive recursion. □

Example 4.3.3. The exponential function $\exp : \mathbb{R} \to \mathbb{R}$ is computable. We use the estimation

$$\exp(x) = \sum_{i=0}^{N} \frac{x^i}{i!} + r_N(x) \ , \text{ where } \ r_N(x) \leq 2 \cdot \frac{|x|^{N+1}}{(N+1)!} \ , \text{ if } |x| \leq 1 + \frac{N}{2} \ .$$

Let M be a Type-2 machine which for any $p = \iota(u_0)\iota(u_1)\ldots \in \mathrm{dom}(\rho_C)$ computes a sequence $q = \iota(v_0)\iota(v_1)\ldots$, where for each n, v_n is determined as follows.
1. M determines the smallest $N_1 \in \mathbb{N}$ with $|\overline{u}_0| + 1 \leq 1 + N_1/2$.
2. M determines the smallest $N \in \mathbb{N}$, $N \geq N_1$, with

$$2 \cdot \frac{|1 + N_1/2|^{N+1}}{(N+1)!} \leq 2^{-n-2} \ .$$

3. M determines the smallest $m \in \mathbb{N}$ with

$$2^{-m} \cdot \sum_{i=1}^{N} \frac{i \cdot (1 + N_1/2)^{i-1}}{i!} \leq 2^{-n-2} \ .$$

4. M determines $v_n \in \Sigma^*$ such that

$$\overline{v}_n = \sum_{i=0}^{N} \frac{\overline{u}_m^{\, i}}{i!} \ .$$

Assume $x = \rho_C(p) = \rho_C(\iota(u_0)\iota(u_1)\ldots)$. Then $|x| \leq 1 + N_1/2$ and $|\overline{u}_m| \leq 1 + N_1/2$. We obtain

$$|\exp(x) - \overline{v}_n| \leq |\sum_{i=0}^{N} \frac{x^i}{i!} - \sum_{i=0}^{N} \frac{\overline{u}_m^{\, i}}{i!}| + |r_N(x)|$$

$$\leq |x - \overline{u}_m| \cdot \sum_{i=1}^{N} \frac{|x^{i-1} + x^{i-2}\overline{u}_m + \ldots + \overline{u}_m^{\, i-1}|}{i!} + 2^{-n-2}$$

$$\leq 2^{-m} \cdot \sum_{i=1}^{N} \frac{i \cdot (1 + N_1/2)^{i-1}}{i!} + 2^{-n-2}$$

$$\leq 2^{-n-2} + 2^{-n-2}$$

$$\leq 2^{-n-1} \ .$$

We obtain furthermore $\exp(x) = \rho_C(\iota(v_0)\iota(v_1)\ldots)$, since for $i < j$,

$$|\overline{v}_i - \overline{v}_j| \le |\overline{v}_i - \exp(x)| + |\exp(x) - \overline{v}_j| \le 2^{-i-1} + 2^{-j-1} \le 2^{-i} .$$

Therfore, f_M realizes the exponential function.   □

The computable real functions are closed under composition (Theorem 3.1.6) and under primitive recursion (Theorem 3.1.7). There are some other useful operations which map computable real functions to computable real functions.

Corollary 4.3.4. If $f, g :\subseteq \mathbb{R}^n \to \mathbb{R}$ are computable functions and $a \in \mathbb{R}$ is a computable number, then $x \mapsto a \cdot f(x)$, $x \mapsto f(x)+g(x)$, $x \mapsto f(x) \cdot g(x)$, $x \mapsto \max(f(x), g(x))$, $x \mapsto \min(f(x), g(x))$ and $x \mapsto 1/f(x)$ are computable functions.

Proof: By Theorem 3.1.6 the computable real functions are closed under composition. Apply Theorem 4.3.2.   □

The join of two computable functions at a computable point is a computable function (Fig. 4.4).

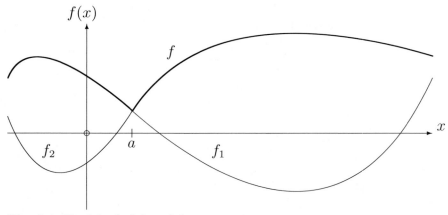

Fig. 4.4. The join f of f_1 and f_2 at a

Lemma 4.3.5 (join of functions). Let $f_1, f_2 :\subseteq \mathbb{R} \to \mathbb{R}$ be computable real functions and let $c \in \mathbb{R}$ be a computable real number. Then the function $f :\subseteq \mathbb{R} \to \mathbb{R}$, defined by

$$f(x) := \begin{cases} f_1(x) & \text{if } x < c, \\ f_2(x) & \text{if } x > c, \\ f_1(a) & \text{if } x = c \text{ and } f_1(a) = f_2(a), \\ \text{div} & \text{otherwise,} \end{cases}$$

is computable.

Proof: First consider the case $c = 0$. We may assume that f_1 and f_2 are (ρ^a, ρ^a)-computable, ρ^a from Lemma 4.1.6. There are Type-2 machines M_1 and M_2 such that f_{M_1} and f_{M_2} realize f_1 and f_2, respectively. For $i = 1, 2$ let $M_i(p, k)$ be the output written by the machine M_i in k steps. For any $w \in \Sigma^*$ let $N(w) := (\lambda$, if w has no subword $\iota(w')$, its rightmost subword $\iota(w')$ otherwise). There is a Type-2 machine M which on input $p = \iota(u_0)\iota(u_1) \ldots \in \mathrm{dom}(\rho^a)$ operates in stages $k = 0, 1, 2 \ldots$ as follows:

Stage k:

- If $0 \in \mathrm{I}^1(u_k)$ and if $N \circ M_1(p, k) = \lambda$ or $N \circ M_2(p, k) = \lambda$, then M goes to the next stage;
- If $0 \in \mathrm{I}^1(u_k)$ and if $N \circ M_1(p, k) = \iota(w_1)$ and $N \circ M_2(p, k) = \iota(w_2)$, then M writes some word $\iota(w)$, such that $\mathrm{I}^1(w)$ is the smallest interval with $\mathrm{I}^1(w_1) \cup \mathrm{I}^1(w_2) \subseteq \mathrm{I}^1(w)$;
- If $\mathrm{I}^1(u_k) < 0$, then M writes $N \circ M_1(p, k)$;
- if $\mathrm{I}^1(u_k) > 0$, then M writes $N \circ M_2(p, k)$.

Suppose that $x = \rho^a(p)$ and $f(x)$ exists. If $x < 0$, then M finally produces the output intervals of M_1 on input p. If $x > 0$, then M finally produces the output intervals of M_2 on input p. If $x = 0$, then M produces a combination of both outputs which converges to $f_1(0) = f_2(0)$. The result is always a ρ^b-name of $f(x)$, ρ^b from Lemma 4.1.6. Therefore, f is (ρ^a, ρ^b)-computable. If $c \neq 0$, apply the above join operation to the functions f_1' and f_2', $f_1'(x) := f_1(x + c)$ and $f_2' := f_2(x + c)$, which are computable by Theorem 4.3.2, and shift the result f': $f(x) := f'(x - c)$. $\qquad \square$

We turn now to functions on infinite sequences of real numbers. We use the representation $[\rho]^\omega$ (Definition 3.3.3) which is equivalent to $[\nu_\mathbb{N} \to \rho]_\mathbb{N}$ by Lemma 3.3.16. Projection and partial summation are computable:

Lemma 4.3.6 (sequences). For sequences $(x_0, x_1, \ldots)$ of real numbers

1. The projection $\mathrm{pr} : \big((x_0, x_1, \ldots), i\big) \mapsto x_i$ is $([\rho]^\omega, \nu_\mathbb{N}, \rho)$-computable;
2. The function $S_0 : \big((x_0, x_1, \ldots), i\big) \mapsto x_0 + x_1 + \ldots + x_i$ is $([\rho]^\omega, \nu_\mathbb{N}, \rho)$-computable;
3. the function $S : (x_0, x_1, \ldots) \mapsto (y_0, y_1, \ldots)$ where $y_i := x_0 + x_1 + \ldots + x_i$, is $([\rho]^\omega, [\rho]^\omega)$-computable.

Proof: 1. The function $(\langle p_0, p_1, \ldots \rangle, w) \mapsto p_{\nu_\mathbb{N}(w)}$ realizes the projection.

2. We apply Theorem 3.1.7 on primitive recursion. Define

$$h\big(0, (x_0, x_1, \ldots)\big) = x_0$$

$$h\big(n + 1, (x_0, x_1, \ldots)\big) = h\big(n, (x_0, x_1, \ldots)\big) + x_{n+1} .$$

Since $f(x_0, x_1, \ldots) := x_0$ and $f'(n, y, (x_0, x_1, \ldots)) := y + x_{n+1}$ are computable by Property 1, h is computable by Theorem 3.1.7. Since $S_0\big((x_0, x_1, \ldots), i\big) = x_0 + x_1 + \ldots + x_i = h\big(i, (x_0, x_1, \ldots)\big)$, S_0 is $([\rho]^\omega, \nu_\mathbb{N}, \rho)$-computable.

3. By Property 2 and Theorem 3.3.15, S is $([\rho]^\omega, [\nu_\mathbb{N} \to \rho]_\mathbb{N})$-computable, and so $([\rho]^\omega, [\rho]^\omega)$-computable. □

The limits of converging sequences and series are computable. We add as a further variable a modulus $e : \mathbb{N} \to \mathbb{N}$ of convergence and use the representation $[\nu_\mathbb{N} \to \nu_\mathbb{N}]_\mathbb{N}$ (Definition 3.3.13).

Theorem 4.3.7 (limit of sequences and series of numbers). For sequences $(x_0, x_1, \ldots)$ of real numbers and modulus functions $e : \mathbb{N} \to \mathbb{N}$, the functions

$$L : \big((x_0, x_1, \ldots), e\big) \mapsto \lim_{i \to \infty} x_i \quad \text{and} \tag{4.1}$$

$$\text{SL} : \big((x_0, x_1, \ldots), e\big) \mapsto \sum_{i \in \mathbb{N}} x_i \tag{4.2}$$

where $((x_0, x_1, \ldots), e) \in \text{dom}(L)$, iff $(\forall j > i \geq e(n))|x_j - x_i| \leq 2^{-n}$, and $((x_0, x_1, \ldots), e) \in \text{dom}(\text{SL})$, iff $(\forall j \geq i \geq e(n))|x_i + \ldots + x_j| \leq 2^{-n}$, are $([\rho]^\omega, [\nu_\mathbb{N} \to \nu_\mathbb{N}]_\mathbb{N}, \rho)$-computable.

Proof: 4.1. We generalize the proof of Theorem 4.2.3. It suffices to show that the function is $([\rho_C]^\omega, [\nu_\mathbb{N} \to \nu_\mathbb{N}]_\mathbb{N}, \rho_C)$-computable. There is a Type-2 machine M which on input $\big(\langle p_0, p_1, \ldots \rangle, q\big)$, $p_i = \iota(u_{i0})\iota(u_{i1}) \ldots \in \text{dom}(\rho_C)$, $e := [\nu_\mathbb{N} \to \nu_\mathbb{N}]_\mathbb{N}(q)$, writes the sequence $q := \iota(u_{e(2)2})\iota(u_{e(3)3})\iota(u_{e(4)4}) \cdots$. Then f_M is a $([\rho_C]^\omega, [\nu_\mathbb{N} \to \nu_\mathbb{N}]_\mathbb{N}, \rho_C)$-realization of L (cf. the proof of Theorem 4.2.3).

4.2. By Lemma 4.3.6, $S : (x_0, x_1, \ldots) \mapsto (y_0, y_1, \ldots)$, $y_i := \sum_{m \leq i} x_i$, is $([\rho]^\omega, [\rho]^\omega)$-computable. If for all $j \geq i \geq e(n)$, $|x_i + \ldots + x_j| \leq 2^{-n}$, then for all $j > i \geq e(n)$, $|y_j - y_i| = |x_{i+1} + \ldots + x_j| \leq 2^{-n}$. Therefore, $\text{SL}\big((x_0, x_1, \ldots), e\big) = L\big(S(x_0, x_1, \ldots), e\big)$, L from 4.1 above, if $|x_i + \ldots + x_j| \leq 2^{-n}$ for all $j \geq i \geq e(n)$, and so SL is $([\rho]^\omega, [\nu_\mathbb{N} \to \nu_\mathbb{N}]_\mathbb{N}, \rho)$-computable. □

We apply Theorem 4.3.7 to show that the uniform limit of a fast converging computable sequence of real-valued functions is computable (Fig 4.5).

Theorem 4.3.8 (limit of sequences and series of functions). Let $\delta :\subseteq \Sigma^\omega \to M$ be a representation, let $X \subseteq M$. Let $(f_i)_{i \in \mathbb{N}}$ with $f_i :\subseteq M \to \mathbb{R}$ and $\text{dom}(f_i) = X$ be a sequence of functions such that $(i, x) \to f_i(x)$ is $(\nu_\mathbb{N}, \delta, \rho)$-computable.

1. If there is a computable function $e : \mathbb{N} \to \mathbb{N}$ with $|f_j(x) - f_i(x)| \leq 2^{-n}$ for all $j > i \geq e(n)$ and $x \in X$, then the function $f :\subseteq M \to \mathbb{R}$, defined by $\text{dom}(f) = X$ and $f(x) = \lim_{i \to \infty} f_i(x)$, is (δ, ρ)-computable.
2. If there is a computable function $e : \mathbb{N} \to \mathbb{N}$ with $|f_i(x) + \ldots + f_j(x)| \leq 2^{-n}$ for all $j \geq i \geq e(n)$ and $x \in X$, then the function $f :\subseteq M \to \mathbb{R}$, defined by $\text{dom}(f) = X$ and $f(x) = \sum_{i \in \mathbb{N}} f_i(x)$, is (δ, ρ)-computable.

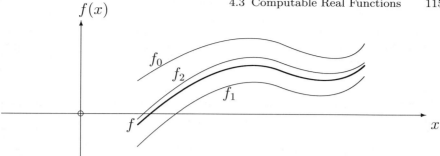

Fig. 4.5. Functions $f_0, f_1, f_2 \ldots$ uniformly converging to f

Proof: 1. Since the function $(x, i) \mapsto f_i(x)$ is $(\delta, \nu_{\mathbb{N}}, \rho)$-computable, by Theorem 3.3.15.2 the function $F : x \mapsto (f_i(x))_{i \in \mathbb{N}}$ is $(\delta, [\nu_{\mathbb{N}} \to \rho]_{\mathbb{N}})$-computable and hence $(\delta, [\rho]^\omega)$-computable by Lemma 3.3.16. We obtain $f(x) = L(F(x), e)$ for all $x \in X$ where L is the limit operator from Theorem 4.3.7,4.1. The function f is (δ, ρ)-computable, since L is $([\rho]^\omega, [\nu_{\mathbb{N}} \to \nu_{\mathbb{N}}]_{\mathbb{N}}, \rho)$-computable and e is $[\nu_{\mathbb{N}} \to \nu_{\mathbb{N}}]_{\mathbb{N}}$-computable.

1. This follows correspondingly from Theorem 4.3.7.4.2.   □

Lemma 4.3.6 and Theorems 4.3.7 and 4.3.8 can be generalized straightforwardly from $\mathbb{R}$ to $\mathbb{R}^n$.

Every complex number $z = x + iy \in \mathbb{C}$ has an absolute value $|z| = \sqrt{x^2 + y^2}$ and a norm $||z|| = \max(|x|, |y|)$ satisfying $||z|| \leq |z| \leq \sqrt{2} \cdot ||z||$. The set $\mathbb{C}$ of complex numbers can be identified with the set $\mathbb{R}^2$ of pairs of real numbers, $x + iy \leftrightarrow (x, y)$, with standard representation $[\rho]^2$. Then $\mathbb{C}^n$ is represented by $[[\rho]^2]^n \equiv [\rho]^{2n} \equiv \rho^{2n}$ (where we assume that the Cartesian product is associative, see Definitions 3.3.3, 4.1.17). We call a point $a \in \mathbb{C}$ computable, iff it is ρ^2-computable (iff it is (ρ, ρ)-computable), a function $f :\subseteq \mathbb{C}^n \to \mathbb{C}$ computable, iff it is (ρ^{2n}, ρ^2)-computable etc.. A complex-valued function is computable, iff its real part and its imaginary part are computable (Lemma 3.3.6).

Theorem 4.3.9 (computable complex functions). The complex functions $z \mapsto a$ (for computable $a \in \mathbb{C}$), $(z_1, z_2) \mapsto z_1 + z_2$, $(z_1, z_2) \mapsto z_1 \cdot z_2$, $z \mapsto 1/z$, $z \mapsto |z|$, $z \mapsto ||z||$, $z \mapsto \mathrm{Re}(z)$ and $z \mapsto \mathrm{Im}(z)$ are computable. Furthermore, every complex polynomial function with computable coefficients and the function $(j, z) \mapsto z^j$ are computable.

Proof: Consider the function $f : z \mapsto 1/z$. By Lemma 3.3.6 it suffices to show that the projections $\mathrm{Re}(f)$ and $\mathrm{Im}(f)$ are (ρ, ρ, ρ)-computable. We have

$$f(x + iy) = \frac{1}{x + iy} = \frac{x - iy}{x^2 + y^2} = \frac{x}{x^2 + y^2} + i\frac{-y}{x^2 + y^2} ,$$

therefore, f is computable by Theorem 4.3.2. Computability of $|z|$ follows from computability of $x \mapsto \sqrt{x}$ for $x \in \mathbb{R}, x \geq 0$, which will be proved below (Example 4.3.13.6). The remaining proofs are left to the reader.   □

Theorem 4.3.10 (sequences of complex numbers). Lemma 4.3.6 and Theorem 4.3.7 hold for sequences of complex numbers accordingly.

Every sequence $(a_j)_{j \in \mathbb{N}}$ of complex numbers defines a power series with co-efficients $a_0, a_1, \ldots$ and a function $f : \subseteq \mathbb{C} \to \mathbb{C}$, defined by

$$f(z) := \sum_{j=0}^{\infty} a_j \cdot z^j \ .$$

The sum $f(z)$ of the series is defined for all z with $|z| < R$ and is not defined for all z with $|z| > R$ where $R := 1/\limsup_{j \to \infty} \sqrt[j]{|a_j|}$ is the *radius of convergence*. By Cauchy's estimate for every number $r < R$ there is a constant M such that

$$|a_j| \leq M \cdot r^{-j} \quad \text{for all} \ \ j \in \mathbb{N} \ .$$

Neither the radius R of convergence nor a function h mapping each $r < R$ to an appropriate constant M for Cauchy's estimate can be computed from $(a_j)_{j \in \mathbb{N}}$ in general. For computing $f(z)$ from z and the sequence $(a_j)_{j \in \mathbb{N}}$ of coefficients further information about this sequence must be available. We will use a radius $r < R$ and a number M such that $|a_j| \leq M \cdot r^{-j}$ for all $j \in \mathbb{N}$.

We represent the set of sequences $j \mapsto a_j$ of complex numbers by $[\nu_{\mathbb{N}} \to \rho^2]_{\mathbb{N}}$ (Definition 3.3.13) or by $[\rho^2]^\omega$ (Definition 3.3.3) which are equivalent by Lemma 3.3.16.

Theorem 4.3.11 (power series). The function

$$P : \left((a_j)_{j \in \mathbb{N}}, r, M, z\right) \longmapsto \sum_{j=0}^{\infty} a_j \cdot z^j$$

defined for arguments with $|z| < r$ and $|a_j| \leq M \cdot r^{-j}$ for all j is

$$\left([\rho^2]^\omega, \nu_{\mathbb{Q}}, \nu_{\mathbb{N}}, \rho^2, \rho^2\right)\text{-computable.}$$

Proof: The multi-valued function $h : \subseteq \mathbb{Q} \times \mathbb{C} \rightrightarrows \mathbb{Q}$ with graph $R_h := \{(r, z, s) \mid |z| < s < r\}$ is $(\nu_{\mathbb{Q}}, [\rho]^2, \nu_{\mathbb{Q}})$-computable.
Now we show that the following variant Q of P which has a further input parameter s,

$$Q : \left((a_j)_{j \in \mathbb{N}}, r, s, M, z\right) \longmapsto \sum_{j=0}^{\infty} a_j \cdot z^j$$

defined for arguments with $|z| < s < r$ and $|a_j| \le M \cdot r^{-j}$ for all j, is

$$\left([\rho^2]^\omega, \nu_\mathbb{Q}, \nu_\mathbb{Q}, \nu_\mathbb{N}, \rho^2, \rho^2\right)\text{-computable.}$$

The function $(j, z) \mapsto z^j$ is $(\nu_\mathbb{N}, \rho^2, \rho^2)$-computable (Theorem 4.3.9). By the complex generalization of Lemma 4.3.6.1, the function $\left((a_j)_{j\in\mathbb{N}}, z, k\right) \mapsto a_k \cdot z^k$ is $([\rho^2]^\omega, \rho^2, \nu_\mathbb{N}, \rho^2)$-computable. Therefore, by Theorem 3.3.15.2 and Lemma 3.3.16,

$$G : \left((a_j)_{j\in\mathbb{N}}, z\right) \mapsto (a_j \cdot z^j)_{j\in\mathbb{N}} \quad \text{is} \quad ([\rho^2]^\omega, \rho^2, [\rho^2]^\omega)\text{-computable.}$$

Next we determine a modulus of convergence. The function

$$H : (r, s, M, n) \mapsto \min\left\{ m \in \mathbb{N} \mid M \cdot \left(\frac{s}{r}\right)^m \cdot \frac{r}{r - s} \le 2^{-n} \right\}$$

$(r, s \in \mathbb{Q}, s < r, M, n \in \mathbb{N})$ is computable, and so by Theorem 3.3.15.2,

$$H' : (r, s, M) \mapsto e, \quad e(n) := H(r, s, M, n) ,$$

is $(\nu_\mathbb{Q}, \nu_\mathbb{Q}, \nu_\mathbb{N}, [\nu_\mathbb{N} \to \nu_\mathbb{N}])$-computable. For $|z| \le s < r$ and $k \ge j \ge e(n)$ we have

$$|a_j \cdot z^j + \ldots + a_k \cdot z^k| \le \textstyle\sum_{k \ge j} |a_k| \cdot |z|^k \le \sum_{k \ge j} M \cdot \left(\frac{s}{r}\right)^k$$
$$= M \cdot \left(\frac{s}{r}\right)^j \cdot \frac{r}{r - s} \le 2^{-n} .$$

Let SL be the complex version of the summation operator from Theorem 4.3.7.4.2. Then

$$\sum_{i=0}^{\infty} a_i \cdot z^i = \text{SL}\left(G((a_j)_{j\in\mathbb{N}}, z), H'(r, s, M)\right) ,$$

if $|z| \le s < r$, and $|a_j| \le M \cdot r^{-j}$ for all j. Therefore, the function Q is computable.

Combining machines for h and Q we can construct a machine computing P. □

For a computable sequence $(a_j)_{j\in\mathbb{N}}$ of complex numbers we obtain the following useful consequences:

Theorem 4.3.12. Let $(a_j)_{j\in\mathbb{N}}$ be a computable sequence of complex numbers and let $R := 1/\limsup_{j\to\infty} \sqrt[j]{|a_j|}$.
1. The function $f : z \mapsto \sum_{i=0}^{\infty} a_i \cdot z^i$ is computable on every closed ball $\{z \in \mathbb{C} \mid |z| \le r\}$ with $r < R$.
2. Let $k \mapsto r_k$, and $k \mapsto M_k$ $(r_k \in \mathbb{Q}, M_k \in \mathbb{N})$ be computable sequences such that $|a_j| \le M_k \cdot r_k^{-j}$ for all j, k.
   Then the function $f : z \mapsto \sum_{i=0}^{\infty} a_i \cdot z^i$ is computable on the open ball $\{z \in \mathbb{C} \mid |z| < \sup_{k\in\mathbb{N}} r_k\}$.

Proof: 1. There is some rational number r' such that $r < r' < R$. There is a Cauchy bound $M \in \mathbb{N}$ for r'. Then for all $|z| < r'$, $f(z) = P((a_j)_{j \in \mathbb{N}}, r', M, z)$, P from Theorem 4.3.11.

2. For input z first find some number k with $|z| < r_k$ and then compute $f(z) = P((a_i)_{i \in \mathbb{N}}, r_k, M_k, z)$, P from Theorem 4.3.11.   □

The above theorems have many applications.

Example 4.3.13.
1. If $(a_j)_{j \in \mathbb{N}}$ is a computable sequence of complex numbers, $z_0 \in \mathbb{C}$ is computable and $r < R := 1/\limsup_{j \to \infty} \sqrt[j]{|a_j|}$, then $g :\subseteq \mathbb{C} \to \mathbb{C}$, defined by

$$g(z) := \sum_{j=0}^{\infty} a_j \cdot (z - z_0)^j \ ,$$

   is computable on the disc $\{z \mid |z - z_0| \le r\}$. For a proof consider the function f from Theorem 4.3.12. Then $g(z) = f(z - z_0)$, hence g is computable by Theorem 4.3.9.
2. The exponential function $\exp : \mathbb{C} \to \mathbb{C}$ can be defined by the power series

$$\exp(z) = \sum_{j=0}^{\infty} \frac{z^j}{j!}$$

   for all $z \in \mathbb{C}$. Since $\limsup_{j \to \infty} \sqrt[j]{1/j!} = 0$, the radius of convergence is $R = \infty$. By Theorem 4.3.12, for every $N \in \mathbb{N}$ the exponential function is computable on the disc $\{z \mid |z| \le N\}$. This means, for every number N there is a machine M_N which computes exp on this disc.
   There is also a single machine computing $\exp(z)$ for all $z \in \mathbb{C}$. To show this we apply Theorem 4.3.12.2. Consider $N \ge 1$. For $j \le N$ we have

$$\frac{1}{j!} \le 1 \le N^{N-j} = N^N \cdot N^{-j} \ ,$$

   and for $j > N$ we have

$$\frac{1}{j!} \le \frac{1}{(N+1) \cdot (N+2) \cdot \ldots \cdot j} \le \frac{1}{N^{j-N}} = N^N \cdot N^{-j} \ .$$

   Define $r_k := k + 1$ and $M_k := r_k^{r_k}$. By Theorem 4.3.12.2, the exponential function is computable on $\mathbb{C}$.
3. The trigonometric functions $\sin : \mathbb{C} \to \mathbb{C}$ and $\cos : \mathbb{C} \to \mathbb{C}$ are computable. For a proof use the identities $\sin(z) = (\exp(iz) - \exp(-iz))/(2i)$ and $\cos(z) = (\exp(iz) + \exp(-iz))/2$ and Example 2. In particular, the real trigonometric functions are computable.

4. For $|z| < 1$ we have

$$\log(1 + z) = \sum_{j=1}^{\infty} (-1)^{j+1} \frac{z^j}{j} \; .$$

The radius of convergence is 1. The function $z \mapsto \log(1+z)$ is computable on the disc $\{z \mid |z| < 1\}$ (Exercise 4.3.12).

5. The real function $x \mapsto \log x$ is computable on the interval $(0; \infty)$: By Example 4.3.13.4 the number $\log 2 = \log(1 + 1/3) - \log(1 - 1/3)$ is computable. There is a machine M which on input x (more precisely, on input p with $\rho(p) = x > 0$) first determines some integer $d \in \mathbb{Z}$ with $1/2 < x \cdot 2^d < 3/2$ and then computes $-d \cdot \log 2 + \log(x \cdot 2^d)$. The function f_M realizes the function $x \mapsto \log x$. (The multi-valued function $g :\subseteq \mathbb{R} \rightrightarrows \mathbb{Z}$ with $\mathrm{R}_g = \{(x, d) \mid 1/2 < x \cdot 2^d < 3/2\}$ is $(\rho, \nu_{\mathbb{Z}})$-computable but has no $(\rho, \nu_{\mathbb{Z}})$-continuous choice function.)

6. The function $(x, y) \mapsto x^y$ for $x, y \in \mathbb{R}$ and $x > 0$ is computable: $x^y = \exp(y \cdot \log x)$. In particular, $x \mapsto \sqrt{x}$ is computable.

7. For real x with $|x| < 1$ we have

$$\arcsin x = x + \frac{1}{2} \cdot \frac{x^3}{3} + \frac{1 \cdot 3}{2 \cdot 4} \cdot \frac{x^5}{5} + \frac{1 \cdot 3 \cdot 5}{2 \cdot 4 \cdot 6} \cdot \frac{x^7}{7} + \dots \; .$$

The function arcsin is computable on $(-1; 1)$ (Exercise 4.3.13).

8. Since computable functions map computable elements to computable ones, and since the rational numbers are computable, numbers like $\sqrt{2} = 2^{1/2}$, $\sqrt[m]{n}$ $(m, n > 1)$, $\mathrm{e} = \exp 1$, $\log 2$, $\log_a b = \log b / \log a$ $(a, b \in \mathbb{Q}, a, b > 0)$, $\pi = 6 \cdot \arcsin(1/2)$ and e^π are computable real numbers. $\qquad \square$

Complex functions $f :\subseteq \mathbb{C} \to \mathbb{C}$ which can be defined by power series are called *analytic* [Ahl66]. We will discuss computability of analytic functions in Sect. 6.5.

Let $f : \mathbb{R} \to \mathbb{R}$ be a (total) computable real function and let $X \subseteq \mathbb{R}$ be a "very complicated" set. Then by definition the restriction $f\rfloor_X$ is also computable. But $f\rfloor_X$ has a computable extension with the simple domain $\mathbb{R}$. There is, however, a computable real function with very complicated domain, which has no computable extension.

Example 4.3.14 (a computable function with inherent G_δ-domain).
Define the function $f :\subseteq \mathbb{R} \to \mathbb{R}$ by

$$f(x) := \begin{cases} \sum \{2^{-i} \mid \mu(i) < x\} & \text{if } x \notin \mathbb{Q} \\ \mathrm{div} & \text{otherwise,} \end{cases}$$

where $\mu\langle i, j, k \rangle := (i - j)/(1 + k)$. Let M be a Type-2 machine which on input $p = \iota(u_0)\iota(u_1)\dots \in \mathrm{dom}(\rho_{\mathrm{C}})$ operates in stages $i = 0, 1, 2, \dots$ as follows. Let $\nu_{\mathbb{Q}}(v_{-1}) = 0$.

Stage i:

M searches for some m with $|\bar{u}_m - \mu(i)| > 2^{-m}$. If no such m exists, then the computation remains in Stage i forever. Otherwise, M prints $\iota(v_i)$ where $\bar{v}_i = \bar{v}_{i-1} + 2^{-i}$, if $\mu(i) < \bar{u}_m - 2^{-m}$, and $v_i = v_{i-1}$ otherwise.

Then, M produces an infinite output $q := \iota(v_0)\iota(v_1)\ldots \in \mathrm{dom}(\rho_C)$, iff $\rho_C(p) \notin \mathbb{Q}$, and in this case $\rho_C(q) = f \circ \rho_C(p)$. Therefore, f_M realizes the function f.

The function f has the domain $\mathrm{dom}(f) = \mathbb{R} \setminus \mathbb{Q}$ which is a G_δ-set, that is, it can be written as an intersection of a sequence of open sets: $\mathbb{R} \setminus \mathbb{Q} = \bigcap_{i\in\mathbb{N}}(\mathbb{R}\setminus\{\mu(i)\})$. $\mathbb{R}\setminus\mathbb{Q}$ is even a "computable" G_δ-set (Exercise 4.3.17). Since for each rational number $a = \nu(k)$, $\lim_{x\to a+} f(x) - \lim_{x\to a-} f(x) = 2^{-k}$, the function f has no proper continuous extension and hence no proper computable extension. $\qquad\square$

Every continuous partial real function has an extension with G_δ-domain [Kur66]. Every (ρ^n, ρ)-computable real function has a strongly (ρ^n, ρ)-computable extension. The domain of each strongly (ρ^n, ρ)-computable real function is a computable G_δ-set (Exercise 3.1.5 and Exercises 4.3.17 and 18).

In Chap. 9 we will discuss several other definitions for computable real functions and compare them with the definitions given here. As a special case of Corollary 3.2.13, every computable or continuous function from $\mathbb{R}^n$ to a discrete space is constant.

Lemma 4.3.15. Let μ be a notation of a set M, $M \neq \emptyset$. If a function $f : \mathbb{R}^n \to M$ is (ρ^n, μ)-continuous, then f is a constant function.

Proof: The final topology $\tau_{\mathbb{R}^n}$ of the admissible representation ρ^n of $\mathbb{R}^n$ is connected. Apply Corollary 3.2.13. $\qquad\square$

Corollary 4.3.16.

1. Every $(\rho^n, \nu_\mathbb{N})$-continuous or -computable function $f : \mathbb{R}^n \to \mathbb{N}$ is constant.
2. Every $(\rho^n, \nu_\mathbb{Q})$-continuous or -computable function $f : \mathbb{R}^n \to \mathbb{Q}$ is constant.
3. Every $(\rho^n, \delta_\mathbb{B})$-continuous or -computable function $f : \mathbb{R}^n \to \mathbb{B}$ is constant ($\delta_\mathbb{B}$ from Definition 3.1.2).

Proof: The first two statements are immediate. Consider $i \in \mathbb{N}$. The function $H_i : h \mapsto h(i)$ is $(\delta_\mathbb{B}, \nu_\mathbb{N})$-computable, hence $H_i \circ f$ is $(\rho^n, \nu_\mathbb{N})$-continuous, and so constant. Therefore, f is constant. $\qquad\square$

Exercises 4.3.

$\diamond$ 1. Let $f : \mathbb{R} \times \mathbb{R} \to \mathbb{R}$ be a computable function and let $c \in \mathbb{R}$ be a computable constant. Define $g(x) := f(x, c)$ for all x. Show that g is computable.

◇ 2. The Gauß staircase is $(\rho, \rho_>)$-computable. Its restriction to $\mathbb{R} \setminus \mathbb{Z}$ is (ρ, ρ)-computable.

◇ 3. Use Exercise 3.1.4 to show that $\emptyset$ and $\mathbb{R}$ are the only ρ-decidable subsets or $\mathbb{R}$.

◇ 4. Show that in Theorem 4.3.2 Property 3 follows from Properties 1,2 and 5.

5. Show that the real functions $x \mapsto |x|$, $x \mapsto ||x||$, $(x, y) \mapsto d(x, y)$ and $(x, y) \mapsto d^e(x, y)$ for $x, y \in \mathbb{R}^n$ are computable.

6. Let $a_0, b_0, \ldots, a_n, b_n \in \mathbb{Q}$ with $a_0 < \ldots < a_n$. Show that the *rational polygon* $f : \mathbb{R} \to \mathbb{R}$, defined by

$$f(x) := \begin{cases} b_0 & \text{if } x < a_0 \\ b_i + (x - a_i)(b_{i+1} - b_i)/(a_{i+1} - a_i) & \text{if } a_i \le x < a_{i+1} \\ b_n & \text{if } a_n < x, \end{cases}$$

is computable.

7. Show that the function

$$((x_i)_{i \in \mathbb{N}}, n) \mapsto \prod_{i=0}^{n} x_i$$

is $([\rho]^\omega, \nu_\mathbb{N}, \rho)$-computable.

8. Show that Lemma 4.3.6, Theorem 4.3.7 and Theorem 4.3.8 hold for $\mathbb{R}^n$ replacing $\mathbb{R}$ (and, in particular, for complex numbers).

9. Let $(f_i)_{i \in \mathbb{N}}$ with $f_i : \mathbb{R} \to \mathbb{R}$ be a sequence of real functions with
    a) $(i, x) \mapsto f_i(x)$ is $(\nu_\mathbb{N}, \rho, \rho)$-computable,
    b) there is a computable function $e : \mathbb{N}^2 \to \mathbb{N}$ with $|f_i(x) - f_j(x)| \le 2^{-n}$ for all $i, j \ge e(n, k)$ and $|x| < k$.
    Show that the sequence converges to a computable function $f : \mathbb{R} \to \mathbb{R}$.

10. Complete the proof of Theorem 4.3.9.

11. Let $a, b \in \mathbb{R}$, $a < b$, a right-computable and b left-computable, a and b not computable. Show that there is a real function $f :\subseteq \mathbb{R} \to \mathbb{R}$ such that f is strictly increasing, $\mathrm{dom}(f) = (0; 1)$, $\mathrm{range}(f) = (a; b)$, f and f^{-1} are computable. The computable function f has a (unique) continuous extension to $[0; 1]$ which is not computable. (Hint: consider Example 4.2.4.3; let f be an infinite polygon.)

12. Prove that $z \mapsto \log(1 + z)$ is computable on the open disc $\{z \mid |z| < 1\}$.

13. Prove that $x \mapsto \arcsin x$ is computable on the open interval $\{x \mid |x| < 1\}$.

14. Show that the real exponential function $x \mapsto e^x$ is $(\rho_<, \rho_<)$-computable.

15. (Sorting real numbers)
    a) Show that the function $f : \mathbb{R}^n \to \mathbb{R}^n$, defined by $f(x_1, \ldots, x_n) := (y_1, \ldots, y_n)$ such that $\{x_1, \ldots, x_n\} = \{y_1, \ldots, y_n\}$ and $y_1 \le y_2 \le \ldots \le y_n$, is computable.
    b) Show that the multi-valued function $f : \mathbb{R}^2 \rightrightarrows \mathbb{N}$, defined by $R_f := \{((x_1, x_2), i) \mid x_i = \min(x_1, x_2)\}$, is not $(\rho^2, \nu_\mathbb{N})$-continuous.

c) Show that the function

$$(x_1, \ldots, x_n) \mapsto \pi, \quad x_i \in \mathbb{R}, \ x_i \neq x_j \text{ for } 1 \leq i, j \leq n,$$

where π is the permutation of $\{1, \ldots, n\}$ such that $x_{\pi(1)} < \cdots < x_{\pi(n)}$, is (ρ^n, ν)-computable (where ν is a canonical notation of the permutations of $\{1, \ldots, n\}$).

16. Let $0 < a < b$ be left-computable real numbers. Show that there is a $(\nu_{\mathbb{N}}, \nu_{\mathbb{Q}})$-computable sequence of rational numbers a_i with $b - a = \lim_{i \to \infty} a_i$.

17. Show that there is some r.e. set $A \subseteq \Sigma^* \times \Sigma^*$ with $\mathbb{R} \setminus \mathbb{Q} = \bigcap_u \bigcup_v \{I_v^1 \mid (u, v) \in A\}$ (that is, $\mathbb{R} \setminus \mathbb{Q}$ is a "computable" G_δ-set).

♦18. [Wei93] Call $X \subseteq \mathbb{R}$ a computable G_δ-set, iff $X = \bigcap_u \bigcup_v \{I_v^1 \mid (u, v) \in A\}$ for some r.e. set $A \subseteq \Sigma^* \times \Sigma^*$.

a) Show that every computable real function $f :\subseteq \mathbb{R} \to \mathbb{R}$ has a strongly (ρ, ρ)-computable extension (Exercise 3.1.5).

b) Show that $\mathrm{dom}(f)$ is a computable G_δ-set for every strongly (ρ, ρ)-computable real function $f : \mathbb{R} \to \mathbb{R}$.

c) Show that for every computable G_δ-set X there is a strongly (ρ, ρ)-computable real function $f : \mathbb{R} \to \mathbb{R}$ with $X = \mathrm{dom}(f)$.

d) Let $X \subseteq \mathbb{N}$ be a Π_2-subset, that is, $X = \{x \in \mathbb{N} \mid (\forall i)(\exists k)(x, i, k) \in B\}$ for some decidable set $B \subseteq \mathbb{N}^3$. Show that X is a computable G_δ-subset of $\mathbb{R}$. (If X is r.e. , then X, $\mathbb{N} \setminus X$ and $\mathbb{R} \setminus X$ are computable G_δ-subsets of $\mathbb{R}$.)

♦19. For a function $f : \mathbb{R}^m \to \mathbb{R}^n$, the following properties are equivalent:

a) f is (ρ^m, ρ^n)-computable,

b) the set $\{(u, v) \mid f[\overline{I}^m(u)] \subseteq I^n(v)\}$ is r.e. ,

c) $f^{-1}[I^n(v)] = \bigcup \{I^m(u) \mid (u, v) \in B\}$ for some r.e. set $B \subseteq \Sigma^* \times \Sigma^*$ (Theorem 3.2.14).

20. Show that every continuous, $(\rho^n, \mathrm{id}_{\Sigma^\omega})$-continuous or $(\rho^n, \mathrm{id}_{\Sigma^\omega})$-computable function $f : \mathbb{R}^n \to \Sigma^\omega$ is constant.

◊21. Let $f : \mathbb{R}^n \to \mathbb{R}^n$ be computable. Show that the function $(n, x) \mapsto f^n(x)$ is $(\nu_{\mathbb{N}}, \rho^n, \rho^n)$-computable. Hint: Apply Theorem 3.1.7.

22. For $c \in \mathbb{C}$ define $f_c : \mathbb{C} \to \mathbb{C}$ by $f_c(z) := z^2 + c$. Show that the function $(n, c) \mapsto f_c^n(0)$ is $(\nu_{\mathbb{N}}, \rho^2, \rho^2)$-computable. Hint: Apply Theorem 3.1.7.

23. Show that every continuous (and hence (ρ^n, ρ)-continuous) function $f : \mathbb{R}^n \to \mathbb{R}$ with $\mathrm{range}(f) \subseteq \mathbb{Q}$ is constant.

5. Computability on Closed, Open and Compact Sets

Since the set $2^{\mathbb{R}}$ of all subsets of $\mathbb{R}$ has a cardinality greater than that of Σ^{ω}, it has no representation. Therefore, the framework of TTE cannot be applied to define computable functions on all subsets of $\mathbb{R}$ like $\sup :\subseteq 2^{\mathbb{R}} \to \mathbb{R}$, which determines the least upper bound for *each* set $X \subseteq \mathbb{R}$ with upper bound. In this section we select three subclasses of subsets of $\mathbb{R}^n$ which have continuum cardinality, the open, the closed and the compact sets. As usual for a subset X of $\mathbb{R}$, we denote the *closure* of X by $\overline{X}$ and the *open kernel* or *interior* of X by X°. We will use the notation I^n of the set $\mathrm{Cb}^{(n)}$ of open rational cubes (balls) in $\mathbb{R}^n$ from Definition 4.1.2. By $\overline{I}^n(w)$ we denote the closure of the open cube $I^n(w)$.

Convention 5.0.1. In the following let $n \geq 1$ be a fixed natural number, let $\mathcal{A}$ be the set of closed, $\mathcal{O}$ the set of open and $\mathcal{K}$ the set of compact subsets of $\mathbb{R}^n$.

In Sect. 5.1 we introduce and discuss three standard representations of the set $\mathcal{A}$ and various characterizations of them. The corresponding computable closed sets are called recursively enumerable, co-recursively enumerable and recursive, respectively. We transfer the concepts to the open subsets by complementation. Various representations of the compact sets are introduced and discussed in Sect. 5.2, in particular, representations via finite open covers and the Hausdorff representation.

5.1 Closed Sets and Open Sets

A closed subset $A \subseteq \mathbb{R}^n$ can be identified by the set of all *open* cubes $J \in \mathrm{Cb}^{(n)}$ intersecting A. The set A can also be identified by the set of all *closed* cubes $\overline{J}$, $J \in \mathrm{Cb}^{(n)}$ not intersecting $A \subseteq \mathbb{R}^n$. (See Exercise 5.1.1) Fig. 5.1 shows a closed subset of $\mathbb{R}^2$ with some rational open cubes intersecting it and some rational closed cubes not intersecting it. From these observations we arrive at two computable topological spaces for the set $\mathcal{A}$ of closed subsets of $\mathbb{R}^n$.

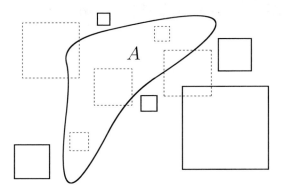

Fig. 5.1. Open cubes intersecting and closed cubes not intersecting a closed set A

Definition 5.1.1 (the representations $\psi_<$, $\psi_>$ and ψ of $\mathcal{A}$).
1. *Define computable topological spaces* $\mathbf{S}_<^{\mathcal{A}} = (\mathcal{A}, \sigma_<^{\mathcal{A}}, \nu_<^{\mathcal{A}})$ *and* $\mathbf{S}_>^{\mathcal{A}} = (\mathcal{A}, \sigma_>^{\mathcal{A}}, \nu_>^{\mathcal{A}})$ *by*

$$A \in \nu_<^{\mathcal{A}}(w) \iff I^n(w) \cap A \neq \emptyset ,$$

$$A \in \nu_>^{\mathcal{A}}(w) \iff \bar{I}^n(w) \cap A = \emptyset$$

for all closed sets $A \in \mathcal{A}$ and $w \in \Sigma^*$.
2. *Let $\psi_<$ and $\psi_>$ be the associated standard representations and $\tau_<^{\mathcal{A}}$ and $\tau_>^{\mathcal{A}}$ the associated topologies on the set $\mathcal{A}$, respectively (Definitions 3.2.1, 3.2.2).*
3. *Let $\psi := \psi_< \wedge \psi_>$ and $\tau^{\mathcal{A}} := \tau_<^{\mathcal{A}} \wedge \tau_>^{\mathcal{A}}$ (Definition 3.3.7).*
4. *A closed set $A \in \mathcal{A}$ is called*

$$
\begin{aligned}
r.e. \quad &:\iff A \text{ is } \psi_<\text{-computable,}\\
co\text{-}r.e. \quad &:\iff A \text{ is } \psi_>\text{-computable,}\\
recursive : &\iff A \text{ is } \psi\text{-computable.}
\end{aligned}
$$

A $\psi_<$-name of a closed set $A \subseteq \mathbb{R}^n$ is a list of all open rational cubes intersecting A, it "enumerates A". For each open rational cube $J \in \mathrm{Cb}^{(n)}$ the set

$$\{A \in \mathcal{A} \mid J \cap A \neq \emptyset\}$$

is a subbase element of the topology $\tau_<^{\mathcal{A}}$ on the set $\mathcal{A}$ of closed subsets of $\mathbb{R}^n$. Similarly, a $\psi_>$-name of a closed set $A \subseteq \mathbb{R}^n$ is a list of all closed rational cubes not intersecting A, it "enumerates the exterior of A". For each closed rational cube $\bar{J}$, $J \in \mathrm{Cb}^{(n)}$, the set

$$\{A \in \mathcal{A} \mid \bar{J} \cap A = \emptyset\}$$

is a subbase element of the topology $\tau_>^{\mathcal{A}}$ on the set $\mathcal{A}$ of closed subsets of $\mathbb{R}^n$. Furthermore, by Lemma 3.3.9, ψ is admissible and a ψ-name of a closed set

$A \subseteq \mathbb{R}^n$ is a combination of a list of all open rational cubes intersecting A and a list of all closed rational cubes not intersecting A, and the set $\sigma_<^{\mathcal{A}} \cup \sigma_>^{\mathcal{A}}$ is a subbase of the final topology $\tau^{\mathcal{A}}$ of ψ. By definition, $\psi \leq \psi_<$ and $\psi \leq \psi_>$, but:

Lemma 5.1.2. $\psi_< \not\leq_t \psi, \quad \psi_> \not\leq_t \psi, \quad \psi_< \not\leq_t \psi_>, \quad \psi_> \not\leq_t \psi_<$.

Proof: By Theorem 3.1.8 it suffices to show $\tau_<^{\mathcal{A}} \not\subseteq \tau_>^{\mathcal{A}}$ and $\tau_>^{\mathcal{A}} \not\subseteq \tau_<^{\mathcal{A}}$.
We have $\emptyset \notin U$, if $U \in \sigma_<^{\mathcal{A}}$. On the other hand, $\emptyset \in V$ for all $V \in \sigma_>^{\mathcal{A}}$. Since $\emptyset \in \mathcal{A}$, $\emptyset \in W$ for all non-empty $W \in \tau_>^{\mathcal{A}}$. Therefore, $\tau_<^{\mathcal{A}} \not\subseteq \tau_>^{\mathcal{A}}$.
The proof of $\tau_>^{\mathcal{A}} \not\subseteq \tau_<^{\mathcal{A}}$ is similar (use $\mathbb{R}^n$ instead of $\emptyset$). □

By definition, a closed set $A \subseteq \mathbb{R}^n$ is recursive, iff it is r.e. and co-r.e. However, in contrast to subsets of $\mathbb{N}$ the complement of the closed set A is open, hence of different type.

Example 5.1.3.   The proofs of the first four examples are very easy and left to Exercise 5.1.2.
1. The empty set and $\mathbb{R}^n$ are recursive closed subsets of $\mathbb{R}^n$.
2. A closed real interval $[a; b]$ is
    a) r.e. , iff a is right-computable and b is left-computable,
    b) co-r.e. , iff a is left-computable and b is right-computable,
    c) recursive, iff a and b are computable.
3. The closed balls $\overline{B}^e(a, r) := \{x \in \mathbb{R}^n \mid |x - a| \leq r\}$ and $\overline{B}(a, r) := \{x \in \mathbb{R}^n \mid ||x - a|| \leq r\}$ with computable $a \in \mathbb{R}^n$ and computable $r \in \mathbb{R}$ are recursive.
4. The closed sets $\{(x, y) \in \mathbb{R}^2 \mid x \leq y\}$ and $\{(x, y) \in \mathbb{R}^2 \mid x = y\}$ are recursive.
5. Assume $n = 1$. Define a function $F :\subseteq \mathbb{R} \times \mathbb{R} \to \mathcal{A}$ by $\text{dom}(F) := \{(x, y) \mid x < y\}$ and $F(x, y) := [x; y]$. Then F is
    • $(\rho_>, \rho_<, \psi_<)$-computable,
    • $(\rho_<, \rho_>, \psi_>)$-computable,
    • (ρ, ρ, ψ)-computable.
    Let $x = \rho_>(p)$ and $y = \rho_<(q)$. Then for any open rational interval $J \subseteq \mathbb{R}$, $J \cap [x; y] \neq \emptyset$, iff $J \cap (a; b) \neq \emptyset$ for some rational number a listed by p and some rational number b listed by q. There is some Type-2 machine M which on input p, q lists all open intervals J with rational endpoints such that $J \cap (a; b) \neq \emptyset$ for some rational number a listed by p and some rational number b listed by q. Then f_M realizes the function F w.r.t. $(\rho_>, \rho_<, \psi_<)$.
    Let $x = \rho_<(p)$ and $y = \rho_>(q)$. Then for any closed rational interval $J \subseteq \mathbb{R}$, $J \cap [x; y] = \emptyset$, iff $J \cap (a; b) = \emptyset$ for some rational number a listed by p and some rational number b listed by q. As above one shows that F is $(\rho_<, \rho_>, \psi_>)$-computable.

From the first two statements we conclude that F is $(\rho, \rho, \psi_<)$-computable and $(\rho, \rho, \psi_>)$-computable. Since $\psi_< \wedge \psi_> \leq \psi$, F is also (ρ, ρ, ψ)-computable.

6. The function $F : x \mapsto \{x\}$ is (ρ^n, ψ)-computable:
The identity function $\mathrm{id}_{\Sigma^\omega}$ on Σ^ω is a $(\rho^n, \psi_<)$-realization of the function F. There is a Type-2 machine N which on input $p \in \mathrm{dom}(\rho^n)$ lists all words w such that $\{\rho^n(p)\} \cap \overline{\mathrm{I}}^n(w) = \emptyset$. Then f_N is a $(\rho^n, \psi_>)$-realization of the function F. Define $g :\subseteq \Sigma^\omega \to \Sigma^\omega$ by $g(p) := \langle p, f_N(p) \rangle$ (Definition 3.3.7). Then for $p \in \mathrm{dom}(\rho^n)$, $F \circ \rho^n(p) = \psi_<(p) = \psi_> \circ f_N(p)$, and so $F \circ \rho^n(p) = \psi_< \wedge \psi_> \langle p, f_N(p) \rangle = \psi \circ g(p)$. Therefore, the function g is a (ρ^n, ψ)-computable realization of F.

7. The function $x \mapsto \overline{B}(x, 1)$ is (ρ^n, ψ)-computable (Exercise 5.1.3).   □

Definition 5.1.1.4 generalizes the definitions of recursively enumerable and recursive subsets of $\mathbb{N}$.

Lemma 5.1.4.   A subset $A \subseteq \mathbb{N}$ is
1. r.e. , iff it is an r.e. closed subset of $\mathbb{R}$ (that is, $\psi_<$-computable),
2. co-r.e. , iff it is a co-r.e. closed subset of $\mathbb{R}$ (that is, $\psi_>$-computable),
3. recursive, iff it is recursive as a closed subset of $\mathbb{R}$ (that is, ψ-computable).

Proof: The statements are trivial for the empty set. Suppose $A \neq \emptyset$. There is a computable function $k \mapsto w_k$ such that $\mathrm{I}^1(w_k) := (k - 1/2 \; ; \; k + 1/2)$.

1. If A is r.e. , then $\{w \mid \mathrm{I}^1(w) \cap A \neq \emptyset\}$ is r.e. and so A is $\psi_<$-computable. On the other hand, let A be a $\psi_<$-computable subset of $\mathbb{R}$. Then $X_A := \{w \mid \mathrm{I}^1(w) \cap A \neq \emptyset\}$ is r.e. Since $A = \{k \mid w_k \in X_A\}$, A is an r.e. set.

2. Assume that $\mathbb{N} \setminus A$ is r.e. Since

$$\overline{\mathrm{I}}^1(w) \cap A = \emptyset \iff \overline{\mathrm{I}}^1(w) \cap \mathbb{N} \text{ is a (finite) subset of } \mathbb{N} \setminus A \, ,$$

the set $\{w \mid \overline{\mathrm{I}}^1(w) \cap A = \emptyset\}$ is r.e. and so A is $\psi_>$-computable. On the other hand, let A be $\psi_>$-computable. Then $Y_A := \{w \mid \overline{\mathrm{I}}^1(w) \cap A = \emptyset\}$ is r.e. and $\mathbb{N} \setminus A = \{k \mid w_k \in Y_A\}$, which is also an r.e. set.

3. The statement for recursive sets follows from Properties 1 and 2.   □

While many closed subsets of $\mathbb{R}^n$ are recursive only trivial subsets of $\mathbb{R}^n$ are ρ^n-decidable (Definition 3.1.3), that is, $[\rho]^n$-decidable or $(\rho, \dots, \rho)$-decidable (Lemmas 3.3.5, 3.3.6).

Theorem 5.1.5. $\emptyset$ and $\mathbb{R}^n$ are the only ρ^n-decidable subsets of $\mathbb{R}^n$.

Proof: Let $X \subseteq \mathbb{R}^n$ be ρ^n-decidable. Then by Exercise 3.1.4 its characteristic function $\mathrm{cf}_X : \mathbb{R}^n \to \mathbb{N}$ is $(\rho^n, \nu_\mathbb{N})$-computable. By Corollary 4.3.16, cf_X is a constant function, hence $X = \emptyset$ or $X = \mathbb{R}^n$.   □

The atomic properties available from a ψ-name of a closed set $A \subseteq \mathbb{R}^2$ are sufficient for plotting it on a screen with arbitrary "metric" precision. Consider the closed set A from Fig. 5.1 and assume $A \subseteq [0;1]^2$. Suppose we have a screen with k rows and k columns. For $0 \leq i,j < k$ define

$$K_{ij} := \left[\frac{i}{k}; \frac{i+1}{k}\right] \times \left[\frac{j}{k}; \frac{j+1}{k}\right] , \quad L_{ij} := \left(\frac{i-1}{k}; \frac{i+2}{k}\right) \times \left(\frac{j-1}{k}; \frac{j+2}{k}\right) .$$

Then for every i,j, $\overline{K}_{ij} \cap A = \emptyset$ or $L_{ij} \cap A \neq \emptyset$ (or both). Let $p \in \Sigma^\omega$ be a ψ-name of the set A. Then p is (more precisely, encodes) a list of all $I \in \mathrm{Cb}^{(2)}$ with $\overline{I} \cap A = 0$ and a list of all $J \in \mathrm{Cb}^{(2)}$ with $J \cap A \neq 0$. For every i,j, K_{ij} is in the first list or L_{ij} is in the second list (or both properties hold).

For every i,j determine the color of the "pixel" at Position (i,j) as follows: Search the lists simultaneously until K_{ij} has been detected in the first list or L_{ij} has been detected in the second list.

$$\text{Print the pixel} \begin{cases} \text{white,} & \text{if } K_{ij} \text{ has been detected} \\ \text{black,} & \text{if } L_{ij} \text{ has been detected.} \end{cases}$$

Every pixel which is outside A and far enough from A will be colored white and every pixel intersecting A will be colored black. Only pixels close to the boundary of A may be colored black or white, depending on the orders of the two lists and the procedure for searching. Fig. 5.2 shows a plot of our set A on a 24×24 pixel screen.

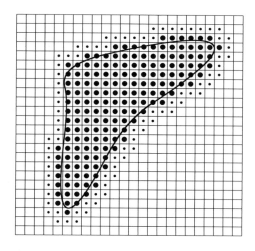

• black pixels
· black or white
  pixels

Fig. 5.2. A a 24×24 pixel plot of a closed set A

By this example it is reasonable to call the closed set $A \subseteq \mathbb{R}^2$ with computable ψ-name *recursive*, even though membership $x \in A$ cannot be decided for $x \in \mathbb{R}^2$ by Theorem 5.1.5.

By Exercise 3.1.4, a set X is ρ^n-decidable, iff its characteristic function $\mathrm{cf}_X : \mathbb{R}^n \to \mathbb{N}$ is $(\rho^n, \nu_\mathbb{N})$-computable. Another generalization of the discrete characteristic functions $\mathrm{cf}_A : \mathbb{N} \to \mathbb{N}$ is the metric distance function. Every closed set $A \subseteq \mathbb{R}^n$ is defined uniquely by its distance function $d_A : \mathbb{R}^n \to \mathbb{R}$, $d_A(x) := d(x, A) := \inf_{y \in A} ||x - y||$, which is continuous. Figure 5.3 shows the maximum distance of a point x and the Euclidean distance of a point y of a closed set $A \subseteq \mathbb{R}^2$.

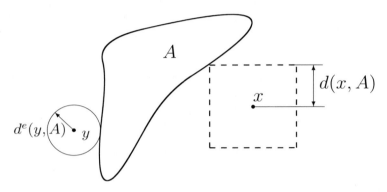

Fig. 5.3. The "maximum distance"-ball with center x touching the set A and the Euclidean ball with center y touching the set A

In the following definition we use names of the functions d_A as names of closed sets A. For including also the empty set $\emptyset \in \mathcal{A}$ we define $= d(x, \emptyset) := \infty$ for all $x \in \mathbb{R}^n$ and use the canonical extensions $\overline{\rho}_<$, $\overline{\rho}_>$ and $\overline{\rho}$ of the representations $\rho_<$, $\rho_>$ and ρ, respectively, to $\overline{\mathbb{R}} := \mathbb{R} \cup \{-\infty, \infty\}$ (Exercise 4.1.21).

Definition 5.1.6. *Define representations* $\psi_<^{\mathrm{dist}}$, $\psi_>^{\mathrm{dist}}$ *and* ψ^{dist} *of the set* $\mathcal{A}$ *of closed subsets of* $\mathbb{R}^n$ *by*

$$\psi_<^{\mathrm{dist}}(p) = A \iff d_A = [\rho^n \to \overline{\rho}_>]_{\mathbb{R}^n}(p) ,$$
$$\psi_>^{\mathrm{dist}}(p) = A \iff d_A = [\rho^n \to \overline{\rho}_<]_{\mathbb{R}^n}(p) ,$$
$$\psi^{\mathrm{dist}}(p) = A \iff d_A = [\rho^n \to \overline{\rho}]_{\mathbb{R}^n}(p)$$

where $d_A(x) := d(x, A) := \inf_{y \in A} ||x - y||$.

Notice the pairings $(\psi_>^{\mathrm{dist}}, \overline{\rho}_<)$ and $(\psi_<^{\mathrm{dist}}, \overline{\rho}_>)$ and remember that by Definition 3.3.13, $f = [\delta \to \delta']_N(p)$, iff $\eta_p^{\omega\omega}$ is a (δ, δ')-realization of f.

Lemma 5.1.7. $\psi_<^{\mathrm{dist}} \equiv \psi_< $, $\psi_>^{\mathrm{dist}} \equiv \psi_> $, $\psi^{\mathrm{dist}} \equiv \psi$.

Proof: 1. $\psi_< \le \psi_<^{\mathrm{dist}}$: For any $x \in \mathbb{R}^n$, $r \in \mathbb{Q}$ and $A \in \mathcal{A}$, $d(x, A) < r$, iff there are balls $B(a, s), B(b, t) \in \mathrm{Cb}^{(n)}$ with rational centers and radii such that

$$x \in B(a, s), \quad B(b, t) \cap A \ne \emptyset, \quad \text{and} \quad d(a, b) + s + t < r .$$

Remember that a ρ^n-name of x is a list of all words $\iota(v)$ with $x \in \mathrm{I}^n(v)$, a $\psi_<$-name of A is a list of all words $\iota(v)$ with $\mathrm{I}^n(w) \cap A \ne \emptyset$ and a $\overline{\rho}_>$-name of $y \in \mathbb{R}$ is a list of all words $\iota(u)$ with $y < \nu_{\mathbb{Q}}(u)$. There is a Type-2 machine M which on inputs $p \in \mathrm{dom}(\rho^n)$ and $q \in \mathrm{dom}(\psi_<)$ operates in stages $k = 0, 1, \ldots$ as follows.

Stage k: M writes the "dummy" symbol 1 and for every subword $\iota(v) \lhd p_{<k}$ and every subword $\iota(w) \lhd q_{<k}$ it writes every word $\iota(u)$ with $u \in \mathrm{dom}(\nu_{\mathbb{Q}})$, $\lg(u) \le k$ and

$$d(a, b) + s + t < \nu_{\mathbb{Q}}(u) ,$$

where $B(a, s) := \mathrm{I}^n(v)$ and $B(b, t) := \mathrm{I}^n(w)$. If $A = \psi_<(q)$, it follows that

$$d_A \circ \rho^n(p) = d(\rho^n(p), \psi_<(q)) = \overline{\rho}_> \circ f_M(p, q) .$$

Notice that $\overline{\rho}_> \circ f_M(p, q) = \overline{\rho}_>(111\ldots) = \infty$, if $A = \emptyset$. By the smn-theorem for $\eta^{\omega\omega}$ (Theorem 2.3.13) there is a computable function $F : \Sigma^\omega \to \Sigma^\omega$ such that $f_M(p, q) = \eta^{\omega\omega}_{F(q)}(p)$. We obtain

$$d_A \circ \rho^n(p) = \overline{\rho}_> \circ \eta^{\omega\omega}_{F(q)}(p) ,$$

and so $d_A = [\rho^n \to \overline{\rho}_>]_{\mathbb{R}^n}(F(p))$, that is, $\psi_<^{\mathrm{dist}}(F(p)) = A$. Therefore, the computable function F translates $\psi_<$ to $\psi_<^{\mathrm{dist}}$.

2. $\psi_> \le \psi_>^{\mathrm{dist}}$: For any closed set $A \subseteq \mathbb{R}^n$, any $x \in \mathbb{R}^n$ and any $r > 0$ we have

$$r < d_A(x) \iff \left(\exists a \in \mathbb{Q}^n, s \in \mathbb{Q}\right)\left(s > 0, \ x \in B(a, s) \text{ and } \overline{B}(a, r+s) \cap A = \emptyset\right).$$

Let $\left((u_i, v_i, w_i, m_i)\right)_{i \in \mathbb{N}}$ be a computable enumeration of all tuples (u, v, w, m) with $u, v \in \mathrm{dom}(\mathrm{I}^n)$, $w \in \mathrm{dom}(\nu_{\mathbb{Q}})$ and $m \in \mathbb{N}$ such that each such tuple occurs infinitely often. There is a Type-2 machine M which on input $(p, q) \in \Sigma^\omega \times \Sigma^\omega$ operates in stages $i = 0, 1, \ldots$ as follows:

Stage i: If $\nu_{\mathbb{Q}}(w_i) \le 0$, or:
$\nu_{\mathbb{Q}}(w_i) > 0$, $\iota(u_i) \lhd p_{m_i}$, $\iota(v_i) \lhd q_{m_i}$ and there are $a \in \mathbb{Q}^n$ and $s \in \mathbb{Q}$ such that $s > 0$, $\mathrm{I}^n(u_i) = B(a, s)$ and $\overline{\mathrm{I}}^n(u_i) = \overline{B}(a, \nu_{\mathbb{Q}}(w_i) + s)$,
then M prints $\iota(w_i)$.

If $x = \rho^n(p)$ and $A = \psi_>(q)$, then $f_M(p, q)$ is a list of all w such that $\nu_{\mathbb{Q}}(w) < d_A(x)$. Therefore, $d_A \circ \rho^n(p) = \overline{\rho}_< \circ f_M(p, q)$. By the smn-theorem for $\eta^{\omega\omega}$ there is a computable function $F : \Sigma^\omega \to \Sigma^\omega$ such that $f_M(p, q) = \eta^{\omega\omega}_{F(q)}(p)$. We obtain

$$d_A \circ \rho^n(p) = \overline{\rho}_< \circ \eta^{\omega\omega}_{F(q)}(p) ,$$

and so $d_A = [\rho^n \to \overline{\rho}_<]_{\mathbb{R}^n}(F(p))$, that is, $\psi_>^{\mathrm{dist}}(F(p)) = A$. Therefore, the computable function F translates $\psi_>$ to $\psi_>^{\mathrm{dist}}$.

3. $\psi \le \psi^{\mathrm{dist}}$: We reduce this case to the cases 1 and 2 above. By definition, $\overline{\rho} = \overline{\rho}_< \wedge \overline{\rho}_>$. By Case 1 there is a computable function $F : \Sigma^\omega \to \Sigma^\omega$ translating $\psi_<$ to $\psi_<^{\mathrm{dist}}$, and by Case 2 there is a computable function $G : \Sigma^\omega \to \Sigma^\omega$ translating $\psi_>$ to $\psi_>^{\mathrm{dist}}$. By the utm-theorem and the smn-theorem there is a computable function $H : \Sigma^\omega \to \Sigma^\omega$ such that

$$\langle \eta_{G(q)}^{\omega\omega}(r), \eta_{F(p)}^{\omega\omega}(r) \rangle = \eta_{H\langle p,q \rangle}^{\omega\omega}(r) .$$

Since for all $p, q \in \Sigma^\omega$,

$$\begin{aligned}
\psi\langle p,q \rangle = A &\iff \psi_<(p) = A \text{ and } \psi_>(q) = A \\
&\iff \psi_<^{\mathrm{dist}} F(p) = A \text{ and } \psi_>^{\mathrm{dist}} G(q) = A \\
&\iff (\forall r)\Big(d_A \circ \rho^n(r) = \overline{\rho}_> \circ \eta_{F(p)}^{\omega\omega}(r) \text{ and} \\
&\qquad d_A \circ \rho^n(r) = \overline{\rho}_< \circ \eta_{G(q)}^{\omega\omega}(r)\Big) \\
&\iff (\forall r)\Big(d_A \circ \rho^n(r) = \overline{\rho}_< \wedge \overline{\rho}_> \langle \eta_{G(q)}^{\omega\omega}(r), \eta_{F(p)}^{\omega\omega}(r) \rangle \Big) \\
&\iff (\forall r)\Big(d_A \circ \rho^n(r) = \overline{\rho}\langle \eta_{G(q)}^{\omega\omega}(r), \eta_{F(p)}^{\omega\omega}(r) \rangle \Big) \\
&\iff (\forall r)\Big(d_A \circ \rho^n(r) = \overline{\rho} \circ \eta_{H\langle p,q \rangle}^{\omega\omega}(r) \Big) \\
&\iff \psi^{\mathrm{dist}} H\langle p,q \rangle = A ,
\end{aligned}$$

the computable function H translates ψ to ψ^{dist}.

4. $\psi_<^{\mathrm{dist}} \le \psi_<$: Each element of Cb^n is a ball $B(a,r)$ with $a \in \mathbb{Q}^n$ and $r \in \mathbb{Q}$. For $A \in \mathcal{A}$ we have

$$B(a,r) \cap A \ne \emptyset \iff d_A(a) < r .$$

There are computable functions f, g such that $\rho^n \circ f(w)$ is the center and $\nu_{\mathbb{Q}} \circ g(w)$ is the radius of $\mathrm{I}^n(w)$. It follows that

$$\begin{aligned}
&\mathrm{I}^n(w) \cap \psi_<^{\mathrm{dist}}(q) \ne \emptyset \\
&\iff \overline{\rho}_> \eta_q^{\omega\omega} f(w) < \nu_{\mathbb{Q}} g(w) \\
&\iff \nu_{\mathbb{Q}}(v) < \nu_{\mathbb{Q}} g(w) \text{ for some subword } \iota(v) \text{ of } \eta_q^{\omega\omega} f(w) .
\end{aligned}$$

By the utm-theorem there is a Type-2 machine M which on input $q \in \mathrm{dom}(\psi_<^{\mathrm{dist}})$ lists all words w such that $\nu_{\mathbb{Q}}(v) < \nu_{\mathbb{Q}} g(w)$ for some subword $\iota(v)$ of $\eta_q^{\omega\omega} f(w)$. The function f_M translates $\psi_<^{\mathrm{dist}}$ to $\psi_<$.

5. $\psi_>^{\mathrm{dist}} \le \psi_>$: For each ball $B(a,r)$ with $a \in \mathbb{Q}^n$ and $r \in \mathbb{Q}$ and each $A \in \mathcal{A}$ we have

$$\overline{B}(a,r) \cap A = \emptyset \iff r < d_A(a) .$$

With the functions f, g from 4 we have

$$\overline{I}^n(w) \cap \psi_>^{\text{dist}}(q) = \emptyset$$
$$\iff \nu_{\mathbb{Q}} g(w) < \overline{\rho}_< \eta_q^{\omega\omega} f(w)$$
$$\iff \nu_{\mathbb{Q}} g(w) < \nu_{\mathbb{Q}}(v) \text{ for some subword } \iota(v) \text{ of } \eta_q^{\omega\omega} f(w) .$$

Continue as in 4 above.

6. $\psi^{\text{dist}} \leq \psi$:

We have $\overline{\rho} \leq \overline{\rho}_>$, therefore, $\psi^{\text{dist}} \leq \psi_>^{\text{dist}} \leq \psi_<$ by 4, and correspondingly, $\psi^{\text{dist}} \leq \psi_>^{\text{dist}} \leq \psi_>$ by 5. Since ψ is the greatest lower bound of $\psi_<$ and $\psi_>$ (Definition 5.1.1, Lemma 3.3.8), $\psi^{\text{dist}} \leq \psi$. $\qquad\square$

Corollary 5.1.8. A closed set $A \subseteq \mathbb{R}^n$ is

> r.e. , iff the distance function d_A is $(\rho^n, \overline{\rho}_>)$-computable ,
>
> co-r.e. , iff the distance function d_A is $(\rho^n, \overline{\rho}_<)$-computable ,
>
> recursive , iff the distance function d_A is $(\rho^n, \overline{\rho})$-computable .

We introduce two further enumeration representations.

Definition 5.1.9. *Define representations $\psi_<^{\text{en}}, \psi_>^{\text{en}} :\subseteq \Sigma^\omega \to \mathcal{A}$ as follows:*
$$\psi_<^{\text{en}}(0^\omega) := \emptyset ,$$
$$\psi_<^{\text{en}}(0^k 1p) := \text{cls} \circ \text{range}\big([\nu_{\mathbb{N}} \to \rho^n]_{\mathbb{N}}(p)\big)$$

for all $k \in \mathbb{N}$ and $p \in \Sigma^\omega$, and

$$p \in \text{dom}(\psi_>^{\text{en}}) : \iff (\forall w \in \Sigma^*)\big(\iota(w) \lhd p \implies w \in \text{dom}(I^n)\big) ,$$
$$\psi_>^{\text{en}}(p) \quad := \quad \mathbb{R}^n \setminus \bigcup \{I^n(w) \mid \iota(w) \lhd p\}$$

for all $p \in \Sigma^\omega$.

For a non-empty closed set A a $\psi_<^{\text{en}}$-name of A is a sequence $0^k 1p$ such that p is a "program" (Definition 3.3.13) of a function $f : \mathbb{N} \to \mathbb{R}^n$ such that $\text{range}(f) = \{f(0), f(1), f(2), \ldots\}$ is dense in A. The sequence 0^ω is the only $\psi_<^{\text{en}}$-name of the empty set. A $\psi_>^{\text{en}}$-name of A is a list of open rational cubes exhausting $\mathbb{R}^n \setminus A$. Notice that not all open cubes in $\mathbb{R}^n \setminus A$ must appear in the list.

Lemma 5.1.10. $\qquad \psi_<^{\text{en}} \equiv \psi_< , \qquad \psi_>^{\text{en}} \equiv \psi_> .$

Proof: 1. $\psi_< \leq \psi_<^{\text{en}}$: Define a multi-valued function $F :\subseteq \mathcal{A} \times \text{Cb}^{(n)} \rightrightarrows \mathbb{R}^n$ by $\text{R}_F := \{((A, J), x) \mid x \in A \cap J\}$. We show that F is $(\psi_<, I^n, \rho^{na})$-computable, where ρ^{na} is the n-dimensional version of ρ^a from Lemma 4.1.6. There is a Type-2 machine M which on input (p, w) such that $\psi_<(p) \cap I^n(w) \neq \emptyset$ determines a sequence $\iota(u_0)\iota(u_1)\ldots$ such that $u_0 = w$ and for all $k \geq 1$, $\iota(u_k) \lhd p$, the diameter of $I^n(u_k)$ is less than 2^{-k} and $\text{cls}(I^n(u_k)) \subseteq I^n(u_{k-1})$. If $\psi_<(p) \cap I^n(w) \neq \emptyset$, then $f_M(p, w)$ is a ρ^{na}-name of a point in $\psi_<(p) \cap I^n(w)$.

There is a Type-2 machine N which on input $(p, v) \in \mathrm{dom}(\psi_<) \times \mathrm{dom}(\nu_\mathbb{N})$ tries to determine the k-th ($k = \nu_\mathbb{N}(v) + 1$) subword $\iota(w)$ of p and tries to produce the output $f_M(p, w)$. Let h be a computable translation from ρ^{na} to ρ^n. By the smn-theorem there is a computable function $g : \Sigma^\omega \to \Sigma^\omega$ with $\eta^{*\omega}_{g(p)}(v) = h \circ f_N(p, v)$. If $\psi_<(p) \neq \emptyset$, then $g(p)$ exists and $\eta^{*\omega}_{g(p)}$ is a $(\nu_\mathbb{N}, \rho)$-realization of an enumeration of a dense subset of $\psi_<(p)$. There is a Type-2 machine L which on input $p \in \mathrm{dom}(\psi)$ prints 0^ω, if $\iota(w) \lhd p$ for no word w, and $0^k 1 g(p)$ for some k, otherwise. Then f_L translates $\psi_<$ to $\psi_\leqslant^{\mathrm{en}}$.

    2. $\psi_\leqslant^{\mathrm{en}} \leq \psi_<$: (See Exercise 5.1.10)

    3. $\psi_> \leq \psi_>^{\mathrm{en}}$: The identity on Σ^ω translates $\psi_>$ to $\psi_>^{\mathrm{en}}$.

    4. $\psi_>^{\mathrm{en}} \leq \psi_>$: Since $\overline{J}$ is compact for $J \in \mathrm{Cb}^{(n)}$, finitely many open balls enumerated by a $\psi_>^{\mathrm{en}}$-name p suffice to cover $\overline{J}$, iff $\overline{J} \cap \psi_>^{\mathrm{en}}(p) = \emptyset$, and so

$$\overline{\mathrm{I}}^n(w) \cap \psi_>^{\mathrm{en}}(p) = \emptyset \iff (\exists u_1, \ldots, u_k)(\iota(u_i) \lhd p \text{ for } 1 \leq i \leq k$$
$$\text{and } \overline{\mathrm{I}}^n(w) \subseteq \mathrm{I}^n(u_1) \cup \ldots \cup \mathrm{I}^n(u_k)).$$

Therefore, there is a Type-2 machine M which on input $p \in \mathrm{dom}(\psi_>^{\mathrm{en}})$ enumerates all $\iota(w)$ such that $\overline{\mathrm{I}}^n(w) \cap \psi_>^{\mathrm{en}}(p) = \emptyset$. Then f_M translates $\psi_>^{\mathrm{en}}$ to $\psi_>$. □

Corollary 5.1.11. A closed set $A \in \mathcal{A}$ is
1. r.e. , iff it is empty or has a dense computable sequence $(x_i)_{i \in \mathbb{N}}$,
2. co-r.e. , iff $\mathbb{R}^n \setminus A = \bigcup_{w \in X} \mathrm{I}^n(w)$ for some r.e. set $X \subseteq \Sigma^*$.

Example 5.1.12.
1. For any point $x \in \mathbb{R}^n$,

    x is computable $\iff \{x\}$ is r.e. $\iff \{x\}$ is co-r.e. $\iff \{x\}$ is recursive.

    By Lemma 4.1.19, x is computable, iff $\{w \mid x \in \mathrm{I}^n(w)\}$ is r.e. , iff $\{w \mid \{x\} \cap \mathrm{I}^n(w) \neq \emptyset\}$ is r.e. , iff $\{x\}$ is $\psi_<$-computable, iff $\{x\}$ is r.e.

    If x is computable then the set of all $J \in \mathrm{Cb}^{(n)}$ with $\{x\} \cap J = \emptyset$ can be enumerated, and so $\{x\}$ is co-r.e.

    On the other hand, assume that $\{x\}$ is co-r.e. , that is, $\psi_>^{\mathrm{en}}$-computable. Then $\mathbb{R}^n \setminus \{x\} = \bigcup_{w \in X} \mathrm{I}^n(w)$ for some r.e. set $X \subseteq \Sigma^*$. For some $k \in \mathbb{N}$, $\|x\| < k$. Then for every $K \in \mathrm{Cb}^{(n)}$ with $x \in K$ the compact set $\overline{B}(0, k) \setminus K$ is covered by already finitely many elements $\mathrm{I}^n(w)$, $w \in X$. Therefore, from an enumeration of X we can find smaller and smaller neighborhoods $K \in \mathrm{Cb}^{(n)}$ of x. Therefore, x is computable.

    By definition, $\{x\}$ is r.e. and co-r.e. , iff it is recursive.

2. A closed set $A \subseteq \mathbb{R}^n$ is co-r.e. , iff there is a computable function $f : \mathbb{R}^n \to \mathbb{R}$ such that $A = f^{-1}[\{0\}]$.

    Let A be co-r.e. If $A = \mathbb{R}^n$, then choose the constant zero function for f. Assume $A \neq \mathbb{R}^n$. Then by Corollary 5.1.11, $\mathbb{R}^n \setminus A = \bigcup_k \mathrm{I}^n \circ g(k)$ for some computable function $g : \mathbb{N} \to \Sigma^*$. For $k \in \mathbb{N}$ define $h_k : \mathbb{R}^n \to \mathbb{R}$ by

$$h_k(x) := 2^{-k} \cdot \max(0, r - d(x, a)) \,,$$

where $I^n \circ g(k) = B(a, r)$. Define $f_k(x) := h_0(x) + \ldots + h_k(x)$. Then $(k, x) \mapsto f_k(x)$ is computable and $|f_i(x) - f_j(x)| \leq 2^{-k}$ for $i, j \geq k$. By Theorem 4.3.8, the function f, defined by $f(x) := \lim_{k \to \infty} f_k(x)$, is computable. Obviously, $A = f^{-1}[\{0\}]$.

On the other hand, let $f : \mathbb{R}^n \to \mathbb{R}$ be computable. There is a Type-2 machine M such that f_M is a (ρ^n, ρ_C)-realization of f. Suppose $f(x) \neq 0$. Then there are sequences $p = \iota(u_0)\iota(u_1)\ldots$ and $q = \iota(v_0)\iota(v_1)\ldots$ and some $j \in \mathbb{N}$ such that $\rho^n(p) = x$, $\rho(q) = f(x)$, $f_M(p) = q$ and $|f(x)| > 2 \cdot 2^{-j}$. There are a number k and words $u_0, u_1, \ldots, u_k, v_0, v_1, \ldots, v_j$ such that $u := \iota(u_0)\iota(u_1)\ldots\iota(u_k)$ is a prefix of p, $v := \iota(v_0)\iota(v_1)\ldots\iota(v_j)$ is a prefix of q and $(u, 0^k, v) \in T_M$, T_M the set from Lemma 2.1.5. Then $x \in I^n(w) \subseteq I^n(u_0) \cap \ldots \cap I^n(u_k)$ for some word w, $|\nu_{\mathbb{Q}}(v_j)| > 2^{-j}$ and so $0 \notin f[I^n(u_0) \cap \ldots \cap I^n(u_k)]$. Now let X be the set of all words w such that there are numbers j, k and words $u_0, u_1, \ldots, u_k, v_0, v_1, \ldots, v_j$ such that $u := \iota(u_0)\iota(u_1)\ldots\iota(u_k)$ is a prefix of some $p \in \text{dom}(\rho^n)$, $v := \iota(v_0)\iota(v_1)\ldots\iota(v_j)$ is a prefix of some $q \in \text{dom}(\rho)$, $(u, 0^k, v) \in T_M$ and $I^n(w) \subseteq I^n(u_0) \cap \ldots \cap I^n(u_k)$. Then X is r.e. by Lemma 2.1.5 and $\{x \mid f(x) \neq 0\} = \bigcup_{w \in X} I^n(w)$. By Corollary 5.1.11, $f^{-1}[\{0\}]$ is co-r.e.

3. For every computable function $f : \mathbb{R}^n \to \mathbb{R}^n$ the function $F : A \mapsto \text{cls}(f[A])$ for $A \in \mathcal{A}$ is $(\psi_<, \psi_<)$-computable:

It suffices to show that f is $(\psi_<^{\text{en}}, \psi_<^{\text{en}})$-computable. If $D \subseteq A \subseteq \mathbb{R}$ is dense in A and $f : \mathbb{R} \to \mathbb{R}$ is continuous, then $f[D]$ is dense in $f[A]$ and also in $\text{cls}(f[A])$. Let $h :\subseteq \Sigma^\omega \to \Sigma^\omega$ be a (ρ^n, ρ^n)-realization of f. By the utm-theorem and the smn-theorem for $\eta^{*\omega}$ there is a computable function $r : \Sigma^\omega \to \Sigma^\omega$ such that $h \circ \eta_p^{*\omega} = \eta_{r(p)}^{*\omega}(w)$. There is a computable function $g :\subseteq \Sigma^\omega \to \Sigma^\omega$ such that $g(0^\omega) = 0^\omega$ and $g(0^k 1 p) = 0^k 1 r(p)$. The function g is a $(\psi_<, \psi_<)$-realization of the function $F : A \mapsto \text{cls}(f[A])$.

$\square$

Theorem 5.1.13 (union and intersection of closed sets).
1. Union $(A, B) \mapsto A \cup B$ on $\mathcal{A}$ is $(\psi_<, \psi_<, \psi_<)$-computable, $(\psi_>, \psi_>, \psi_>)$-computable and (ψ, ψ, ψ)-computable.
2. Intersection $(A, B) \mapsto A \cap B$ on $\mathcal{A}$ is $(\psi_>, \psi_>, \psi_>)$-computable.
3. The function $A \mapsto A \cap \{0\}$ is not $(\psi, \psi_<)$-continuous, and so intersection $(A, B) \mapsto A \cap B$ on $\mathcal{A}$ is not $(\psi, \psi, \psi_<)$-continuous.

Proof: 1. For any $J \in \text{Cb}^{(n)}$, $(A \cup B) \cap J \neq \emptyset \iff (A \cap J \neq \emptyset \text{ or } B \cap J \neq \emptyset)$. There is a Type-2 machine M which on input (p, q) produces a list of all words "$\iota(w)$" such that "$\iota(w)$" is a subword of p or a subword of q. If $A = \psi_<(p)$ and $B = \psi_<(q)$, then $A \cup B = \psi_< \circ f_M(p, q)$. Therefore, union is $(\psi_<, \psi_<, \psi_<)$-computable.

For any $J \in \text{Cb}^{(n)}$, $(A \cup B) \cap \overline{J} = \emptyset \iff (A \cup \overline{J} = \emptyset \text{ and } B \cup \overline{J} = \emptyset)$. There is a Type-2 machine M which on input (p, q) produces a list of all words "$\iota(w)$" such that "$\iota(w)$" is a subword of p and a subword of q. If $A = \psi_>(p)$ and $B = \psi_>(q)$, then $A \cup B = \psi_> \circ f_M(p, q)$. Therefore, union is $(\psi_>, \psi_>, \psi_>)$-computable.

Since $\psi = \psi_< \wedge \psi_>$, union is also (ψ, ψ, ψ)-computable.

2. If $\mathbb{R}^n \setminus A = \bigcup_{w \in X} I^n(w)$ and $\mathbb{R}^n \setminus B = \bigcup_{w \in Y} I^n(w)$, then

$$\mathbb{R}^n \setminus (A \cap B) = \mathbb{R}^n \setminus A \ \cup \ \mathbb{R}^n \setminus B \ = \bigcup_{w \in X \cup Y} I^n(w) .$$

There is a Type-2 machine M which on input (p, q) produces a list of all words "$\iota(w)$" such that "$\iota(w)$" is a subword of p or a subword of q. If $A = \psi_>^{en}(p)$ and $B = \psi_>^{en}(q)$, then $A \cap B = \psi_>^{en} \circ f_M(p, q)$. Therefore, intersection is $(\psi_>^{en}, \psi_>^{en}, \psi_>^{en})$-computable. Apply Lemma 5.1.10.

3. Assume that $f : \mathcal{A} \to \mathcal{A}$, defined by $f(A) := \{0\} \cap A$, is $(\psi, \psi_<)$-continuous, and so $(\tau^{\mathcal{A}}, \tau_<^{\mathcal{A}})$-continuous by Theorem 3.2.11. Define $U \in \sigma_<^{\mathcal{A}}$ by $U := \{A \in \mathcal{A} \mid A \cap B(0, 1) \neq \emptyset\}$. Since f is continuous in $\{0\} \in \mathcal{A}$ and $f(\{0\}) = \{0\} \in U$, there is an open set $V \in \tau^{\mathcal{A}}$ such that $\{0\} \in V$ and $f[V] \subseteq U$. Since $\sigma_<^{\mathcal{A}} \cup \sigma_>^{\mathcal{A}}$ is a subbase of $\tau^{\mathcal{A}}$, we may choose $V = A_1 \cap \ldots \cap A_k \cap B_1 \cap \ldots \cap B_m$ where $A_i \in \sigma_<^{\mathcal{A}}$ and $B_i \in \sigma_>^{\mathcal{A}}$. Since $\{0\} \in A_i$, there is some $r_i > 0$ such that $\{x\} \in A_i$, whenever $|x| < r_i$ ($i = 1, \ldots, k$). Since $\{0\} \in B_i$, there is some $s_i > 0$ such that $\{x\} \in B_i$, whenever $|x| < s_i$ ($i = 1, \ldots, m$). Choose $x \in \mathbb{R}^n$ with $0 < |x| < \min(\min_{i \leq k} r_i, \min_{i \leq m} s_i)$. Then $\{x\} \in V$ and $f(\{x\}) = \emptyset \notin U$ (contradiction). $\qquad \square$

As in many examples, also for intersection of closed sets discontinuity is connected with computational irregularity: There are two recursive closed sets the intersection of which is not r.e. (Exercise 5.1.15).

In Definition 5.1.1 for a closed set $A \subseteq \mathbb{R}^n$, we have considered atomic properties

$$J \cap A \neq \emptyset \quad \text{and} \quad \overline{J} \cap A = \emptyset$$

with $J \in \mathrm{Cb}^{(n)}$, but not the other two variants

$$\overline{J} \cap A \neq \emptyset \quad \text{and} \quad J \cap A = \emptyset .$$

Only the first two cases induce robust definitions (cf. Theorem 4.1.11).

Theorem 5.1.14 ($\psi_<$ and $\psi_>$ are robust). Consider the representations $\psi_<$ and $\psi_>$ (Definition 5.1.1) as functions of the notation $\nu_{\mathbb{Q}}$, that is, $\psi_< = E_<(\nu_{\mathbb{Q}})$ and $\psi_> = E_>(\nu_{\mathbb{Q}})$. Let $E'_<$ be the modification of $E_<$ where "$J \cap A \neq \emptyset$" is replaced by "$\overline{J} \cap A \neq \emptyset$", and let $E'_>$ be the modification of $E_>$ where "$\overline{J} \cap A = \emptyset$" is replaced by "$J \cap A = \emptyset$". For every notation $\nu_Q :\subseteq \Sigma^* \to Q$ of a dense subset of $\mathbb{R}^n$,

1. $\psi_< = E_<(\nu_{\mathbb{Q}}) \equiv_t E_<(\nu_Q)$ and $\psi_> = E_>(\nu_{\mathbb{Q}}) \equiv_t E_>(\nu_Q)$,
2. $\psi_< = E_<(\nu_{\mathbb{Q}}) \equiv E_<(\nu_Q)$ and $\psi_> = E_>(\nu_{\mathbb{Q}}) \equiv E_>(\nu_Q)$, if $\nu_{\mathbb{Q}}$ and ν_Q are r.e.-related.
3. $E'_<(\nu_{\mathbb{Q}}) \not\equiv_t E'_<(\nu_Q)$ and $E'_>(\nu_{\mathbb{Q}}) \not\equiv_t E'_>(\nu_Q)$ in general.

Proof : 1. and 2. Let ν_Q be a notation of a dense subset Q of $\mathbb{R}$. Define

$$I_Q(\iota(v_1)\ldots\iota(v_n)\iota(w)) := B((\nu_Q(v_1),\ldots,\nu_Q(v_n)),\nu_Q(w))$$

(cf. Definition 4.1.2). Define a notation ν of a subbase of a topology τ_Q on $\mathcal{A}$ by

$$A \in \nu(w) : \Longleftrightarrow I_Q(w) \cap A \neq \emptyset$$

(cf. Definition 5.1.1) and let $\delta_Q := E_<(\nu_Q)$ be the induced representation of $\mathcal{A}$. From the definitions

$$\begin{aligned}
A \in \nu^{\mathcal{A}}_{\lessgtr}(w) &\Longleftrightarrow I^n(w) \cap A \neq \emptyset \\
&\Longleftrightarrow (\exists v)(\overline{I}_Q(v) \subseteq I^n(w) \text{ and } I_Q(v) \cap A \neq \emptyset) \\
&\Longleftrightarrow (\exists v)(\overline{I}_Q(v) \subseteq I^n(w) \text{ and } A \in \nu(v)) .
\end{aligned}$$

Let $p \in \Sigma^\omega$ be a list of all (v,w) with $\overline{I}_Q(v) \subseteq I^n(w)$. There is a Type-2 machine M which on input $(p,q) \in \Sigma^\omega \times \Sigma^\omega$ operates in stages $0,1,\ldots$ as follows. In Stage $\langle k,n\rangle$, M searches for the k-th subword "$\iota(v)$" in q with $v \in \mathrm{dom}(I_Q)$ and the n-th pair (v_n,w_n) from the list p and prints "$\iota(w_n)$", if $v_n = v$. If q is a δ_Q-name of A, that is, a list of all v with $A \in \nu(v)$, then $f_M(p,q)$ is a list of all w with $A \in \nu^{\mathcal{A}}_{\lessgtr}(w)$. Therefore, the continuous function $q \mapsto f_M(p,q)$ translates δ_Q to $\psi_<$. If $\nu_{\mathbb{Q}}$ and ν_Q are recursively related, then there is a computable sequence $p \in \Sigma^\omega$ and the translation becomes computable.

The properties $\psi_< \leq_t \delta_Q$ and $\psi_< \leq \delta_Q$, respectively, can be proved in the same way. This proves the first parts of Properties 1 and 2.

For the case of $\psi_>$ the proof is only slightly more complicated. If $A \in \nu^{\mathcal{A}}_{\gtrless}(w)$, that is, $\overline{I}^n(w) \cap A = \emptyset$, then for some $\varepsilon > 0$, $d(\overline{I}^n(w),A) \geq \varepsilon$, since $\overline{I}^n(w)$ is compact (where $d(X,Y) := \inf\{d(x,y) \mid x \in X, y \in Y\}$).

Again since $\overline{I}^n(w)$ is compact, it can be covered by the union of finitely many open cubes $I_Q(v_1), I_Q(v_2), \ldots$ of diameter $< \varepsilon/2$ intersecting it. Then $\overline{I}_Q(v_i) \cap A = \emptyset$ for each i. We obtain $A \in \nu^{\mathcal{A}}_{\gtrless}(w)$, iff there are words $v_1,\ldots,v_k$ such that

$$\overline{I}^n(w) \subseteq I_Q(v_1) \cup \ldots \cup I_Q(v_k) \quad \text{and} \quad (\forall i)A \in \nu_Q(v_i) .$$

The rest of the proof is similar to that of the first case. Notice that $\{\iota(w)\iota(v_1)\ldots\iota(v_k) \mid \overline{I}^n(w) \subseteq I_Q(v_1) \cup \ldots \cup I_Q(v_k)\}$ is r.e. , if $\nu_{\mathbb{Q}}$ and ν_Q are r.e.-related.

3. We construct counter examples for the space $\mathbb{R}^1$ which can be generalized easily to the case $\mathbb{R}^n$ for $n > 1$. We show that we obtain non-equivalent representations, if $\mathbb{Q}$ is replaced by the dense set $\mathbb{Q}_{10} := \{z/10^m \mid z \in \mathbb{Z}, m \in \mathbb{N}\}$ of finite decimal fractions. Let I_{10} be a standard notation of the set of all open intervals with endpoints in $\mathbb{Q}_{10}$.

Define computable topological spaces on $\mathcal{A}$ by the notations ν and ν_{10} of subsets of $\mathcal{A}$ by

$$A \in \nu(w) : \iff \bar{\mathrm{I}}^1(w) \cap A \neq \emptyset ,$$

$$A \in \nu_{10}(w) : \iff \bar{\mathrm{I}}_{10}(w) \cap A \neq \emptyset .$$

Let δ and δ_{10} be the induced standard representations of $\mathcal{A}$ and let τ and τ_{10} be the induced topologies on $\mathcal{A}$, respectively (Definitions 3.2.1, 3.2.2).
We prove $\tau_{10} \subset \tau$.
Since range(ν_{10}) $\subseteq$ range(ν), $\tau_{10} \subseteq \tau$.
There is some $w \in \Sigma^*$ with $\mathrm{I}^1(w) = (0; 1/3)$ and so $[1/3; 1] \in \nu(w) \in \tau$.
Assume $\nu(w) \in \tau_{10}$. Since range(ν_{10}) is a subbase of τ_{10}, there are words $v_1, \ldots, v_k$ such that

$$[1/3; 1] \in \nu_{10}(v_1) \cap \ldots \cap \nu_{10}(v_k) \subseteq \nu(w) .$$

It follows that $[1/3; 1] \cap \bar{\mathrm{I}}_{10}(v_i) \neq \emptyset$ for all i. Since the $\bar{\mathrm{I}}_{10}(v_i)$ are intervals with endpoints in $\mathbb{Q}_{10}$ and $1/3 \notin \mathbb{Q}_{10}$, there is some m such that $a \notin (1/3 - 2^{-m}; 1/3 + 2^{-m})$ for all endpoints a of the intervals $\bar{\mathrm{I}}_{10}(v_i)$. It follows that $[1/3 + 2^{-m-1}; 1] \in \nu_{10}(v_i)$ for all i but $[1/3 + 2^{-m-1}; 1] \notin \nu(w)$, and so $\nu_{10}(v_1) \cap \ldots \cap \nu_{10}(v_k) \not\subseteq \nu(w)$ (contradiction). Therefore, $\tau \not\subseteq \tau_{10}$ and by Theorem 3.1.8 $\delta_{10} \not\leq_t \delta$ and $\delta_{10} \not\leq \delta$.

For the second case "$A \in \nu(w) : \iff \mathrm{I}^n(w) \cap A = \emptyset$" the proof is similar. Choose the closed set $[1/3 - 2^{-m-1}; 1]$ as a counter-example. $\square$

Also in this situation topological irregularities are correlated with computational ones (Exercise 5.1.18). We will not consider the non-robust variants, since they do not seem to be relevant.

If $\psi_>^{\mathrm{en}}(p) = A$, then p is a list of open balls $J \in \mathrm{Cb}^{(n)}$ which exhaust $\mathbb{R}^n \setminus A$ (Definition 5.1.11). If $E'_>(\nu_{\mathbb{Q}})(p) = A$, then p is a list of all open balls $J \in \mathrm{Cb}^{(n)}$ with $J \subseteq \mathbb{R}^n \setminus A$. Since $E'_>(\nu_{\mathbb{Q}})$ is a restriction of $\psi_>^{\mathrm{en}}$, $E'_>(\nu_{\mathbb{Q}}) \leq \psi_>^{\mathrm{en}}$, however, $\psi_>^{\mathrm{en}} \not\leq_t E'_>(\nu_{\mathbb{Q}})$ (Exercise 5.1.16). Since a set $U \subseteq \mathbb{R}^n$ is open, iff its complement $\mathbb{R}^n \setminus U$ is closed, each representation δ of the set $\mathcal{A}$ of closed sets induces a representation δ' of the set $\mathcal{O}$ of open sets: $\delta'(p) := \mathbb{R}^n \setminus \delta(p)$.

Definition 5.1.15 (the representations $\theta_<$, $\theta_>$, θ and $\theta_<^{\mathrm{en}}$ of $\mathcal{O}$).
1. Define representations of the set of open subsets $\mathcal{O}$ of $\mathbb{R}^n$ by

$$\theta_<(p) := \mathbb{R}^n \setminus \psi_>(p), \quad \theta_>(p) := \mathbb{R}^n \setminus \psi_<(p), \quad \theta(p) := \mathbb{R}^n \setminus \psi(p) .$$

2. Define topologies on $\mathcal{O}$ by

$$U \in \tau_<^{\mathcal{O}} \iff \{\mathbb{R}^n \setminus B \mid B \in U\} \in \tau_>^{\mathcal{A}},$$

$$U \in \tau_>^{\mathcal{O}} \iff \{\mathbb{R}^n \setminus B \mid B \in U\} \in \tau_<^{\mathcal{A}},$$

$$U \in \tau^{\mathcal{O}} \iff \{\mathbb{R}^n \setminus B \mid B \in U\} \in \tau^{\mathcal{A}}.$$

3. An open set $U \in \mathcal{O}$ is called

$$\begin{aligned}
r.e. &\quad : \iff U \text{ is } \theta_<\text{-computable,} \\
co\text{-}r.e. &\quad : \iff U \text{ is } \theta_>\text{-computable,} \\
recursive &\quad : \iff U \text{ is } \theta\text{-computable.}
\end{aligned}$$

4. Define the representation $\theta_<^{\mathrm{en}}$ by

$$\theta_<^{\mathrm{en}}(p) := \mathbb{R}^n \setminus \psi_>^{\mathrm{en}}(p) = \bigcup \{ I^n(w) \mid \iota(w) \lhd p \} \ .$$

A $\theta_<$-name of an open set $U \subseteq \mathbb{R}^n$ is a list of all closed rational cubes contained in U, it "enumerates the inside" of U. For each rational cube $J \in \mathrm{Cb}^{(n)}$ the set

$$\{ U \in \mathcal{O} \mid \overline{J} \subseteq U \}$$

is a subbase element of the topology $\tau_<^{\mathcal{O}}$ on the set $\mathcal{O}$ of open subsets of $\mathbb{R}^n$. Similarly, a $\theta_>$-name of an open set $U \subseteq \mathbb{R}^n$ is a list of all open rational cubes not contained in U, it "enumerates the outside" of U. For each open rational cube $J \in \mathrm{Cb}^{(n)}$, the set

$$\{ U \in \mathcal{O} \mid J \not\subseteq U \}$$

is a subbase element of the topology $\tau_>^{\mathcal{O}}$ on the set $\mathcal{O}$ of open subsets of $\mathbb{R}^n$. A $\theta_<^{\mathrm{en}}$-name p of U is a list of open (not necessarily all) rational balls which exhausts the set U.

We summarize some obvious properties of our representations of the open subsets of $\mathbb{R}^n$ (cf. Lemmas 5.1.2 and 5.1.10).

Corollary 5.1.16.
1. $\theta_<$, $\theta_>$ and θ are admissible with final topologies $\tau_<^{\mathcal{O}}$, $\tau_>^{\mathcal{O}}$ and $\tau^{\mathcal{O}}$, respectively,
2. $\theta = \theta_> \wedge \theta_<$, in particular, $\theta \leq \theta_<$ and $\theta \leq \theta_>$,
3. $\theta_< \not\leq_t \theta$, $\theta_> \not\leq_t \theta$, $\theta_< \not\leq_t \theta_>$ and $\theta_> \not\leq_t \theta_<$,
4. $\theta_< \equiv \theta_<^{\mathrm{en}}$.

Example 5.1.17 (r.e. , co-r.e. and recursive open sets).
1. The empty set and $\mathbb{R}^n$ are recursive open subsets of $\mathbb{R}^n$.
2. A real interval $(a; b)$ is
    a) r.e. , iff a is right-computable and b is left-computable,
    b) co-r.e. , iff a is left-computable and b is right-computable,
    c) recursive, iff a and b are computable.
3. The open balls $B^e(a, r) := \{ x \in \mathbb{R}^n \mid |x - a| < r \}$ and $B(a, r) := \{ x \in \mathbb{R}^n \mid ||x - a|| < r \}$ with computable $a \in \mathbb{R}^n$ and computable $r \in \mathbb{R}$ are recursive.
4. The open sets $\{ (x, y) \in \mathbb{R}^2 \mid x < y \}$ and $\{ (x, y) \in \mathbb{R}^2 \mid x \neq y \}$ are recursive.
5. The set U_N from Theorem 4.2.8 containing all computable real numbers is r.e. open.

The easy proofs are left to Exercise 5.1.19. □

From Theorem 5.1.13 we obtain immediately:

Corollary 5.1.18 (union and intersection of open sets).
1. Intersection $(U, V) \mapsto U \cap V$ on $\mathcal{O}$ is $(\theta_<, \theta_<, \theta_<)$-computable, $(\theta_>, \theta_>, \theta_>)$-computable and (θ, θ, θ)-computable.
2. Union $(U, V) \mapsto U \cup V$ on $\mathcal{O}$ is $(\theta_<, \theta_<, \theta_<)$-computable but not $(\theta, \theta, \theta_>)$-continuous.

Therefore, union is not $(\theta, \theta, \theta_>)$-computable and not $(\theta_<, \theta, \theta_>)$-continuous etc., since $\theta \leq \theta_<$ and $\theta \leq \theta_>$. There are two recursive open sets the union of which is not co-r.e. (Exercise 5.1.15).

Example 5.1.19.
1. Let $[\theta_<]^\omega$ be the infinite product of the representation $\theta_<$ according to Definition 3.3.3. Then infinite union $F : \mathcal{O}^\mathbb{N} \to \mathcal{O}$ defined by $F(U_0, U_1, \ldots) := U_0 \cup U_1 \cup \ldots$ is $([\theta_<]^\omega, \theta_<)$-computable. It suffices to show that the function F is $([\theta_<^{\mathrm{en}}]^\omega, \theta_<^{\mathrm{en}})$-computable. There is a Type-2 machine M which on input $\langle p_0, p_1, \ldots \rangle$ prints a list of all $\iota(w)$ such that $\iota(w) \lhd p_i$ for some $i \in \mathbb{N}$. Obviously, the function f_M is a $([\theta_<^{\mathrm{en}}]^\omega, \theta_<^{\mathrm{en}})$-realization of F.
2. This example is a special case of Theorem 6.2.4.1 we will prove later. By definition, a function $f : \mathbb{R}^n \to \mathbb{R}^n$ is continuous, iff $f^{-1}[A]$ is closed for any closed subset $A \subseteq \mathbb{R}^n$, iff $f^{-1}[U]$ is open for any open subset $U \subseteq \mathbb{R}^n$. As a computational version of this fact we show:
For every computable function $f : \mathbb{R}^n \to \mathbb{R}^n$ the function

$$U \mapsto f^{-1}[U] \text{ for } U \in \mathcal{U}$$

is $(\theta_<, \theta_<)$-computable. In general this function is not $(\theta, \theta_>)$-computable. (See Exercise 5.1.6.) We show that the function is $(\theta_<^{\mathrm{en}}, \theta_<^{\mathrm{en}})$-computable. By Theorem 3.2.14 there is an r.e. set $C \subseteq \Sigma^* \times \Sigma^*$ such that $f^{-1}[\mathrm{I}^n(v)] = \bigcup_{(v,w) \in C} \bigcap \{ \mathrm{I}^n(u) \mid \iota(u) \lhd w \}$. Since

$$\left\{ (u', w) \mid \mathrm{I}^n(u') \subseteq \bigcap \{ \mathrm{I}^n(u) \mid \iota(u) \lhd w \} \right\}$$

is r.e. ,

$$f^{-1}[\mathrm{I}^n(v)] = \bigcup_{(v,w) \in D} \mathrm{I}^n(w)$$

for some r.e. set $D \subseteq \Sigma^* \times \Sigma^*$. There is a Type-2 machine M which maps any $p \in \Sigma^\omega$ to a list of all $\iota(w)$ such that $\iota(v) \lhd p$ and $(v, w) \in D$ for some $v \in \Sigma^*$. The function f_M is an $(\theta_<^{\mathrm{en}}, \theta_<^{\mathrm{en}})$-realization of the function $U \mapsto f^{-1}[U]$, which, therefore, has a computable $(\theta_<, \theta_<)$-realization h. Since $f^{-1}[\psi_>(p)] = f^{-1}[\mathbb{R}^n \setminus \theta_<(p)] = \mathbb{R}^n \setminus f^{-1}[\theta_<(p)] = \mathbb{R}^n \setminus \theta_< \circ h(p) = \psi_> \circ h(p)$, the function

$$A \mapsto f^{-1}[A] \text{ for } A \in \mathcal{A}$$

is $(\psi_>, \psi_>)$-computable. $\qquad \square$

By Theorem 3.2.10, for all representations $\delta :\subseteq \Sigma^\omega \to \mathcal{O}$ of the open subsets of $\mathbb{R}^n$,

$$\{(J,U) \in \text{Cb}^{(n)} \times \mathcal{O} \mid \overline{J} \subseteq U\} \text{ is } (I^n, \delta)\text{-r.e.} \iff \delta \leq \theta_< ,$$

that is, $\theta_<$ is complete in the set of all representations δ of $\mathcal{O}$, for which the property $\overline{J} \subseteq U$ is (I^n, δ)-r.e. The equivalence class of $\theta_<$ can be characterized also as follows: for all representations $\delta :\subseteq \Sigma^\omega \to \mathcal{O}$ of the open subsets of $\mathbb{R}^n$,

$$\{(x,U) \in \mathbb{R}^n \times \mathcal{O} \mid x \in U\} \text{ is } (\rho^n, \delta)\text{-r.e.} \iff \delta \leq \theta_< ,$$

that is, θ is, up to equivalence, the poorest representation of the open sets such that membership $x \in U$ can be proved (Exercise 5.1.25).

In this section we have considered representations of subset spaces of $\mathbb{R}^n$ for fixed dimension n. We will use the same names $\psi, \theta_<$ and so on, if the dimension is given in advance. If necessary, we add it as a superscript, for example, $\psi_{\geq}^2$ is a representation of the non-empty closed subsets of $\mathbb{R}^2$.

The definitions and some characterizations of the recursive and recursively enumerable subsets of $\mathbb{R}^n$ are from [Zho96]. Further representations of the closed subsets of $\mathbb{R}^n$ equivalent to $\psi_<$, $\psi_>$ or ψ are discussed in [BW99]. Further examples of non-effective set operations can be found in [Bra99a].

Exercises 5.1.

1. Show that for closed sets A, B, each of the following conditions implies $A = B$:
   - $\{J \in \text{Cb}^{(n)} \mid J \cap A \neq \emptyset\} = \{J \in \text{Cb}^{(n)} \mid J \cap B \neq \emptyset\}$,
   - $\{J \in \text{Cb}^{(n)} \mid \overline{J} \cap A = \emptyset\} = \{J \in \text{Cb}^{(n)} \mid \overline{J} \cap B = \emptyset\}$,
   - $\{J \in \text{Cb}^{(n)} \mid J \cap A = \emptyset\} = \{J \in \text{Cb}^{(n)} \mid J \cap B = \emptyset\}$,
   - $\{J \in \text{Cb}^{(n)} \mid \overline{J} \cap A \neq \emptyset\} = \{J \in \text{Cb}^{(n)} \mid \overline{J} \cap B \neq \emptyset\}$.

   Show that there are closed sets $A, B \subseteq \mathbb{R}^n$ such that $\{a \in \mathbb{Q}^n \mid a \in A\} = \{a \in \mathbb{Q}^n \mid a \in B\}$ but $A \neq B$. Hint: Choose one-element sets.

2. Prove the first four statements in Example 5.1.3.

3. Show that
   a) the function $x \mapsto \overline{B}(x, 1)$ is (ρ^n, ψ)-computable,
   b) the function $J \mapsto J$ is (I^n, θ)-computable,
   c) the function $J \mapsto \text{cls}(J)$ is (I^n, ψ)-computable.

4. Construct an r.e. closed set $A \subseteq \mathbb{R}^n$ which is not co-r.e. and a co-r.e. closed set $B \subseteq \mathbb{R}^n$ which is not r.e.

$\Diamond$ 5. Show that the closed set $\{0\} \cup \{2^{-n} \mid n \in A\} \subseteq \mathbb{R}$ is r.e., iff $A \subseteq \mathbb{N}$ is r.e.

6. Show that there are a computable function $f : \mathbb{R}^n \to \mathbb{R}^n$ and a recursive closed set $A \subseteq \mathbb{R}^n$ such that $f^{-1}[A]$ is not recursively enumerable. Hint: Example 5.1.12.2.

7. In Definition 5.1.1, replace the notation I^n of $Cb^{(n)}$ by the notation I_e^n of rational Euclidean balls, defined by $I_e^n(\iota(v_1)\iota(v_2)\ldots\iota(v_n)\iota(w)) := \{x \in \mathbb{R}^n \mid |x - (\nu_\mathbb{Q}(v_1), \ldots, \nu_\mathbb{Q}(v_n))| < \nu_\mathbb{Q}(w)\}$. Show that the resulting representations are equivalent to the original ones.

8. Prove $\psi_>^{\text{dist}} \leq \psi_>$ in detail.

9. Replace the maximum distance in Definition 5.1.6 by the Euclidean distance and show that Lemma 5.1.7 holds accordingly for the Euclidean distance.

10. Prove $\psi_<^{\text{en}} \leq \psi_<$.

11. Show that a closed set $A \subseteq \mathbb{R}^n$ is recursive, iff there is a recursive subset $Z \subseteq \mathbb{N} \times \mathbb{Q}^n \times \mathbb{Q}$ such that

$$(k, a, b) \in Z \Longrightarrow d(a, A) < b + 2^{-k} ,$$
$$(k, a, b) \notin Z \Longrightarrow d(a, A) > b - 2^{-k}$$

for all $k \in \mathbb{N}$, $a \in \mathbb{Q}^n$ and $b \in \mathbb{Q}$.

12. Define topologies $\tau_<^t$ and $\tau_>^t$ on the set $\mathcal{A}$ of closed subsets of $\mathbb{R}^n$ by (uncountable) subbases:
    a) $\tau_<^t$: all $\{A \in \mathcal{A} \mid A \cap U \neq \emptyset\}$ with open $U \subseteq \mathbb{R}^n$,
    b) $\tau_>^t$: all $\{A \in \mathcal{A} \mid A \cap K = \emptyset\}$ with compact $K \subseteq \mathbb{R}^n$.
    Then $\tau_<^t = \tau_<^{\mathcal{A}}$ and $\tau_>^t = \tau_>^{\mathcal{A}}$. The topologies are discussed in [Bee93].

13. Define the multi-valued function $f :\subseteq \mathcal{A} \rightrightarrows \mathbb{R}^n$ by
    $R_f := \{(A, x) \mid x \in A\}$. Show:
    a) f has no (ψ, ρ^n)-continuous choice function;
    b) f is $(\psi_<, \rho^n)$-computable;
    c) f is not $(\psi_>, \rho^n)$-continuous.
    Therefore, a point $x \in A$ can be determined from a $\psi_<$-name of A but not from a $\psi_>$-name of A. In particular, every r.e. (that is, $\psi_<$-computable) closed non-empty set has a computable point.
    Define the multi-valued function $g :\subseteq \mathcal{O} \rightrightarrows \mathbb{R}^n$ by
    $R_g := \{(O, x) \mid x \in O\}$. Show:
    a) g has no (θ, ρ^n)-continuous choice function;
    b) g is $(\theta_<, \rho^n)$-computable;
    c) g is not $(\theta_>, \rho^n)$-continuous.
    Therefore, a point $x \in U$ can be determined from a $\theta_<$-name of U but not from a $\theta_>$-name of U. In particular, every r.e. (that is, $\theta_<$-computable) open non-empty set has a computable point.
    Hint for a): consider sets $\{x\}$, $\{y\}$ and $\{x, y\}$ ($x \neq y$). See Example 3.1.4.13.

♦14. Let δ and δ' be representations of the set of closed subsets of $\mathbb{R}^n$. Show that intersection $(A, B) \mapsto A \cap B$ is not $(\delta, \delta', \psi_<)$-continuous. (Hint: Exercise 4.1.19)

15. Show that there are recursive closed sets $X, Y \subseteq \mathbb{R}$ such that $X \cap Y$ is not r.e. Hint: Let $f : \mathbb{N} \to \mathbb{N}$ be a computable enumeration of an r.e. set $C \subseteq \mathbb{N}$ which is not recursive. Let $X := \{2^{-n} \mid n \in \mathbb{N}\} \cup \{0\}$ and let Y be the closure of the set $Z := \{(1 + 2^{-k-2}) \cdot 2^{-n} \mid f(i) \neq n \text{ for all } i < k\}$.

◆16. Define a representation ψ' of $\mathcal{A}$ by $\psi'(p) = A$, iff p is a list of all $J \in \mathrm{Cb}^{(n)}$ such that $J \cap A = \emptyset$ (that is, $J \subseteq \mathbb{R}^n \setminus A$). Show:

   a) $\psi' \leq \psi_>^{\mathrm{en}}$ and $\psi_>^{\mathrm{en}} \not\leq_t \psi'$,

   b) There is a ψ-computable (that is, recursive) closed set $A \subseteq \mathbb{R}$ which is not ψ'-computable. Hint: Let $f : \mathbb{N} \to \mathbb{N}$ be a computable enumeration of an r.e. set C which is not recursive, define

$$A := \mathbb{R} \setminus \bigcup \left\{ \left((1 + 2^{-k-2}) \cdot 2^{-n} \; ; \; \frac{3}{2} \cdot 2^{-n} \right) \mid f(i) \neq n \text{ for all } i < k \right\}.$$

17. There is a recursive closed set $Y \subseteq \mathbb{R}$ such that $\{w \in \Sigma^* \mid \overline{\mathrm{I}}^1(w) \cap Y \neq \emptyset\}$ is not r.e. Hint: Exercise 5.1.15.

◆18. Consider Theorem 5.1.14 and let ν_Q be the restriction of $\nu_{\mathbb{Q}}$ to the set $Q := \mathbb{Q} \setminus \{2^{-n} \mid n \in \mathbb{N}\}$.

   a) Show that the set Y from Exercise 5.1.15 is $E'_<(\nu_Q)$-computable but not $E'_<(\nu_{\mathbb{Q}})$-computable.

   b) Show that the set A from Exercise 5.1.16 is $E'_>(\nu_Q)$-computable but not $E'_>(\nu_{\mathbb{Q}})$-computable.

19. Prove the statements from Example 5.1.17.

20. Show that the function $N \mapsto U_N$ from Theorem 4.2.8 is $(\nu_{\mathbb{N}}, \theta_<)$-computable. Generalize the theorem to n dimensions.

21. Let $\mathrm{cls} : \mathcal{O} \to \mathcal{A}$ be the closure operator. Show:

   a) cls is $(\theta_<, \psi_<)$-computable,

   b) cls is not $(\theta, \psi_>)$-continuous.

   c) $\mathrm{cls}(U)$ is not $\psi_>$-computable for some θ-computable open set $U \subseteq \mathbb{R}$. (Hint: $U := \mathbb{R} \setminus A$, A from 16 above)

22. Define a function $\chi : \{0,2\}^\omega \to \mathbb{R}$ by $\chi(a_0 a_1 \ldots) := \sum_{i \in \mathbb{N}} a_i \cdot 3^{-i}$.

   a) Show that χ is injective.

   b) Show that χ is a homeomorphism from $\{0,2\}^\omega$ to $\mathrm{range}(\chi)$ (that is, χ is $(\tau_C, \tau_{\mathbb{R}})$-continuous and χ^{-1} is $(\tau_{\mathbb{R}}, \tau_C)$-continuous, where τ_C is the Cantor topology on $\{0,2\}^\omega$).

   c) Show that the "Cantor discontinuum" $\mathrm{CD} := \mathrm{range}(\chi)$ is a recursive closed set of real numbers.

◆23. Show that for every continuous function $f : \mathbb{R}^n \to \mathbb{R}^n$ the function

$$A \mapsto f^{-1}[A] \text{ for } A \in \mathcal{A}$$

is $(\psi_>, \psi_>)$-continuous.

◆24. A set $X \subseteq \mathbb{R}^n$ is convex, iff $x, y \in X$ implies $x + (y - x) \cdot t \in X$ for $0 \leq t \leq 1$. Let $\mathcal{U}$ be the set of all convex open subsets of $\mathbb{R}^n$. Define computable topological spaces $\mathbf{S}_< = (\mathcal{U}, \sigma_<, \nu_<)$ and $\mathbf{S}_> = (\mathcal{U}, \sigma_>, \nu_>)$ by $U \in \nu_<(w) : \iff [\nu_{\mathbb{Q}}]^n(w) \in U$ and $U \in \nu_>(w) : \iff [\nu_{\mathbb{Q}}]^n(w) \notin U$. Show that $\mathbf{S}_<$ and $\mathbf{S}_>$ are computable topological spaces. Roughly speaking, $\delta_{\mathbf{S}_<}(p) = U$ ($\delta_{\mathbf{S}_>}(p) = U$), iff p is a list of all rational points $a \in U$ ($a \notin U$). Show:

a) $\delta_{\mathbf{S}_<} \not\leq_t \delta_{\mathbf{S}_>}$, $\delta_{\mathbf{S}_>} \not\leq_t \delta_{\mathbf{S}_<}$,
b) $\theta_< |^{\mathcal{U}} \equiv \delta_{\mathbf{S}_<}$,
c) $\delta_{\mathbf{S}_>} \leq \theta_> |^{\mathcal{U}}$,
d) $\theta_> |^{\mathcal{U}} \not\leq_t \delta_{\mathbf{S}_>}$,
e) there is a θ-computable (that is, recursive) convex open subset of $\mathbb{R}^2$ which is not $\delta_{\mathbf{S}_>}$-computable.

♦25. a) Show that for all representations $\delta :\subseteq \Sigma^\omega \to \mathcal{O}$ of the open subsets of $\mathbb{R}^n$,

$$\{(x, U) \in \mathbb{R}^n \times \mathcal{O} \mid x \in U\} \text{ is } (\rho^n, \delta)\text{-r.e.} \iff \delta \leq \theta_< ,$$

$$\{(a, U) \in \mathbb{Q}^n \times \mathcal{O} \mid a \in U\} \text{ is } (\rho^n, \delta)\text{-r.e.} \iff \delta \leq \theta_< .$$

b) The set $\{(a, U) \in \mathbb{Q}^n \times \mathcal{O} \mid a \in U\}$ is not $([\nu_{\mathbb{Q}}]^n, \theta_<)$-r.e. Hint: use the set $U := \mathbb{R}^n \setminus A$ from Exercise 5.1.16 as a counter-example.
c) Let $\mathcal{U}$ be the set of all convex open subsets of $\mathbb{R}^n$. Show that for all representations $\delta :\subseteq \Sigma^\omega \to \mathcal{U}$ of the convex open subsets of $\mathbb{R}^n$,

$$\{(a, U) \in \mathbb{Q}^n \times \mathcal{U} \mid a \in U\} \text{ is } ([\nu_{\mathbb{Q}}]^n, \delta)\text{-r.e.} \iff \delta \leq \theta_< |^{\mathcal{U}} .$$

Hint: Exercise 24 above.

26. For $X \subseteq \mathbb{R}^n$ let $\partial X := \text{cls}(X) \cap \text{cls}(\mathbb{R}^n \setminus X)$ be the boundary of X. Find a recursive open set $U \subseteq \mathbb{R}$ such that ∂U is not co-r.e. closed.
27. Define $\text{pr} : \mathbb{R}^2 \to \mathbb{R}$ by $\text{pr}(x, y) := x$. For every open set $U \subseteq \mathbb{R}^2$ the projection $\text{pr}[U] \subseteq \mathbb{R}$ is open. Show: the function $U \mapsto \text{pr}[U]$ for open $U \subseteq \mathbb{R}^2$ is
a) $(\theta_<^2, \theta_<^1)$-computable,
b) not $(\theta_>^2, \theta_>^1)$-continuous.
28. Show that the Cartesian product $(U, V) \mapsto U \times V$ for open sets $U, V \subseteq \mathbb{R}$ is $(\theta_<^1, \theta_<^1, \theta_<^2)$-computable, $(\theta_>^1, \theta_>^1, \theta_>^2)$-computable and $(\theta^1, \theta^1, \theta^2)$-computable.
29. For $A, B \subseteq \mathbb{R}^n$ define the set distance by

$$d(A, B) := \inf\{d(a, b) \mid a \in A, b \in B\} .$$

For which triples of representations $(\delta_1, \delta_2, \delta_3)$, $\delta_1, \delta_2 \in \{\psi_<, \psi_>, \psi, \theta_<, \theta_>, \theta\}$, $\delta_3 \in \{\rho_<, \rho_>, \rho\}$, is the set distance $(\delta_1, \delta_2, \delta_3)$-computable?
30. Let $f :\subseteq \mathbb{R}^n \to \mathbb{R}$, $U, V \subseteq \text{dom}(f)$, U, V r.e. open. Show that f is computable on $U \cup V$, if f is computable on U and on V.
♦31. Show that a real function $f : \mathbb{R} \to \mathbb{R}$ is computable, iff its graph $\{(x, y) \mid y = f(x)\}$ is an recursive closed set.
32. (Julia sets, Mandelbrot set)
a) Let $f : \mathbb{N} \times \mathbb{R}^n \to \mathbb{R}$ be computable. Show that

$$\{x \in \mathbb{R}^n \mid (\exists k) \, f(k, x) > 0\}$$

is an r.e. open set.

b) The *filled Julia set* of a polynomial function $f : \mathbb{C} \to \mathbb{C}$ [Bar93] is defined by

$$J_f := \{z \in \mathbb{C} \mid \text{the sequence } (f^k(z))_{k \in \mathbb{N}} \text{ is bounded}\} .$$

Show that J_f is a co-r.e. closed set for all computable polynomial functions f of degree greater than 1. Hint: Exercise 4.3.21.[1]

c) The *Mandelbrot set* [Bar93] is defined by

$$M := \{c \in \mathbb{C} \mid (\forall k)|f_c^k(0)| \le 2\} ,$$

where $f_c : \mathbb{C} \to \mathbb{C}$ is defined by $f_c(z) := z^2 + c$. Show that M is a co-r.e. closed subset of $\mathbb{C}$. Hint: Exercise 4.3.22. [2]

5.2 Compact Sets

A set $K \subseteq \mathbb{R}^n$ is compact, iff it is closed and bounded. By the Heine/Borel theorem, K is compact, iff for every set $\alpha \subseteq \tau_{\mathbb{R}^n}$ of open sets with $K \subseteq \bigcup \alpha$ there is a finite subset $\alpha_0 \subseteq \alpha$ such that $K \subseteq \bigcup \alpha_0$ ("every open cover has a finite subcover"). There is another well-known characterization: $K \subseteq \mathbb{R}^n$ is compact, iff every sequence $(x_i)_{i \in \mathbb{N}}$ in K has a subsequence $(x_{i_j})_{j \in \mathbb{N}}$ converging to some $x \in K$. We will deduce representations from the first two characterizations. Remember that $\mathcal{K}$ is the set of all compact subsets of $\mathbb{R}^n$.

Definition 5.2.1 (the representations $\kappa_<$, $\kappa_>$ and κ of $\mathcal{K}$). *Define representations $\kappa_<$, $\kappa_>$ and κ of $\mathcal{K}$ by*
1. $\kappa_< \langle p, w \rangle = K :\iff \psi_<(p) = K$ *and* $K \subseteq B(0, \nu_{\mathbb{N}}(w))$,
2. $\kappa_> \langle p, w \rangle = K :\iff \psi_>(p) = K$ *and* $K \subseteq B(0, \nu_{\mathbb{N}}(w))$,
3. $\kappa \langle p, w \rangle = K :\iff \psi(p) = K$ *and* $K \subseteq B(0, \nu_{\mathbb{N}}(w))$.

Therefore, a κ-name of a compact set K consists of a ψ-name of K and a name of a bound of K. Obviously, $\kappa_< \le \psi_<$, $\kappa_> \le \psi_>$ and $\kappa \le \psi$.

Lemma 5.2.2.
1. $\kappa_< \le \psi_<$, $\kappa_> \le \psi_>$, $\kappa \le \psi$,
2. $\psi_<|^\kappa \not\le_t \kappa_<$, $\psi_>|^\kappa \not\le_t \kappa_>$, $\psi|^\kappa \not\le_t \kappa$,
3. $\kappa \equiv \kappa_< \wedge \kappa_> \equiv \psi_< \wedge \kappa_> \equiv \kappa_< \wedge \psi_>$,
4. $\kappa_< \not\le_t \kappa$, $\kappa_> \not\le_t \kappa$, $\kappa_< \not\le_t \kappa_>$, $\kappa_> \not\le_t \kappa_<$.

Proof: 1. Immediate.
2. It is impossible to find continuously a bound of $\psi(p)$ $(\psi_<(p), \psi_>(p))$ from p. See Exercise 5.2.4.
3. This follows from $\psi \equiv \psi_< \wedge \psi_>$.
4. Similar to the proof of Lemma 5.1.2. $\qquad \square$

---

[1] Many Julia sets are even recursive [Zho98].
[2] It is an open problem whether the Mandelbrot set is recursive ([Pen89], p.124; [BCSS98]; Sect. 9.7 in this book).

Remark 5.2.3. If we replace $\psi_<$ ($\psi_>$, ψ) in Definition 5.2.1 by $\psi_<^{\mathrm{dist}}$ ($\psi_>^{\mathrm{dist}}$, ψ^{dist}), we obtain a representation $\kappa_<^{\mathrm{dist}}$ ($\kappa_>^{\mathrm{dist}}$, κ^{dist}) equivalent to $\kappa_<$ ($\kappa_>$, κ).

We introduce two further representations by means of finite open covers $T \in \mathrm{E}(\mathrm{Cb}^{(n)})$, where $\mathrm{E}(\mathrm{Cb}^{(n)})$ is the set of finite subsets of $\mathrm{Cb}^{(n)}$.

Definition 5.2.4 (representations via covers). *Define a representation κ_{c} by covers and a representation κ_{mc} by minimal covers of the compact subsets of $\mathbb{R}^n$ by computable topological spaces as follows.*

1. $\mathbf{S}_{\mathrm{c}} := (\mathcal{K}, \sigma_{\mathrm{c}}, \nu_{\mathrm{c}})$, *where*

$$\left(K \in \nu_{\mathrm{c}}(w), \text{ iff } \mathrm{I}^n(v) \text{ exists for } \iota(v) \lhd w \text{ and } K \subseteq \bigcup \{\mathrm{I}^n(v) \mid \iota(v) \lhd w\} \right),$$

and $\kappa_{\mathrm{c}} := \delta_{\mathbf{S}_{\mathrm{c}}}$,

2. $\mathbf{S}_{\mathrm{mc}} := (\mathcal{K}, \sigma_{\mathrm{mc}}, \nu_{\mathrm{mc}})$, *where*

$$\left(K \in \nu_{\mathrm{mc}}(w), \text{ iff } \mathrm{I}^n(v) \text{ exists and } \mathrm{I}^n(v) \cap K \neq \emptyset \text{ for all } \iota(v) \lhd w \text{ and } K \subseteq \bigcup \{\mathrm{I}^n(v) \mid \iota(v) \lhd w\} \right),$$

and $\kappa_{\mathrm{mc}} := \delta_{\mathbf{S}_{\mathrm{mc}}}$.

If, in particular, $\kappa_{\mathrm{c}}(p) = \emptyset$, then also words w are listed by p such that $\iota(v) \lhd w$ for no word v (the empty cover). And if $\kappa_{\mathrm{mc}}(p) = \emptyset$, then only words w are listed such that $\iota(v) \lhd w$ for no word v. Since every compact set K can be identified by the set of $J \in \mathrm{Cb}^{(n)}$ such that $J \cap K \neq \emptyset$, the sets σ_{c} and σ_{mc} identify points (that is, compact sets), and so $\mathbf{S}_{\mathrm{c}}$ and $\mathbf{S}_{\mathrm{mc}}$ are in fact computable topological spaces. As a refinement of the Heine/Borel characterization, a set $K \subseteq \mathbb{R}^n$ is compact, if for all $\alpha \subseteq \mathrm{Cb}^{(n)}$ (*not* $\alpha \subseteq \mathcal{O} = \tau_{\mathbb{R}^n}$) $K \subseteq \bigcup \alpha$ implies $K \subseteq \bigcup \beta$ for some finite subset $\beta \subseteq \alpha$ (Exercise 5.2.2).

Lemma 5.2.5.	$\kappa_{\mathrm{c}} \equiv \kappa_>$,	$\kappa_{\mathrm{mc}} \equiv \kappa$.

Proof: $\kappa_{\mathrm{c}} \leq \kappa_>$: Let $p \in \Sigma^\omega$ be a list of all finite covers $T \in \mathrm{E}(\mathrm{Cb}^{(n)})$ of K. The first element gives a bound i such that $K \subseteq B(0, i)$. Furthermore, for any $J \in \mathrm{Cb}^{(n)}$, $\overline{J} \cap K = \emptyset$, iff $\overline{J} \cap \bigcup T = \emptyset$ for some set T enumerated by p. Therefore, some Type-2 machine translates κ_{c} to $\kappa_>$.

$\kappa_> \leq \kappa_{\mathrm{c}}$: Assume $K = \kappa_> \langle p, w \rangle$. Then $K \subseteq L := \overline{B}(0, \nu_{\mathbb{N}}(w))$ and $K = \psi_>(p)$. Let S be the set of cubes $J \in \mathrm{Cb}^{(n)}$ listed by p. Then $T \in \mathrm{E}(\mathrm{Cb}^{(n)})$ covers K, iff $T \cup S$ covers L. Since L is compact, $T \in \mathrm{E}(\mathrm{Cb}^{(n)})$ covers K, iff $T \cup S_0$ covers L for some finite subset $S_0 \subseteq S$. Therefore, a list of all $T \in \mathrm{E}(\mathrm{Cb}^{(n)})$ with $K \subseteq \bigcup T$ can be computed from p.

$\kappa_{\mathrm{mc}} \leq \kappa$: For any $J \in \mathrm{Cb}^{(n)}$, $J \cap K \neq \emptyset$ iff $J \in T$ for some minimal finite cover $T \in \mathrm{E}(\mathrm{Cb}^{(n)})$ of K. Therefore, a list of all $J \in \mathrm{Cb}^{(n)}$ with $J \cap K \neq \emptyset$ can be computed from a list of all minimal finite covers of K. This shows $\kappa_{\mathrm{mc}} \leq \psi_<$. The statement follows from $\kappa_{\mathrm{mc}} \leq \kappa_{\mathrm{c}} \leq \kappa_>$ (see above) and $\psi_< \wedge \kappa_> \leq \kappa$.

$\kappa \leq \kappa_{\mathrm{mc}}$: A cover $T \in \mathrm{E}(\mathrm{Cb}^{(n)})$ of K is minimal, iff $J \cap K \neq \emptyset$ for all $J \in T$. Let $\kappa(p) = K$. Since $\kappa \leq \kappa_> \leq \kappa_{\mathrm{c}}$, from p we can get a list of all covers T of K. Since $\kappa \leq \kappa_<$, we can get a list of all $J \in \mathrm{Cb}^{(n)}$ intersecting K. Therefore, we can produce a list of all minimal covers T of K, and so $\kappa \leq \kappa_{\mathrm{mc}}$. $\qquad\square$

Every compact subset $K \subseteq \mathbb{R}$ has a maximum. By the following lemma it can be computed.

Lemma 5.2.6 (maximum of compact sets). Consider the special case $n = 1$. The maximum function $\max :\subseteq \mathcal{K} \to \mathbb{R}$, $\mathrm{dom}(\max) = \mathcal{K} \setminus \{\emptyset\}$, on the non-empty compact subsets of $\mathbb{R}$ is
1. $(\psi_< |^{\mathcal{K}}, \rho_<)$-computable,
2. $(\kappa_>, \rho_>)$-computable,
3. (κ, ρ)-computable.

Proof: 1. If $\psi_<(p) = K$, then p is a list of all open rational intervals J such that $K \cap J \neq \emptyset$. For every such interval J, $\inf(J)$ is a lower bound of $\max(K)$, and since arbitrarily short intervals are listed by p, these lower bounds approach $\max(K)$ arbitrarily closely. There is a Type-2 machine which on input $p \in \mathrm{dom}(\psi_<)$ produces a list of all $\inf(J)$ (more precisely, of all words $\iota(w)$ with $\nu_{\mathbb{Q}}(w) = \inf(J)$) such that J is listed by p. Then f_M is a $(\psi_<, \rho_<^b)$-realization ($\rho_<^b$ from Lemma 4.1.8) of the function $\max$.

2. We show that $\max$ is $(\kappa_{\mathrm{c}}, \rho_>^b)$-computable ($\rho_>^b$ from Lemma 4.1.8). If $\kappa_{\mathrm{c}}(p) = K$, then p is a list of all finite covers of K with rational open intervals. The supremum of the union of each such cover is a rational upper bound of K, and these upper bounds approach $\max(K)$ arbitrarily closely. There is a Type-2 machine which on input $p \in \mathrm{dom}(\kappa_{\mathrm{c}})$ produces a list of rational numbers containing the supremum of the union of each finite cover listed by p. Then f_M is a $(\kappa_{\mathrm{c}}, \rho_>^b)$-realization of $\max$.

3. Since $\kappa \leq \psi_<$, it follows from Property 1. that $\max$ is $(\kappa, \rho_<)$-computable. Since $\kappa \leq \kappa_>$, it follows from Property 2. that $\max$ is $(\kappa, \rho_>)$-computable. The final statement follows from $\rho_< \wedge \rho_> \leq \rho$. $\qquad\square$

By symmetry, the minimum function is $(\psi_<, \rho_>)$-computable, $(\kappa_>, \rho_<)$-computable and (κ, ρ)-computable. However, the function $\max$ on $\mathcal{K} \setminus \{\emptyset\}$ is not $(\psi_>, \rho_>)$-continuous, since no upper bound of $K = \psi_>(p)$ can be determined from a prefix of p (see Exercise 5.2.5 on negative results). An arbitrary upper bound given additionally by a $\kappa_>$-name allows to approximate the maximum from above.

The Heine/Borel definition of compactness gives rise to another representation of the compact sets. A name of a compact set K will be a program which for every countable open cover of K determines a finite subcover.

Theorem 5.2.7 (computable Heine/Borel theorem). Call $f :\subseteq \Sigma^\omega \to \Sigma^*$ an HB-function for $K \in \mathcal{K}$, iff for all $p = \langle p_0, p_1, \ldots \rangle \in \Sigma^\omega$, $p_i \in \mathrm{dom}(\theta^{\mathrm{en}}_<)$,

$$K \subseteq \bigcup \left\{ \theta^{\mathrm{en}}_<(p_i) \mid i \in \mathbb{N} \right\} \implies K \subseteq \bigcup \left\{ \theta^{\mathrm{en}}_<(p_i) \mid i < \nu_\mathbb{N} \circ f(p) \right\},$$
$$K \not\subseteq \bigcup \left\{ \theta^{\mathrm{en}}_<(p_i) \mid i \in \mathbb{N} \right\} \implies f(p) = \mathrm{div}$$

Define the Heine/Borel representation $\kappa_{\mathrm{HB}} :\subseteq \Sigma^\omega \to \mathcal{K}$ by

$$\kappa_{\mathrm{HB}}(p) = K :\iff \eta^{\omega*}_p \text{ is an HB-function for } K.$$

Then $\kappa_{\mathrm{HB}} \equiv \kappa_> .$

Proof: By Lemma 5.2.5 it suffices to prove $\kappa_{\mathrm{HB}} \equiv \kappa_c$.

$\kappa_{\mathrm{HB}} \leq \kappa_c$: There is a computable function $g :\subseteq \Sigma^* \to \Sigma^\omega$ with $\bigcup \{ \mathrm{I}^n(v) \mid \iota(v) \lhd w \} = \theta^{\mathrm{en}}_< \circ g(w)$. Let $K = \kappa_{\mathrm{HB}}(p)$. Then

$$K \subseteq \bigcup \left\{ \mathrm{I}^n(v) \mid \iota(v) \lhd w \right\} \iff \eta^{\omega*}_p \langle g(w), g(w), \ldots \rangle \text{ exists.}$$

Use standard techniques to construct a Type-2 machine M which on input $p \in \mathrm{dom}(\kappa_{\mathrm{HB}})$ computes a list of all $\iota(w)$ such that $\eta^{\omega*}_p \langle g(w), g(w), \ldots \rangle$ exists. Then f_M translates κ_{HB} to κ_c.

$\kappa_c \leq \kappa_{\mathrm{HB}}$: Assume $\kappa_c(q) = K$. Assume w.l.o.g. that by compactness of K, $K \subseteq \bigcup_{i\in\mathbb{N}} \theta^{\mathrm{en}}_<(p_i)$, iff there are numbers $i_1, \ldots, i_k$ and words $w_{i_1}, \ldots, w_{i_k}$ such that

$$\iota(w_{i_1}) \lhd p_{i_1}, \ldots, \iota(w_{i_k}) \lhd p_{i_k} \text{ and } K \subseteq \mathrm{I}^n(w_{i_1}) \cup \ldots \cup \mathrm{I}^n(w_{i_k}) .$$

Notice that $K \subseteq \mathrm{I}^n(w_{i_1}) \cup \ldots \cup \mathrm{I}^n(w_{i_k})$, iff there is some $w \in \Sigma^*$ such that $\iota(w) \lhd q$ and $\mathrm{I}^n(w_{i_1}) \cup \ldots \cup \mathrm{I}^n(w_{i_k}) = \bigcup \{ \mathrm{I}^n(v) \mid \iota(v) \lhd w \}$.

Use standard techniques to construct a Type-2 machine which on input (q, p), $q \in \mathrm{dom}(\kappa_c)$ and $p = \langle p_0, p_1, \ldots \rangle \in \Sigma^\omega$, tries to find numbers and words with the above properties and in case of success prints some word u with $\nu_\mathbb{N}(u) = 1 + \max\{i_1, \ldots, i_k\}$. By the smn-theorem (Theorem 2.3.13) there is a computable function $f : \Sigma^\omega \to \Sigma^\omega$ such that $f_M(q, p) = \eta^{\omega*}_{f(q)}(p)$. The function f translates κ_c to κ_{HB}.   □

The *Hausdorff metric* is a useful concept for studying computability on the compact sets. In the following we will consider the set $\mathcal{K}^* := \mathcal{K} \setminus \{\emptyset\}$ of non-empty compact subsets of $\mathbb{R}^n$. By definition, $d(x, A) := \inf_{y \in A} d(x, y)$ is the distance from $x \in \mathbb{R}^n$ to $A \subseteq \mathbb{R}^n$. For $s > 0$ define the *s-neighborhood* of $A \in \mathcal{K}^*$ by $A_s := \{ x \in \mathbb{R}^n \mid d(x, A) < s \}$. The Hausdorff distance $d_\mathrm{H} : \mathcal{K}^* \times \mathcal{K}^* \to \mathbb{R}$ is defined by

$$d_\mathrm{H}(A, B) := \inf \{ s > 0 \mid A \subseteq B_s \text{ and } B \subseteq A_s \} .$$

The Hausdorff distance of A and B, $A, B \in \mathcal{K}^*$, is less than s, iff $A \subseteq B_s$ and $B \subseteq A_s$. An equivalent definition is

$$d_{\mathrm{H}}(A, B) := \max(\sup_{y \in B} d(y, A), \sup_{x \in A} d(x, B))$$

(see Exercise 5.2.8). The Hausdorff distance is a metric on the set $\mathcal{K}^*$ of non-empty compact sets, that is,

- $d_{\mathrm{H}}(A, B) \geq 0$,
- $d_{\mathrm{H}}(A, B) = 0 \iff A = B$,
- $d_{\mathrm{H}}(A, B) = d_{\mathrm{H}}(B, A)$,
- $d_{\mathrm{H}}(A, C) \leq d_{\mathrm{H}}(A, B) + d_{\mathrm{H}}(B, C)$

for all $A, B, C \in \mathcal{K}^*$. The countable set $\mathrm{E}^*(\mathbb{Q}^n)$ of the non-empty finite subsets of $\mathbb{Q}^n$ is dense in $\mathcal{K}^*$, that is, for each compact set K and each $r > 0$ there is some finite set $L \subseteq \mathbb{Q}^n$ such that $d_{\mathrm{H}}(K, L) < r$. The proof is easy: Since $\mathbb{Q}^n$ is dense in $\mathbb{R}^n$, $K \subseteq \bigcup \{B(a, r) \mid a \in \mathbb{Q}^n\}$. Since K is compact, finitely many such balls each of which intersects K are sufficient. Let L be the set of their centers. Fig. 5.4 shows a finite cover of a compact set $K \subseteq \mathbb{R}^2$ with balls of radius r.

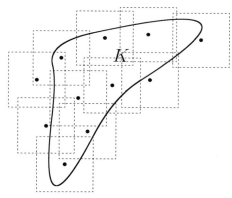

Fig. 5.4. Finitely many open balls of radius r cover K; $d_{\mathrm{H}}(K, L) < r$ for the set L of their centers

For the set L of centers, $d_{\mathrm{H}}(K, L) < r$. Therefore, for each compact set K there is some sequence $(L_i)_{i \in \mathbb{N}}$ of finite subsets of $\mathbb{Q}^n$ such that $d_{\mathrm{H}}(K, L_i) < 2^{-i-1}$, and so this sequence converges to K and satisfies $d_{\mathrm{H}}(L_i, L_j) \leq d_{\mathrm{H}}(L_i, K) + d_{\mathrm{H}}(K, L_j) \leq 2^{-i}$ for $i < j$. On the other hand, let $(L_i)_{i \in \mathbb{N}}$ be a sequence of finite subsets of $\mathbb{Q}^n$ such that $d_{\mathrm{H}}(L_i, L_k) \leq 2^{-i}$ for $i < k$. Let

$$K := \{\lim_{i \to \infty} a_i \mid (a_i)_{i \in \mathbb{N}} \text{ is a sequence with}$$
$$a_i \in L_i \text{ and } d_{\mathrm{H}}(a_i, a_{i+1}) \leq 2^{-i} \text{ for all } i \in \mathbb{N}\} .$$

An easy consideration shows that every Cauchy sequence in K converges to an element of K, and so K is closed. Furthermore, $d_{\mathrm{H}}(K, L_i) \leq 2^{-i}$ for all i.

As usual, we call K the limit of the sequence $(L_i)_{i\in\mathbb{N}}$, $K = \lim_{i\to\infty} L_i$. From the above results it follows easily, that $(\mathcal{K}^*, d_\mathrm{H})$ is a complete metric space with countable dense subset $\mathrm{E}^*(\mathbb{Q}^n)$.

As for the real metric space $(\mathbb{R}, d)$ we can define a representation of $(\mathcal{K}^*, d_\mathrm{H})$ by fast converging Cauchy sequences (Definition 4.1.5).

Definition 5.2.8 (Cauchy representation of compact sets). *Let EQ^n be a standard notation of the set $\mathrm{E}^*(\mathbb{Q}^n)$ of non-empty finite subsets of $\mathbb{Q}^n$. Define the Cauchy representation $\kappa_\mathrm{H} :\subseteq \Sigma^\omega \to \mathcal{K}^*$ by*
$\kappa_\mathrm{H}(p) = K$, *iff* $p = \iota(w_0)\iota(w_1)\dots$, *such that* $d_\mathrm{H}(\mathrm{EQ}^n(w_i), \mathrm{EQ}^n(w_k)) \le 2^{-i}$ *for $i < k$ and $K = \lim_{i\to\infty} \mathrm{EQ}^n(w_i)$.*

Fig. 5.5 shows the first four finite sets of a Cauchy sequence converging fast to a compact set K.

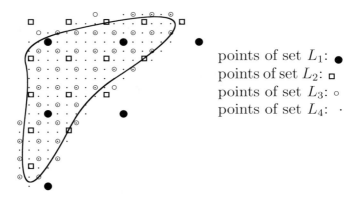

points of set L_1: ●
points of set L_2: □
points of set L_3: ○
points of set L_4: ·

Fig. 5.5. The first four finite sets of a Cauchy sequence converging to a compact set K

By our previous considerations the Cauchy representation κ_H is well-defined and surjective. As for the definition of the Cauchy representation ρ of the real numbers (Definition 4.1.5), replacement of the condition "$d_\mathrm{H}(\mathrm{EQ}^n(w_i), \mathrm{EQ}^n(w_k)) \le 2^{-i}$" in Definition 5.2.8 by one of the conditions "$d_\mathrm{H}(\mathrm{EQ}^n(w_i), \mathrm{EQ}^n(w_k)) < 2^{-i}$", "$d_\mathrm{H}(\mathrm{EQ}^n(w_i), K) < 2^{-i}$" or "$d_\mathrm{H}(\mathrm{EQ}^n(w_i), K) \le 2^{-i}$" gives a representation equivalent to κ_H.

Theorem 5.2.9. $\kappa_\mathrm{H} \equiv \kappa|^{\mathcal{K}^*} \equiv \kappa^{\mathrm{dist}}|^{\mathcal{K}^*}$.

Proof: $\kappa_\mathrm{H} \le \kappa$: Suppose the sequence $L_0, L_1, \dots$ converges to K and $d_\mathrm{H}(L_i, L_k) \le 2^{-i}$ for $i < k$. A bound of K can be obtained from L_0. For $J \in \mathrm{Cb}^{(n)}$ we have

$$J \cap K \ne \emptyset \iff (\exists i)(\exists a \in L_i)B(a, 2^{-i+1}) \subseteq J ,$$

$$\overline{J} \cap K = \emptyset \iff (\exists i)\overline{J} \cap (L_i)_{2^{-i+1}} = \emptyset$$

(where $(L_i)_{2^{-i+1}}$ is the 2^{-i+1}-neighborhood of L_i), and so a Type-2 machine can determine a ψ-name of K from (a code of) $L_0, L_1, \ldots$. Therefore, $\kappa_{\mathrm{H}} \leq \kappa$.

$\kappa_{\mathrm{mc}}|^{\mathcal{K}^*} \leq \kappa_{\mathrm{H}}$: Let $S_0, S_1, \ldots$ be a list of all minimal finite covers of K with balls from $\mathrm{Cb}^{(n)}$. A machine M can compute a Cauchy name $L_0, L_1, \ldots$ of K as follows. For $i \in \mathbb{N}$, M searches for the smallest number j such that the radius of the ball J is less than 2^{-i-1} for each $J \in S_j$ and writes (a code of) L_i, where L_i is the set of centers of the elements J of S_j. Then machine M translates $\kappa_{\mathrm{mc}}|^{\mathcal{K}^*}$ to κ_{H}.
By Lemma 5.2.5, $\kappa_{\mathrm{H}} \equiv \kappa|^{\mathcal{K}^*}$. $\qquad\square$

The equivalence classes of $\kappa_{>}$ and κ can be defined via computable topological spaces.

Lemma 5.2.10. Define a computable topological space $\mathbf{S}_{>} = (\mathcal{K}, \sigma_{>}, \nu_{>})$ by

$$K \in \nu_{>}(0w) :\iff \overline{\mathrm{I}}^n(w) \cap K = \emptyset,$$
$$K \in \nu_{>}(1w) :\iff K \subseteq B(0, \nu_{\mathbb{Q}}(w)).$$

Then: $\delta_{\mathbf{S}_{>}} \equiv \kappa_{>}$; $\psi_{<} \wedge \delta_{\mathbf{S}_{>}} \equiv \kappa$.

Proof: $\delta_{\mathbf{S}_{>}} \leq \kappa_{>}$ is immediate.

$\kappa_{>} \leq \delta_{\mathbf{S}_{>}}$: By Lemma 5.2.5 it suffices to show $\kappa_{\mathrm{c}} \leq \delta_{\mathbf{S}_{>}}$. Suppose, $K = \kappa_{\mathrm{c}}(p)$. Then for $a \in \mathbb{Q}$,

$$K \subseteq B(0, a) \iff (\exists w)\Big(\iota(w) \lhd p \text{ and } \bigcup \{\mathrm{I}^n(v) \mid \iota(v) \lhd w\} \subseteq B(0, a)\Big).$$

Therefore, there is a Type-2 machine M, mapping every κ_{c}-name of a compact set K to a list of all words u with $K \subseteq B(0, \nu_{\mathbb{Q}}(u))$. Since $\kappa_{\mathrm{c}} \leq \psi_{>}$, $\kappa_{\mathrm{c}} \leq \delta_{\mathbf{S}_{>}}$.
$\psi_{<} \wedge \delta_{\mathbf{S}_{>}} \equiv \kappa$: $\psi_{<} \wedge \delta_{\mathbf{S}_{>}} \equiv \psi_{<} \wedge \kappa_{>} \equiv \kappa$ by Lemma 5.2.2. $\qquad\square$

While a $\kappa_{>}$-name of a compact set K contains a single integer bound of it, a $\delta_{\mathbf{S}_{>}}$-name of K contains *all* positive rational bounds of K. The equivalences mean that a single bound suffices to find all other rational bounds. Since $\psi_{<}$ and $\delta_{\mathbf{S}_{>}}$ are admissible, also κ is admissible by Lemma 3.3.9.

As we will prove later, the final topology of a Cauchy representation of a metric space is the topology generated by the open balls of the space. Since $\kappa_{\mathrm{H}} \equiv \psi_{<} \wedge \delta_{\mathbf{S}_{>}}|^{\mathcal{K}^*}$, the sets $\{K \in \mathcal{K}^* \mid J \cap K \neq \emptyset\}$, $\{K \in \mathcal{K}^* \mid \overline{J} \cap K = \emptyset\}$ and $\{K \in \mathcal{K}^* \mid K \subseteq B(0, r)\}$ for $J \in \mathrm{Cb}^{(n)}$ and $r \in \mathbb{Q}$ are a subbase of the metric topology on $\mathcal{K}^*$.

Example 5.2.11. For every computable function $f : \mathbb{R}^n \to \mathbb{R}^n$ the function $F : \mathcal{K} \to \mathcal{K}$, defined by

$$F(K) := f[K]$$

is $(\kappa_>, \kappa_>)$-computable and (κ, κ)-computable.

First we show that F is (κ_c, κ_c)-computable. By Theorem 3.2.14 and Exercise 4.3.19, there is an r.e. set $B \subseteq \Sigma^* \times \Sigma^*$ such that $f^{-1}[\mathrm{I}^n(v)] = \bigcup \{\mathrm{I}^n(u) \mid (u, v) \in B\}$. Suppose, $K = \kappa_c(p)$. Then for any $w \in \Sigma^*$,

$$f[K] \subseteq \bigcup \{\mathrm{I}^n(v) \mid \iota(v) \vartriangleleft w\}$$
$$\Longleftrightarrow (\forall x \in K)(\exists v)\Big(f(x) \in \mathrm{I}^n(v), \iota(v) \vartriangleleft w\Big)$$
$$\Longleftrightarrow (\forall x \in K)(\exists u, t, v)\Big(x \in \mathrm{I}^n(u), (u, t) \in B, \mathrm{I}^n(t) \subseteq \mathrm{I}^n(v), \iota(v) \vartriangleleft w\Big) .$$

Since K is compact, finitely many $\mathrm{I}^n(u)$ suffice to cover K, and so the last property is equivalent to

$$(\exists w', \iota(w') \vartriangleleft p)(\forall u, \iota(u) \vartriangleleft w')(\exists t, v)\Big((u, t) \in B, \mathrm{I}^n(t) \subseteq \mathrm{I}^n(v), \iota(v) \vartriangleleft w\Big) .$$

Since B is r.e. , there is a Type-2 machine M which on input $p \in \mathrm{dom}(\kappa_c)$ prints all words $\iota(w)$ such that $F(K) \subseteq \bigcup \{\mathrm{I}^n(v) \mid \iota(v) \vartriangleleft w\}$. We obtain $F \circ \kappa_c(p) = \kappa_c \circ f_M(p)$, and so F is (κ_c, κ_c)-computable and $(\kappa_>, \kappa_>)$-computable.

We show that F is (κ, κ)-computable. Since F is $(\kappa_>, \kappa_>)$-computable, it is $(\kappa, \kappa_>)$-computable. By Example 5.1.12.3, $A \mapsto \mathrm{cls}(f[A])$ is $(\psi_<, \psi_<)$-computable. Since $\kappa \leq \psi_<$ and $f[K]$ is , hence closed, for compact K, $K \mapsto f[K]$ is $(\kappa, \psi_<)$-computable. Combining the two results we obtain that F is (κ, κ)-computable, since $\kappa_< \wedge \psi_> \leq \kappa$ by Lemma 5.2.2. $\square$

Further representations of the compact subsets of $\mathbb{R}^n$ are discussed in [BW99].

Exercises 5.2.

$\diamond$ 1. Show that a compact set $X \subseteq \mathbb{R}^n$ is recursive, iff it is κ-computable.

$\blacklozenge$ 2. Show that $K \subseteq \mathbb{R}^n$ is compact, iff for all $\alpha \subseteq \mathrm{Cb}^{(n)}$ with $K \subseteq \bigcup \alpha$ there are $J_1, \ldots, J_k \in \alpha$ such that $K \subseteq J_1 \cup \ldots \cup J_k$. (The "if" is the important case!)

3. Let $\kappa'_<$, $\kappa'_>$ and κ' be the restrictions of $\kappa_<$, $\kappa_>$ and κ, respectively to the set $\{\{x\} \mid x \in \mathbb{R}^n\}$. Show $\kappa'_< \equiv \kappa'_> \equiv \kappa' \equiv \rho^n$.

4. Show $\kappa \leq \psi$. Show that the multi-valued function $F : \mathcal{K} \rightrightarrows \mathbb{N}$, defined by its graph $\mathrm{R}_F := \{(K, i) \mid K \subseteq B(0, i)\}$ is $(\kappa, \nu_\mathbb{N})$-computable but not $(\psi|^{\mathcal{K}}, \nu_\mathbb{N})$-continuous and conclude $\psi|^{\mathcal{K}} \not\leq_t \kappa$.

5. Consider the special case $n = 1$. Show that the maximum function $\max :\subseteq \mathcal{K} \rightarrow \mathbb{R}$, $\mathrm{dom}(\max) = \mathcal{K} \setminus \{\emptyset\}$, on the non-empty compact subsets of $\mathbb{R}$ is not
   a) $(\psi|^{\mathcal{K}}, \rho_>)$-continuous,
   b) $(\kappa_>, \rho_<)$-continuous,
   c) $(\kappa_<, \rho_>)$-continuous.

6. Call $f :\subseteq \Sigma^\omega \to \Sigma^*$ a bHB-function for $K \in \mathcal{K}$, iff

$$K \subseteq \theta_<^{\text{en}}(p) \implies K \subseteq \bigcup \{I^n(w) \mid \iota(w) \lhd p_{<\nu_{\mathbb{N}} \circ f(p)}\} \text{ and}$$
$$K \not\subseteq \theta_<^{\text{en}}(p) \implies f(p) = \text{div} .$$

Define $\kappa' :\subseteq \Sigma^\omega \to \mathcal{K}$ by

$$\kappa'(p) = K : \iff \eta_p^{\omega*} \text{ is a bHB-function for } K.$$

Prove $\kappa_c \equiv \kappa'$. (A κ'-name of K is a program of a function which determines for every cover with base elements $J \in \text{Cb}^{(n)}$ a finite subcover.)

7. Call $f :\subseteq \Sigma^\omega \to \Sigma^*$ an mbHB-function for $K \in \mathcal{K}$, iff

$$K \subseteq \theta_<^{\text{en}}(p) \implies K \subseteq \bigcup \{I^n(w) \mid \iota(w) \lhd f(p)\} \text{ and}$$
$$\big((\iota(w) \lhd p \text{ and } I^n(w) \cap K \neq \emptyset), \text{ if } \iota(w) \lhd f(p)\big) ,$$
$$K \not\subseteq \theta_<^{\text{en}}(p) \implies f(p) = \text{div} .$$

Define $\kappa'' :\subseteq \Sigma^\omega \to \mathcal{K}$ by

$$\kappa''(p) = K : \iff \eta_p^{\omega*} \text{ is an mbHB-function for } K.$$

Prove $\kappa \equiv \kappa''$. (A κ''-name of K is a program of a function which determines for every cover with base elements $J \in \text{Cb}^{(n)}$ a minimal finite subcover.)

8. Show

$$\inf\{s > 0 \mid A \subseteq B_s \text{ and } B \subseteq A_s\} = \max(\inf_{A \subseteq B_s} s, \inf_{B \subseteq A_s} s)$$

and $\inf\{s \mid B \subseteq A_s\} = \sup\{d(y, A) \mid y \in B\}$ and conclude that the two given definitions of the Hausdorff distance are equivalent.

9. Show that the function d_{H} is in fact a metric on $\mathcal{K}^*$.

10. Let $(K_i)_{i \in \mathbb{N}}$ be a $(\nu_{\mathbb{N}}, \kappa_{\text{H}})$-computable sequence of non-empty compact sets such that $d_{\text{H}}(K_i, K_j) \leq 2^{-k}$, if $i, j \geq e(k)$ for some computable function $e : \mathbb{N} \to \mathbb{N}$. Show that the sequence converges to a κ_{H}-computable compact set K.

11. a) Show that intersection $(K, L) \mapsto K \cap L$ on $\mathcal{K}$ is not $(\kappa, \kappa, \kappa_<)$-continuous.

  b) Find two κ-computable sets $A, B \subseteq \mathbb{R}$ such that $A \cap B$ is not $\kappa_<$-r.e.

12. Define a computable topological space $\mathbf{S}'_> = (\mathcal{K}, \sigma_>, \nu_>)$ by
$K \in \nu_>(0w) : \iff \bar{I}^n(w) \cap K = \emptyset, \quad K \in \nu_>(1w) : \iff K \subseteq B(0, \nu_{\mathbb{N}}(w))$. Then: $\delta_{\mathbf{S}'_>} \equiv \kappa_>$. (cf. Lemma 5.2.10)

13. Generalize Example 5.2.11 to computable functions $f : \mathbb{R}^m \to \mathbb{R}^n$.

14. For $A, B \subseteq \mathbb{R}^n$ define the set distance by

$$d(A, B) := \inf\{d(a, b) \mid a \in A, b \in B\} .$$

For which triples of representations $(\delta_1, \delta_2, \delta_3)$, $\delta_1 \in \{\kappa_<, \kappa_>, \kappa\}$, $\delta_2 \in \{\kappa_<, \kappa_>, \kappa, \psi_<, \psi_>, \psi, \theta_<, \theta_>, \theta\}$, $\delta_3 \in \{\rho_<, \rho_>, \rho\}$ is the set distance $(\delta_1, \delta_2, \delta_3)$-computable?

15. For dimension $n = 1$ let δ be the restriction of the representation κ to the set $E(\mathbb{N})$ of the finite subsets of $\mathbb{N} \subseteq \mathbb{R}$. Define a standard notation ν of $E(\mathbb{N})$ by

$$\nu\big(\iota(w_1)\ldots\iota(w_n)\big) := \{\nu_{\mathbb{N}}(w_1),\ldots,\nu_{\mathbb{N}}(w_n)\} \ .$$

Show $\nu \equiv \delta$.

16. Show that every computable function $f : \mathbb{R}^m \to \mathbb{R}^m$ has a computable *rate of growth*, that is, a computable function $g : \mathbb{N} \to \mathbb{N}$ such that

$$||x|| \leq n \implies ||f(x)|| \leq g(n)$$

for all $x \in \mathbb{R}^m$ and all $n \in \mathbb{N}$. (Hint: use Example 5.2.11)

6. Spaces of Continuous Functions

This chapter is devoted to representations of continuous functions and to applications of the concepts introduced so far. In Sect. 6.1 we define and discuss several representations of spaces of continuous real functions, in particular, representations via names of realizing programs, the "compact-open" representations and the representations by uniform approximation with rational polygons. In Sect. 6.2 we prove computability of many standard operations on functions, closed, open and compact sets. In particular, we prove a computable version of Urysohn's lemma for closed subsets of $\mathbb{R}^n$. Computability of zero-finding for real functions under various restrictions is discussed in Sect. 6.3. Sect. 6.4 is devoted to computability problems of differentiation and integration, and Sect. 6.5 contains some further results on analytic functions.

6.1 Various representations

In this section let $m, n \geq 1$ be fixed natural numbers. We will introduce and compare representations of spaces of real functions $f :\subseteq \mathbb{R}^m \to \mathbb{R}^n$. For any subset $A \subseteq \mathbb{R}^m$ let

$$C(A, \mathbb{R}^n) := \{f :\subseteq \mathbb{R}^m \to \mathbb{R}^n \mid f \text{ continuous and } \operatorname{dom}(f) = A\}$$

be the set of continuous partial functions from $\mathbb{R}^m$ to $\mathbb{R}^n$ with domain A. Usually, $C(A, \mathbb{R}^n)$ denotes the set of all *total* continuous functions $f : A \to \mathbb{R}^n$, but the above definition is more convenient in our context. As usual we will abbreviate

$$C(A) := C(A, \mathbb{R}) \quad \text{and} \quad C[a; b] := C([a; b], \mathbb{R}) .$$

For finding a "natural" representation of the set $C(A, \mathbb{R}^n)$ we remember the main theorem for admissible representations (Theorem 3.2.11), by which a real function $f :\subseteq \mathbb{R}^m \to \mathbb{R}^n$ is continuous, iff it is (ρ^m, ρ^n)-continuous (Definitions 4.1.17 and 3.3.13, Fig. 6.1).

Definition 6.1.1 (standard representation of $C(A, \mathbb{R}^n)$). *Define a representation* $\delta^A_\to :\subseteq \Sigma^\omega \to C(A, \mathbb{R}^n)$ *of* $C(A, \mathbb{R}^n)$ *by*

$$\delta^A_\to := [\rho^m \to \rho^n]_A .$$

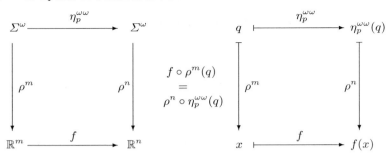

Fig. 6.1. $f = [\rho^m \to \rho^n]_A(p) = \delta^A_\to(p)$

The following lemma summarizes some immediate consequences.

Lemma 6.1.2.
1. A real function $f :\subseteq \mathbb{R}^m \to \mathbb{R}^n$ with $\text{dom}(f) = A$ is (ρ^m, ρ^n)-computable, iff it is $\delta^A_\to$-computable.
2. The function $(f, x) \mapsto f(x)$ for $f \in C(A, \mathbb{R}^n)$ and $x \in \mathbb{R}^m$ is $(\delta^A_\to, \rho^m, \rho^n)$-computable.
3. For every representation δ of a subset of $C(A, \mathbb{R}^n)$, the function $(f, x) \mapsto f(x)$ for $f \in C(A, \mathbb{R}^n)$ and $x \in \mathbb{R}^m$ is (δ, ρ^m, ρ^n)-computable, iff $\delta \leq \delta^A_\to$.

Proof: 1. Immediate from the definitions.
2., 3. This follows from Lemma 3.3.14.   □

Therefore, $\delta^A_\to$ is up to equivalence the weakest or "poorest" representation δ of $C(A, \mathbb{R}^n)$ for which the evaluation functionis (δ, ρ^m, ρ^n)-computable. Furthermore, a real function $f \in C(A, \mathbb{R}^n)$ is computable, iff it has a computable (ρ^m, ρ^n)-realization, iff it is $\delta^A_\to$-computable. The representation $\delta^A_\to$ is tailor-made for evaluation (w.r.t. ρ-names).

For open or compact sets $A \subseteq \mathbb{R}^m$ the sets $C(A, \mathbb{R}^n)$ of continuous functions $f : A \to \mathbb{R}^n$ are of particular importance in analysis. We will introduce other representations of $C(A, \mathbb{R}^n)$ for r.e. open and for r.e. closed sets $A \subseteq \mathbb{R}^m$ (Definitions 5.1.1 and 5.1.15). Important examples are $C(\mathbb{R})$ and $C[0; 1]$. The next representation of $C(A, \mathbb{R}^n)$ is defined by means of a computable topological space generated by "compact-open boxes".

Definition 6.1.3 (admissible representation by "box-properties").
Let $A \subseteq \mathbb{R}^m$ be an r.e. open or r.e. closed set. Define a computable topological space $\mathbf{S}_{\text{co}} = (C(A, \mathbb{R}^n), \sigma_{\text{co}}, \nu_{\text{co}})$, such that

$$\sigma_{\text{co}} := \begin{cases} \{R(J, L) \mid J \in \text{Cb}^{(m)}, L \in \text{Cb}^{(n)}, \overline{J} \subseteq A\}, & \text{if } A \text{ is open,} \\ \{R(J, L) \mid J \in \text{Cb}^{(m)}, L \in \text{Cb}^{(n)}, J \cap A \neq \emptyset\}, & \text{if } A \text{ is closed,} \end{cases}$$

where

$$R(J, L) := \{f \in C(A, \mathbb{R}^n) \mid f[\overline{J}] \subseteq L\} \quad and$$
$$\nu_{\mathrm{co}}(\iota(u)\iota(v)) := R(\mathrm{I}^m(u), \mathrm{I}^n(v)) \ .$$

We will call $\delta_{\mathrm{co}}^A := \delta_{\mathbf{S}_{\mathrm{co}}}$ *the compact-open representation of* $C(A, \mathbb{R}^n)$.

Convention 6.1.4. In the following let $A \subseteq \mathbb{R}^m$ be a fixed r.e. open set (that is, A is open and $\{w \in \Sigma^* \mid \overline{\mathrm{I}}^m(w) \subseteq A\}$ is r.e.) or a fixed r.e. closed set (that is, A is closed and $\{w \in \Sigma^* \mid \mathrm{I}^m(w) \cap A \neq \emptyset\}$ is r.e.). See Definitions 5.1.1 and 5.1.15.

For $\overline{J} \subseteq A$, if A is open, and for $J \cap A \neq \emptyset$, if A is closed, a function f has the "box-property" $R(J, L)$, iff it maps the *compact* cube $\overline{J} \subseteq \mathbb{R}^m$ into the *open* cube L. A δ_{co}^A-name of a continuous real function $f \in C(A, \mathbb{R}^n)$ is a list of all of its box-properties. Fig. 6.2 is an illustration for the case $m = n = 1$.

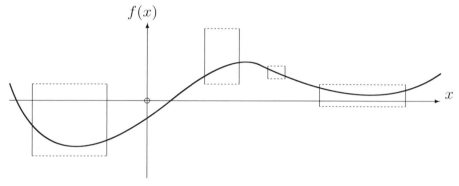

$f(x)$

x

Fig. 6.2. Some box-properties of a continuous function $f \in C(\mathbb{R})$

By definition [Eng89], on the space $C(A, B)$ of continuous functions from A to B, the *compact-open topology* is generated by the set of all $T(C, U)$ as a subbase where C is compact and U is open, and $T(C, U)$ is the set of all continuous functions f such that $f[C] \subseteq U$. In our example, the final topology τ_{co}^A of δ_{co}^A is not generated by the set of all $T(C, U)$ but by the properly smaller set σ_{co}. However, the generated topology is the same:

Lemma 6.1.5. The final topology $\tau_{\mathrm{co}}^A := \tau_{\mathbf{S}_{\mathrm{co}}}$ of the admissible representation δ_{co}^A is the compact-open topology on $C(A, \mathbb{R}^n)$.

Proof: This follows from Proposition 2 below. Let (M, d) be a locally compact metric space and let (N, τ) be a topological space. The compact-open topology τ_{co} on the set $C(M, N)$ of continuous functions $f : M \to N$ is defined by the subbase of all $T(K, U) := \{f \in C(M, N) \mid f[K] \subseteq U\}$ such that

$K \subseteq M$ is compact and $U \subseteq N$ is open. Let β be a base of (M, d) such that $\overline{I} := \mathrm{cls}(I)$ is compact for every $I \in \beta$ and let γ be a base of (N, τ).

Proposition 1:
Let $f : M \to N$ be continuous and let $U \subseteq N$ be open. Then $x \in f^{-1}[U]$, iff there are $I \in \beta$ and $J \in \gamma$ such that $x \in I$ and $f[\overline{I}] \subseteq J \subseteq U$.

Proof 1: There is some $J \in \gamma$ such that $f(x) \in J \subseteq U$ and by continuity of f there is some $\varepsilon > 0$ such that $f[B(x, \varepsilon)] \subseteq J$. Since β is a base, there is some $I \in \beta$ such that $x \in I \subseteq B(x, \varepsilon/2)$. Then $x \in \overline{I}$ and $f[\overline{I}] \subseteq J \subseteq U$.

Proposition 2:
The set of all $T(\overline{I}, J)$ such that $I \in \beta$ and $J \in \gamma$ is a subbase of the compact open-topology τ_{co} on $\mathrm{C}(M, N)$.

Proof 2:
Let $K \subseteq M$ be compact, $U \subseteq N$ open and $f : M \to N$ continuous. By Proposition 1 and compactness of K we obtain:

$\quad f \in T(K, U)$,

iff $f[K] \subseteq U$,

iff $(\forall x \in K)(\exists I_x \in \beta, J_x \in \gamma)(x \in I_x$ and $f[\overline{I}_x] \subseteq J_x \subseteq U)$,

iff there are families $(I_x)_{x \in K}$ and $(J_x)_{x \in K}$ $(I_x \in \beta$ and $J_x \in \gamma)$ such that

$\quad x \in I_x$ and $f[\overline{I}_x] \subseteq J_x \subseteq U$ for all $x \in K$,

iff there are $I_1, \ldots, I_m \in \beta$ and $J_1, \ldots, J_m \in \gamma$ such that

$\quad K \subseteq I_1 \cup \ldots \cup I_m$ and $f[\overline{I}_k] \subseteq J_k \subseteq U$ for $1 \leq k \leq m$,

iff there are $I_1, \ldots, I_m \in \beta$ and $J_1, \ldots, J_m \in \gamma$ such that

$\quad K \subseteq I_1 \cup \ldots \cup I_m$, and $f \in T(\overline{I}_k, J_k)$ and $J_k \subseteq U$ for $1 \leq k \leq m$,

iff $f \in \bigcup\{\bigcap_{1 \leq k \leq m} T(\overline{I}_k, J_k) \mid m \geq 1, \ I_1, \ldots, I_m \in \beta, \ J_1, \ldots, J_m \in \gamma,$

$\quad (\forall k) J_k \subseteq U, \ K \subseteq I_1 \cup \ldots \cup I_m\}$. $\qquad \square$

In Definition 6.1.3 we have considered atomic properties $f[\overline{J}] \subseteq L$ where J and L are balls with a rational center and radius.

The definition is robust, that is, one can replace the balls $\overline{J}$ and L by other appropriate compact and open sets, respectively, without changing the induced computability concept (c.f. Theorems 4.1.11, 5.1.14). For an example see Exercise 6.1.1. More general robustness properties can be obtained from the formula derived for $T(K, U)$ in the proof of Lemma 6.1.5. If in Definition 6.1.3 the condition "$f[\overline{J}] \subseteq L$" is replaced by one of the conditions "$f[\overline{J}] \subseteq \overline{L}$", "$f[J] \subseteq L$" or "$f[J] \subseteq \overline{L}$", the resulting definitions are very sensitive (Exercise 6.1.2). However, if not *all* properties "$f[J] \subseteq L$" must be listed but only sufficiently many of them, we obtain another useful representation of $\mathrm{C}(A, \mathbb{R}^n)$ (cf. Definition 5.1.15 of the representations $\theta_<$ and $\theta_<^{\mathrm{en}}$).

Definition 6.1.6. *Define a representation δ_{oo}^A of $\mathrm{C}(A, \mathbb{R}^n)$ as follows. $\delta_{\mathrm{oo}}^A(p) = f$, iff 1. and 2. hold:*
1. *If $\iota(z) \lhd p$, then there are words $u \in \mathrm{dom}(\mathrm{I}^m), v \in \mathrm{dom}(\mathrm{I}^n)$ such that $z = \iota(u)\iota(v)$, $f[\mathrm{I}^m(u)] \subseteq \mathrm{I}^n(v)$ and*
   $\quad \overline{\mathrm{I}}^m(u) \subseteq A$ *for open A and $\mathrm{I}^m(u) \cap A \neq \emptyset$ for closed A.*
2. *For all $x \in A$ and $k \in \mathbb{N}$ there are words u, v such that*
   $$x \in \mathrm{I}^m(u), \quad \iota(\iota(u)\iota(v)) \lhd p \quad \text{and} \quad \mathrm{radius}(\mathrm{I}^n(v)) < 2^{-k} .$$

Lemma 6.1.7. The three representations $\delta_{\rightarrow}^A$, δ_{co}^A and δ_{oo}^A are equivalent.

Proof: We prove the special case $m = n = 1$. The general case can be proved analogously. We assume that $A \subseteq \mathbb{R}$ is open. For closed A the proof is similar.

$\delta_{\rightarrow}^A \leq \delta_{\mathrm{oo}}^A$: Since $\rho \equiv \rho^a$ (ρ^a from Lemma 4.1.6), $\delta_{\rightarrow}^A = [\rho \to \rho]_A \equiv [\rho^a \to \rho^a]_A$. We prove $[\rho^a \to \rho^a]_A \leq \delta_{\mathrm{oo}}^A$. By Definition 3.3.13, $[\rho^a \to \rho^a]_A(p) \circ \rho^a(q) = \rho^a \circ \eta_p^{\omega\omega}(q)$. Let M be a Type-2 machine computing the universal function of the representation $\eta^{\omega\omega}$ (Theorem 2.3.13). The set

$$L := \{\iota(u_0)\ldots\iota(u_k) \mid \iota(u_0)\ldots\iota(u_k) \sqsubseteq q \text{ for some } q \in \mathrm{dom}(\rho^a)\}$$

of "complete" prefixes of ρ^a-names is a recursive subset of Σ^*. Let T_M be the recursive set from Lemma 2.1.5. Since A is r.e. open, there is a Type-2 machine N which on input p prints a list of all $\iota(\iota(u)\iota(v))$ for which there are words $w, v_1 \in \Sigma^*$, $u_1 \in L$ and a number $t \in \mathbb{N}$ such that $\overline{\mathrm{I}}^1(u) \subseteq A$, $(w, u_1, 0^t, v_1) \in T_M$, $w \sqsubseteq p$, $\iota(u)$ is a suffix of u_1 and $\iota(v) \lhd v_1$.

Consider $f = [\rho^a \to \rho^a]_A(p)$. We show $f = \delta_{\mathrm{oo}}^A \circ f_N(p)$. Let $\iota(\iota(u)\iota(v)) \lhd f_N(p)$. Then there are words $w, v_1 \in \Sigma^*$, $u_1 \in L$ and a number $t \in \mathbb{N}$ such that $\overline{\mathrm{I}}^1(u) \subseteq A$, $(w, u_1, 0^t, v_1) \in T_M$, $w \sqsubseteq p$, $\iota(u)$ is a suffix of u_1 and $\iota(v) \lhd v_1$. Therefore, $f_M(p, u_1 q) \in v_1 \Sigma^\omega$ for all $q \in \Sigma^\omega$ such that $(p, u_1 q) \in \mathrm{dom}(f_M)$, and so $f[\mathrm{I}^1(u)] = f \circ \rho^a[u_1 \Sigma^\omega] \subseteq \rho^a \circ \eta_p^{\omega\omega}[u_1 \Sigma^\omega] \subseteq \rho^a[v_1 \Sigma^\omega] \subseteq \mathrm{I}^1(v)$. This proves the first condition from Definition 6.1.6.

Let $x \in A$ and $k \in \mathbb{N}$. There is some q such that $\rho^a(q) = x$. Since $f \circ \rho^a(q) = \rho^a \circ f_M(p, q)$, there are words $w, v_1, u, v \in \Sigma^*$, $u_1 \in L$ and a number $t \in \mathbb{N}$ such that $(w, u_1, 0^t, v_1) \in T_M$, $w \sqsubseteq p$, $u_1 \sqsubseteq q$, $\iota(u)$ is a suffix of u_1, $\iota(v) \lhd v_1$, $\overline{\mathrm{I}}^1(u) \subseteq A$ and $\mathrm{radius}(\mathrm{I}^n(v)) < 2^{-k}$. Then $x \in \mathrm{I}^1(u)$ and $\iota(\iota(u)\iota(v)) \lhd f_N(p)$. This proves the second condition from Definition 6.1.6.

$\delta_{\mathrm{oo}}^A \leq \delta_{\mathrm{co}}^A$: We apply Corollary 5.1.16 for the case of open subsets of $\mathbb{R}$ ($n = 1$). Suppose $f := \delta_{\mathrm{oo}}^A(p)$. Then p is a list of words $\iota(\iota(u)\iota(v))$ such that $\overline{\mathrm{I}}^1(u) \subseteq A$, $f[\mathrm{I}^1(u)] \subseteq \mathrm{I}^1(v)$ and for any $x \in A$ and $k \in \mathbb{N}$ the list p contains a word $\iota(\iota(u)\iota(v))$ such that $x \in \mathrm{I}^1(u)$ and $\mathrm{radius}(\mathrm{I}^1(v)) < 2^{-k}$. Therefore,

$$f^{-1}[\mathrm{I}^1(w)] = \bigcup \{\mathrm{I}^1(u) \mid \iota(\iota(u)\iota(v)) \lhd p \text{ and } \mathrm{I}^1(v) \subseteq \mathrm{I}^1(w) \text{ for some } v\}.$$

There is a Type-2 machine M which on input $(p, w) \in \Sigma^\omega \times \Sigma^*$ lists all $\iota(u)$ such that $\iota(\iota(u)\iota(v)) \lhd p$ and $\mathrm{I}^1(v) \subseteq \mathrm{I}^1(w)$ for some v. If $f = \delta_{\mathrm{oo}}^A(p)$, then $f^{-1}[\mathrm{I}^1(w)] = \theta_<^{\mathrm{en}} \circ f_M(p, w)$. Let M_1 be a Type-2 machine translating $\theta_<^{\mathrm{en}}$ to $\theta_<$ (Corollary 5.1.16). Combining M and M_1 we get a machine M_2 mapping (p, w) to a $\theta_<$-name of $f^{-1}[\mathrm{I}^1(w)]$. From M_2 one can construct a machine N mapping $p \in \mathrm{dom}(\delta_{\mathrm{oo}}^A)$ to a list of all $\iota(\iota(u)\iota(v))$, such that $\overline{\mathrm{I}}^1(u) \subseteq A$ and $f[\overline{\mathrm{I}}^1(u)] \subseteq \mathrm{I}^1(v)$. The function f_N translates δ_{oo}^A to δ_{co}^A. $\delta_{\mathrm{co}}^A \leq [\rho \to \rho]_A$: By Lemma 4.1.6 it suffices to prove $\delta_{\mathrm{co}}^A \leq [\rho^a \to \rho^b]_A$. If $\delta_{\mathrm{co}}^A(p) = f$, then p is a list of all $\iota(\iota(u)\iota(v))$, such that $f[\overline{\mathrm{I}}^1(u)] \subseteq \mathrm{I}^1(v)$ and $\overline{\mathrm{I}}^1(u) \subseteq A$. If $\rho^a(q) = x$, then q encodes a nested sequence of closed rational balls converging to x. There is a Type-2 machine M which on input (p, q), $p \in \mathrm{dom}(\delta_{\mathrm{co}}^A)$, $q \in \mathrm{dom}(\rho^a)$, produces a list of all $\iota(v)$ for which there is some word $\iota(u)$ listed by q such that $\iota(\iota(u)\iota(v)) \lhd p$. If $\delta_{\mathrm{co}}^A(p) = f$ and $\rho^a(q) = x$, then $\rho^b \circ f_M(p, q) = f(x)$, that is, $\delta_{\mathrm{co}}^A(p) \circ \rho^a(q) = \rho^b \circ f_M(p, q)$ for all $p \in \mathrm{dom}(\delta_{\mathrm{co}}^A)$ and $q \in \mathrm{dom}(\rho^a)$. By the smn-theorem for $\eta^{\omega\omega}$ there is some computable function $h : \Sigma^\omega \to \Sigma^\omega$ such that $f^M(p, q) = \eta_{h(p)}^{\omega\omega}(q)$. We obtain $\delta_{\mathrm{co}}^A(p) = [\rho^a \to \rho^b]_A \circ h(p)$ for all $p \in \mathrm{dom}(\delta_{\mathrm{co}}^A)$, hence $\delta_{\mathrm{co}}^A \leq [\rho^a \to \rho^b]_A$. $\square$

Example 6.1.8.

1. Consider $m = n = 1$. For $a \in \mathbb{R}$ define $H(a) : \mathbb{R} \to \mathbb{R}$ by $H(a)(x) := a \cdot x$. Then the function $H : \mathbb{R} \to C(\mathbb{R})$ is $(\rho, \delta_\to^{\mathbb{R}})$-computable. This follows immediately from Theorem 3.3.15.

2. Consider $m = n$. Composition on $C(\mathbb{R}^n, \mathbb{R}^n)$ is $(\delta_\to^{\mathbb{R}^n}, \delta_\to^{\mathbb{R}^n}, \delta_\to^{\mathbb{R}^n})$-computable. For a proof, assume $f = \delta_\to^{\mathbb{R}^n}(p)$ and $g = \delta_\to^{\mathbb{R}^n}(q)$. Then

$$g \circ f \circ \rho^n(r) = g \circ \rho^n \circ \eta_p^{\omega\omega}(r) = \rho^n \circ \eta_q^{\omega\omega} \circ \eta_p^{\omega\omega}(r) \ .$$

   Applying the utm-theorem and the smn-theorem we get a computable function $h : \Sigma^\omega \to \Sigma^\omega$ such that $\eta_q^{\omega\omega} \circ \eta_p^{\omega\omega}(r) = \eta_{h(q,p)}^{\omega\omega}(r)$. Therefore, $\delta_\to^{\mathbb{R}^n}(q) \circ \delta_\to^{\mathbb{R}^n}(p) = \delta_\to^{\mathbb{R}^n} \circ h(q, p)$, and so composition is $(\delta_\to^{\mathbb{R}^n}, \delta_\to^{\mathbb{R}^n}, \delta_\to^{\mathbb{R}^n})$-computable.

3. Consider $m = n = 1$. We call a function $\mathrm{moc} : \mathbb{N} \to \mathbb{N}$ a modulus of continuity of the function $f : \mathbb{R} \to \mathbb{R}$ in $x \in \mathbb{R}$, iff $|f(y) - f(x)| \leq 2^{-k}$ whenever $|y - x| \leq 2^{-\mathrm{moc}(k)}$. If $f \in C(\mathbb{R})$ is continuous, it has a modulus of continuity at every point $x \in \mathbb{R}$. We consider the representation $\delta_\mathbb{B}$ of the set $\mathbb{B}$ of all functions $f : \mathbb{N} \to \mathbb{N}$ from Definition 3.1.2. Can a modulus of continuity of f in 0 be computed from f? We formalize the problem by means of the multi-valued function $F : C(\mathbb{R}) \rightrightarrows \mathbb{N}^\mathbb{N}$, defined by $(f, \mathrm{moc}) \in R_F$, iff the function $\mathrm{moc} : \mathbb{N} \to \mathbb{N}$ is a modulus of continuity of f in 0.
   We show that F is $(\delta_{\mathrm{oo}}^\mathbb{R}, \delta_\mathbb{B})$-computable:
   If $\delta_{\mathrm{oo}}^\mathbb{R}(p) = f$ then p is a list of "sufficiently many" $\iota(\iota(u)\iota(v))$ such that $f[\mathrm{I}^1(u)] \subseteq \mathrm{I}^1(v)$ etc. (Definition 6.1.6). There is a Type-2 machine M which on input $p \in \mathrm{dom}(\delta_{\mathrm{oo}}^\mathbb{R})$ prints a sequence $0^{n_0} 1 0^{n_1} 1 \ldots$, where n_i is determined as follows. M searches in p a subword $\iota(\iota(u)\iota(v))$ such that $0 \in \mathrm{I}^1(u)$ and $\mathrm{I}^1(v)$ has a diameter less than 2^{-i}. Then let n_i be the smallest number $k \in \mathbb{N}$ such that $[-2^{-k}; 2^{-k}] \subseteq \mathrm{I}^1(u)$. Now, if $|y - 0| \leq 2^{-n_i}$, then $y \in \mathrm{I}^1(u)$, hence $f(0), f(y) \in \mathrm{I}^1(v)$ and so $|f(y) - f(0)| \leq 2^{-i}$. Therefore, the function f_M realizes the multi-valued function F.
   Assume that F has a $(\delta_\to^\mathbb{R}, \delta_\mathbb{B})$-continuous choice function G. Then the function $a \mapsto (G \circ H(a))(0)$, H from Example 1 above, is $(\rho, \nu_\mathbb{N})$-continuous. By Corollary 4.3.16, this function has a constant value $k \in \mathbb{N}$. Therefore, for all $a \in \mathbb{R}$, $|a \cdot y| \leq 2^{-0} = 1$, whenever $|y| \leq 2^{-k}$, which cannot be true. $\square$

For any closed finite interval $A \subseteq \mathbb{R}$ the maximum metric on $C(A)$ is defined by

$$d(f, g) := \max_{x \in A} |f(x) - g(x)| \, .$$

Notice that the maximum $d(f, g) \in \mathbb{R}$ exists since the functions f and g are continuous and the interval A is compact. By the Weierstraß approximation theorem the set of polynomials with rational coefficients restricted to A form a dense subset of $C[A]$. The polygon functions with rational vertices are another dense subset. Fig. 6.3 shows a function $f \in C[0; 1]$ and a polygon function with rational vertices close to it.

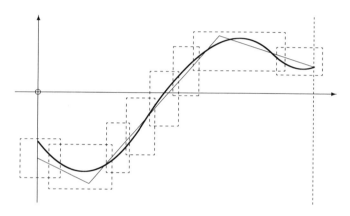

Fig. 6.3. A function $f \in C[0; 1]$, a rational polygon close to f and a set of open-open squares narrowly enclosing $\text{graph}(f)$

In the following let $A = [a_l; a_r]$ be a recursive closed interval. Remember that the endpoints of A are computable real numbers. We introduce the Cauchy representation δ_C^A of $C(A)$, where a name of a function f is a sequence of rational polygons converging fast to it.

Definition 6.1.9 (Cauchy representation of $C(A)$). *Let $A = [c; d]$ be a recursive interval. A rational polygon (on A) is a function $f \in C(A)$ for which there are rational numbers $a_0, b_0, \ldots, a_k, b_k$ such that $a_0 < a_1 < \ldots < a_k$, $a_0 \leq c$, $d \leq a_k$ and for all $x \in A$,*

$$f(x) = b_{i-1} + (x - a_{i-1})(b_i - b_{i-1})/(a_i - a_{i-1}) \quad \text{if} \quad a_{i-1} \leq x \leq a_i$$

for $i = 1, \ldots, k$. Let ν_{Pg} be a standard notation of the set Pg of rational polygons on A. The Cauchy representation δ_C^A of $C(A)$ is defined as follows:

$$\delta_C^A(p) = f : \iff p = \iota(w_0)\iota(w_1) \ldots \text{ such that}$$
$$d(\nu_{\text{Pg}}(w_i), \nu_{\text{Pg}}(w_k)) \leq 2^{-i} \text{ for } i < k$$
$$\text{and } f = \lim_{i \to \infty} \nu_{\text{Pg}}(w_i) \, .$$

If $\delta^A_C(\iota(w_0)\iota(w_1)\ldots) = f$, then for each n, f is in the ball with center $\nu_{\mathrm{Pg}}(w_n)$ and radius 2^{-n} which is presented graphically by a band around $\nu_{\mathrm{Pg}}(w_n)$ of width $2 \cdot 2^{-n}$ (Fig. 1.5).

Lemma 6.1.10. For any recursive closed finite interval A, $\delta^A_C \equiv \delta^A_\to$.

Proof: By Lemma 6.1.7 it suffices to prove $\delta^A_C \leq \delta^A_{\mathrm{co}}$ and $\delta^A_{\mathrm{oo}} \leq \delta^A_C$.

$\delta^A_C \leq \delta^A_{\mathrm{co}}$: Assume $\delta^A_C(p) = f$ and $p = \iota(w_0)\iota(w_1)\ldots$. Then $f \in R(J,L)$ (Definition 6.1.3), iff $\nu_{\mathrm{Pg}}(w_n) - 2^{-n} \in R(J,L)$ and $\nu_{\mathrm{Pg}}(w_n) + 2^{-n} \in R(J,L)$ for some $n \in \mathbb{N}$. Therefore, there is some Type-2 machine M which on input $p = \iota(w_0)\iota(w_1)\ldots \in \mathrm{dom}(\delta^A_C)$ prints a list of all pairs of intervals (J,L) such that $f \in R(J,L)$. The function f_M translates δ^A_C to δ^A_{co}.

$\delta^A_{\mathrm{oo}} \leq \delta^A_C$: Assume $\delta^A_{\mathrm{oo}}(p) = f$ and $n \in \mathbb{N}$. Then by Definition 6.1.6 for each $x \in A$ there is a pair (J_x, L_x) of intervals listed by p, such that $x \in J_x$, $f[J_x] \subseteq L_x$ and the diameter of L_x is less than 2^{-n-1}. Since A is compact and $A \subseteq \bigcup\{J_x \mid x \in A\}$,

$$A \subseteq \bigcup\{J_x \mid x \in X_n\} \text{ for some finite set } X_n \subseteq A$$

(finitely many open-open boxes suffice to cover the interval A, see Fig. 6.3). Let $g \in \mathrm{Pg}$ be any rational polygon running through all of these open-open boxes. Then $d(f,g) \leq 2^{-n-1}$. Let M be a Type-2 machine which on input $p \in \mathrm{dom}(\delta^A_{\mathrm{oo}})$ prints a sequence $q = \iota(w_0)\iota(w_1)\ldots \in \Sigma^\omega$, where for each n the word w_n is determined as follows: M searches in p finitely many pairs of intervals (J_i, L_i), $i = 1, \ldots, k$, such that the diameter of L_i is less than 2^{-n-1} for all i and $A \subseteq J_1 \cup \ldots \cup J_k$. Then M determines some word w_n such that the rational polygon $\nu_{\mathrm{Pg}}(w_n)$ runs through the k open-open boxes. Since

$$\begin{aligned} d(\nu_{\mathrm{Pg}}(w_n), \nu_{\mathrm{Pg}}(w_m)) &\leq d(\nu_{\mathrm{Pg}}(w_n), f) + d(f, \nu_{\mathrm{Pg}}(w_m)) \\ &\leq 2^{-n-1} + 2^{-m-1} \\ &< 2^{-n} \end{aligned}$$

for $m > n$, the sequence $q = \iota(w_0)\iota(w_1)\ldots \in \Sigma^\omega$ is a δ^A_C-name of $\delta^A_{\mathrm{oo}}(p)$. Therefore, f_M translates δ^A_{oo} to δ^A_C. $\qquad\square$

The set of open balls $\{B(p,r) \mid p \in \mathrm{Pg}, r \in \mathbb{Q}, r > 0\}$, $B(p,r) := \{f \in C(A) \mid d(p,f) < r\}$, is a base of the final topology of δ^A_C (Exercise 6.1.12), which is the compact-open topology on $C(A)$, since $\delta^A_C \equiv \delta^A_{\mathrm{co}}$. Therefore, the compact-open topology is the topology induced by the metric d.

The rational polygons in Definition 6.1.9 can be replaced by the set Pn of *rational polynomials*, that is, polynomial functions with rational coefficients, restricted to A, which are known to be dense in $C(A)$. Let ν_{Pn} be a standard notation of Pn and let δ^A_{Cp} be the derived Cauchy representation according to Definition 6.1.9. Since $d|_{\mathrm{Pg}\times\mathrm{Pn}}$ is $(\nu_{\mathrm{Pg}}, \nu_{\mathrm{Pn}}, \rho)$-computable, the two Cauchy representations δ^A_{Cp} and δ^A_C are equivalent (Exercise 6.1.7). Lemma 6.1.10

with a polynomial Cauchy representation can be generalized to $C(A, \mathbb{R}^n)$ for κ-computable $A \subseteq \mathbb{R}^m$. The property

$$\delta_{Cp}^A \equiv \delta_{\to}^A$$

can be called the "effective Weierstraß theorem".

Exercises 6.1.

1. Replace $\mathbb{Q}$ in Definitions 4.1.2 by the finite decimal fractions $\mathbb{Q}_{10}$, and let τ and (via Definition 4.1.17) δ be the resulting topology and representation, respectively in Definition 6.1.3. Show that $\tau_{co}^A = \tau$ and $\delta_{co}^A \equiv \delta$.

♦ 2. Let τ' be the topology generated by the set of all $R'(J, L) := \{f \in C(\mathbb{R}) \mid f[J] \subseteq L\}$ $(J, L \in \mathrm{Cb}^{(1)})$ as a subbase, and let τ'' be the topology generated by the set of all $R''(J, L) := \{f \in C(\mathbb{R}) \mid f[J] \subseteq L\}$ such that the center and radius of J and L are finite decimal fractions as a subbase. Show that $\tau_{co}^{\mathbb{R}}$, τ' and τ'' are three different topologies.

♦ 3. A function $f : \mathbb{R} \to \mathbb{R}$ is called *lower semi-continuous*, iff it is $(\tau_\rho, \tau_{\rho_<})$-continuous (Lemma 4.1.3). Define two representations on the set $\mathrm{LSC}(\mathbb{R})$ of lower semi-continuous functions $f : \mathbb{R} \to \mathbb{R}$ as follows.

   a) $\delta_{\to}^< := [\rho \to \rho_<]_{\mathbb{R}}$ (cf. Definition 6.1.1).

   b) Let $\delta_{co}^<$ be the canonical representation for the computable topological space $\mathbf{S}_{co}^< = (\mathrm{LSC}(\mathbb{R}), \sigma_{co}^<, \nu_{co}^<)$, such that $\sigma_{co}^< := \{R_{abc} \mid a, b, c \in \mathbb{Q}, \ a < b\}$ where $R_{abc} := \{f \in \mathrm{LSC}(\mathbb{R}) \mid c < f[[a;b]]\}$ and $\nu_{co}^<$ is a canonical notation of $\sigma_{co}^<$ (cf. Definition 6.1.3).
   Show $\delta_{\to}^< \equiv \delta_{co}^<$.

4. Prove Lemma 6.1.7 for r.e. closed sets A.

5. Show that the function $(a, b) \mapsto f_{ab}$, $a, b, \in \mathbb{R}$, where $f_{ab}(x) := a \cdot x + b$, is $(\rho, \rho, \delta_{\to}^{\mathbb{R}})$-computable.

6. Let $A = [a_l; a_r]$ be a recursive closed interval. Prove:

   a) The restriction $R : C(\mathbb{R}) \to C(A)$,

   $$R(f) := f\rfloor_A ,$$

   is $(\delta_{\to}^{\mathbb{R}}, \delta_{\to}^A)$-computable.

   b) The extension $E : C(A) \to C(\mathbb{R})$,

   $$E(g)(x) := \begin{cases} g(a_l), & \text{if } x < a_l , \\ g(x), & \text{if } a_l \leq x \leq a_r , \\ g(a_r), & \text{if } x > a_r , \end{cases}$$

   is $(\delta_{\to}^A, \delta_{\to}^{\mathbb{R}})$-computable. Hint: consider the representations $\delta_{oo}^{\mathbb{R}}$ and δ_{oo}^A and enrich names.

♦ 7. Let $d : C[0; 1] \times C[0; 1] \to \mathbb{R}$ be the maximum metric on $C[0; 1]$. Let ν_{Pn} be a standard notation of the set of rational polynomials restricted to $[0; 1]$ and let and let $\delta_{Cp}^{[0;1]}$ be the Cauchy representation of $C[0; 1]$ derived from it. Show that

   a) $d|_{\mathrm{Pg} \times \mathrm{Pg}}$ is $(\nu_{\mathrm{Pg}}, \nu_{\mathrm{Pg}}, \rho)$-computable,
   b) d is $(\delta_C^{[0;1]}, \delta_C^{[0;1]}, \rho)$-computable,
   c) $d|_{\mathrm{Pg} \times \mathrm{Pn}}$ is $(\nu_{\mathrm{Pg}}, \nu_{\mathrm{Pn}}, \rho)$-computable,
   d) $\delta_C^{[0;1]} \equiv \delta_{Cp}^{[0;1]}$.

◆ 8. A real function $f : \mathbb{R} \to \mathbb{R}$ is continuous, iff its graph is closed. Define a representation δ_g of $C(\mathbb{R})$ as follows:

$$\delta_g(p) = f : \iff \psi(p) = \mathrm{graph}(f)$$

where ψ is the representation of the closed subsets of $\mathbb{R}^2$ from Definition 5.1.1. Show $\delta_g \equiv \delta_{\to}^{\mathbb{R}}$.

9. Let $A, B \subseteq \mathbb{R}$ be intervals, let $h :\subseteq \mathbb{R} \to \mathbb{R}$ be increasing such that $\mathrm{dom}(h) = A$ and $\mathrm{range}(h) = B$. Define a transformation $T_h : C(B) \to C(A)$ by $T_h(f) := f \circ h$.
   a) Let $A := [a_l; a_r]$, $a_l < a_r$, be a recursive closed interval. Define the real function $h :\subseteq \mathbb{R} \to \mathbb{R}$ by $\mathrm{dom}(h) := A$ and $h(x) := (x - a_l)/(a_r - a_l)$. Show:

$$T_h \circ \delta_{\to}^{[0;1]} \equiv \delta_{\to}^A .$$

   b) Find a computable increasing real function h such that

$$T_h \circ \delta_{\to}^{(0;1)} \equiv \delta_{\to}^{\mathbb{R}} .$$

   c) Let A be an r.e. open interval. Find a computable increasing real function h such that $T_h \circ \delta_{\to}^{\mathbb{R}} \equiv \delta_{\to}^A$ (Exercise 4.3.11).

10. Show that the function $h : a \mapsto f_a$ where $f_a(x) := x^3 - 3 \cdot x + a$ is $(\rho, \delta_{\to}^{\mathbb{R}})$-computable and that its inverse is $(\delta_{\to}^{\mathbb{R}}, \rho)$-computable

◆11. In this chapter we have studied the representation $\delta_{\to}^A$ (and equivalent ones) of the set $C(A, \mathbb{R}^n)$ of all continuous functions $f :\subseteq \mathbb{R}^m \to \mathbb{R}^n$ with explicitly given domain $A \subseteq \mathbb{R}^m$. There are various useful representations of sets C of continuous real functions which have no common domain. Let $R(f, p)$ abbreviate "$\eta_p^{\omega\omega}$ is a (ρ^m, ρ^n)-realization of f" .
   a) The function $[\rho^m \to \rho^n]_s$ (Remark 3.3.17.1) represents the set of *strongly* (ρ^m, ρ^n)-continuous functions. A function $f :\subseteq \mathbb{R}^m \to \mathbb{R}^n$ is strongly (ρ^m, ρ^n)-continuous, iff it is continuous and its domain is a G_δ-subset of $\mathbb{R}^m$ (cf. Definition 2.3.6, [Wei93]). Notice that $R(f, p)$, if $f = [\rho^m \to \rho^n]_s(p)$.
   b) $[\rho^m \to \rho^n]_m(p)$ is the continuous function f with maximal domain such that $R(f, p)$ (Remark 3.3.17.2).
   c) $[\rho^m \to \rho^n]_{\mathrm{set}}(p)$ is the set of all continuous functions f such that $R(f, p)$ (Remark 3.3.17.3).
For the following representations the name of a function contains explicitly a name of its domain (Definition 6.1.3).

$$f = \delta_1 \langle p, q \rangle : \iff R(f, p) \text{ and } \mathrm{dom}(f) = \theta_<(q) ,$$

$$f = \delta_2 \langle p, q \rangle \; : \Longleftrightarrow \; R(f, p) \;\; \text{and} \;\; \text{dom}(f) = \psi_<(q) \; ,$$
$$f = \delta_3 \langle p, q \rangle \; : \Longleftrightarrow \; R(f, p) \;\; \text{and} \;\; \text{dom}(f) = \kappa(q) \; .$$

Generalize Lemma 6.1.7 to the representations δ_1, δ_2 and δ_3 from above, where the set A is no longer fixed but another parameter.

12. Show that the set of open balls $\{B(p, r) \mid p \in \text{Pg}, r \in \mathbb{Q}, r > 0\}$, $B(p, r) :=$ $\{f \in C(A) \mid d(p, f) < r\}$, is a base of the final topology of δ_C^A (Definition 6.1.9).
13. Let $J_1, \ldots, J_m \subseteq \mathbb{R}$ be intervals with computable endpoints and let J be the rectangle $J_1 \times \ldots \times J_m$. Show that there is a $(\delta_{\rightarrow}^J, \delta_{\rightarrow}^{\mathbb{R}^m})$-computable function E extending every continuous function $f \in C(J, \mathbb{R}^n)$ to a continuous total function $E(f) : \mathbb{R}^m \to \mathbb{R}^n$ (Sect. 0.4 in [PER89]).

6.2 Computable Operators on Functions, Sets and Numbers

We are now able to formulate and prove computability of many interesting operators on real numbers, open, closed or compact subsets of real numbers and continuous real functions. In this section, for $m, n \geq 1$ define $\delta_{\rightarrow}^{mn} := [\rho^m \to \rho^n]_{\mathbb{R}^m}$. Our first theorem generalizes Corollary 4.3.4.

Theorem 6.2.1.   Consider $k, m, n \geq 1$.

1. The function H with $H(a)(x) := a$ for $a \in \mathbb{R}^n$ and $x \in \mathbb{R}^m$ is $(\rho^n, \delta_{\rightarrow}^{mn})$-computable;
2. the function H with $H(f, g)(x) := f(x) \text{ op } g(x)$ for $f, g \in C(\mathbb{R}^m)$, where $\text{op} \in \{+ , \cdot , \max , \min\}$, is $(\delta_{\rightarrow}^{m1}, \delta_{\rightarrow}^{m1}, \delta_{\rightarrow}^{m1})$-computable;
3. the function H with $H(f)(x) := 1/f(x)$ for all $f \in C(\mathbb{R}^m)$ with $(\forall x) f(x) \neq 0$ is $(\delta_{\rightarrow}^{m1}, \delta_{\rightarrow}^{m1})$-computable;
4. the function $(f, g) \mapsto g \circ f$ for $f \in C(\mathbb{R}^k, \mathbb{R}^m)$ and $g \in C(\mathbb{R}^m, \mathbb{R}^n)$ is $(\delta_{\rightarrow}^{km}, \delta_{\rightarrow}^{mn}, \delta_{\rightarrow}^{kn})$-computable;
5. the function $(i, f) \mapsto f^i$ for $i \in \mathbb{N}$ and $f \in C(\mathbb{R}^n, \mathbb{R}^n)$ is $(\nu_{\mathbb{N}}, \delta_{\rightarrow}^{nn}, \delta_{\rightarrow}^{nn})$-computable.

Proof: 1. By the smn-theorem there are computable functions $g : \Sigma^\omega \times \Sigma^\omega \to \Sigma^\omega$ and $f : \Sigma^\omega \to \Sigma^\omega$ such that

$$H(\rho^n(p)) \circ \rho^m(q) = \rho^n(p) = \rho^n \circ g(p, q) = \rho^n \circ \eta_{f(p)}^{\omega\omega}(q) \; .$$

We obtain $H(\rho^n(p)) = \delta_{\rightarrow}^{mn} \circ f(p)$, and so H is $(\rho^n, \delta_{\rightarrow}^{mn})$-computable.

2. As an example we consider addition. By Theorem 4.3.2, addition has a computable (ρ, ρ, ρ)-realization h. By the utm-theorem and the smn-theorem there is a computable function $s : \Sigma^\omega \times \Sigma^\omega \to \Sigma^\omega$ such that

$$H(\delta_\to^{m1}(p), \delta_\to^{m1}(q)) \circ \rho_m(r) = \delta_\to^{m1}(p) \circ \rho_m(r) + \delta_\to^{m1}(q) \circ \rho_m(r)$$
$$= \rho \circ \eta_p^{\omega\omega}(r) + \rho \circ \eta_q^{\omega\omega}(r)$$
$$= \rho \circ h(\eta_p^{\omega\omega}(r), \eta_q^{\omega\omega}(r))$$
$$= \rho \circ \eta_{s(p,q)}^{\omega\omega}(r) ,$$

and so $H(\delta_\to^{m1}(p), \delta_\to^{m1}(q)) = \delta_\to^{m1} \circ s(p,q)$. Therefore, H is $(\delta_\to^{m1}, \delta_\to^{m1}, \delta_\to^{m1})$-computable.

3. and 4 can be proved similarly (cf. Exercise 3.3.11).

5. is a special case of Theorem 3.1.7. $\square$

Appropriate generalizations to $C(A)$ or $C(A, \mathbb{R}^n)$ for $A \subseteq \mathbb{R}^m$ are straightforward. Also some fundamental operations on infinite sequences of functions are computable.

Theorem 6.2.2. Let $\delta_\to^{m1} := [\rho^m \to \rho]_{\mathbb{R}^m}$. For sequences $(f_i)_{i\in\mathbb{N}}$, $f_i \in C(\mathbb{R}^m)$ for $i \in \mathbb{N}$,

1. the operation $(f_i)_{i\in\mathbb{N}} \mapsto (h_i)_{i\in\mathbb{N}}$,

$$h_i(x) := f_0(x) + \ldots + f_i(x) ,$$

is $([\delta_\to^{m1}]^\omega, [\delta_\to^{m1}]^\omega)$-computable,

2. the operation

$$L' : (f_i)_{i\in\mathbb{N}} \mapsto f \text{ where } f(x) := \lim_{i\to\infty} f_i(x)$$

and $(f_i)_{i\in\mathbb{N}} \in \text{dom}(L')$, iff $|f_i(x) - f_j(x)| \leq 2^{-i}$ for all $x \in \mathbb{R}^m$ and $i < j$, is $([\delta_\to^{m1}]^\omega, \delta_\to^{m1})$-computable.

Proof: 1. The function $((f_i)_i, j) \mapsto f_j$ is $([\delta_\to^{m1}]^\omega, \nu_\mathbb{N}, \delta_\to^{m1})$-computable, hence $([\delta_\to^{m1}]^\omega, \nu_\mathbb{N}, [\rho^m \to \rho]_{\mathbb{R}^m})$-computable. By Theorem 3.3.15 the function

$$((f_i)_i, j, x) \mapsto f_j(x) \quad \text{is } ([\delta_\to^{m1}]^\omega, \nu_\mathbb{N}, \rho^m, \rho)\text{-computable,}$$

and so $((f_i)_i, x, j) \mapsto f_j(x)$ is $([\delta_\to^{m1}]^\omega, \rho^m, \nu_\mathbb{N}, \rho)$-computable. Again by Theorem 3.3.15 the function

$$((f_i)_i, x) \mapsto (f_j(x))_j \quad \text{is } ([\delta_\to^{m1}]^\omega, \rho^m, [\nu_\mathbb{N} \to \rho]_\mathbb{N})\text{-computable,}$$

and so $([\delta_\to^{m1}]^\omega, \rho^m, [\rho]^\omega)$-computable by Lemma 3.3.16. By composition with the function S from Theorem 4.3.6.3 we obtain that

$$((f_i)_i, x) \mapsto (h_i(x))_i \quad \text{is } ([\delta_\to^{m1}]^\omega, \rho^m, [\rho]^\omega)\text{-computable,}$$

and so $([\delta_\to^{m1}]^\omega, \rho^m, [\nu_\mathbb{N} \to \rho]_\mathbb{N})$-computable by Lemma 3.3.16. By Theorem 3.3.15,

$$((f_i)_i, j, x) \mapsto h_j(x) \quad \text{is } ([\delta_\to^{m1}]^\omega, \nu_\mathbb{N}, \rho^m, \rho)\text{-computable,}$$

$$((f_i)_i, j) \mapsto h_j \quad \text{is } ([\delta_\rightarrow^{m1}]^\omega, \nu_\mathbb{N}, [\rho^m \rightarrow \rho]_{\mathbb{R}^m})\text{-computable},$$

hence $([\delta_\rightarrow^{m1}]^\omega, \nu_\mathbb{N}, \delta_\rightarrow^{m1})$-computable and

$$(f_i)_i \mapsto (h_i)_i \quad \text{is } ([\delta_\rightarrow^{m1}]^\omega, [\nu_\mathbb{N} \rightarrow \delta_\rightarrow^{m1}]_\mathbb{N})\text{-computable},$$

and so $([\delta_\rightarrow^{m1}]^\omega, [\delta_\rightarrow^{m1}]^\omega)$-computable by Lemma 3.3.16.

2. From Part 1 above we know that

$$((f_i)_i, x) \mapsto \big(f_j(x)\big)_j \quad \text{is } ([\delta_\rightarrow^{m1}]^\omega, \rho^m, [\rho]^\omega)\text{-computable}.$$

Applying the limit operator on L from Theorem 4.3.7 to $\big((f_0(x), f_1(x), \ldots), \mathrm{id}_\mathbb{N}\big)$, we obtain by composition that

$$((f_i)_i, x) \mapsto \lim_{i \rightarrow \infty} f_i(x) = f(x) \text{ is } ([\delta_\rightarrow^{m1}]^\omega, \rho^m, \rho)\text{-computable}.$$

By Theorem 3.3.15, $L' : (f_i)_i \mapsto f$ is $([\delta_\rightarrow^{m1}]^\omega, [\rho^m \rightarrow \rho]_{\mathbb{R}^m})$-computable, hence $([\delta_\rightarrow^{m1}]^\omega, \delta_\rightarrow^{m1})$-computable. $\qquad \square$

The theorem can be generalized easily to functions from $C(A)$ with $A \in \mathbb{R}^m$. So far we have considered effectivity on spaces of functions with a common domain. The following example of a more general type of effectivity generalizes Theorem 6.2.1.3.

Example 6.2.3. For any function $f :\subseteq \mathbb{R} \rightarrow \mathbb{R}$ define $H(f) :\subseteq \mathbb{R} \rightarrow \mathbb{R}$ by $H(f)(x) := 1/f(x)$. We show that there is a computable function $h : \Sigma^\omega \rightarrow \Sigma^\omega$ such that $\eta_{h(p)}^{\omega\omega}$ is a (ρ, ρ)-realization of $H(f)$, if $\eta_p^{\omega\omega}$ is a (ρ, ρ)-realization of $f :\subseteq \mathbb{R} \rightarrow \mathbb{R}$. Notice that nothing is said about $\mathrm{dom}(f)$.

Since $x \mapsto 1/x$ is (ρ, ρ)-computable, there is a computable function $g :\subseteq \Sigma^\omega \rightarrow \Sigma^\omega$ such that $1/\rho(q) = \rho \circ g(q)$. By the utm-theorem and the smn-theorem for $\eta^{\omega\omega}$, there is a computable function $h : \Sigma^\omega \rightarrow \Sigma^\omega$ such that $g \circ \eta_p^{\omega\omega}(q) = \eta_{h(p)}^{\omega\omega}(q)$. Suppose, $\eta_p^{\omega\omega}$ is a (ρ, ρ)-realization of $f :\subseteq \mathbb{R} \rightarrow \mathbb{R}$. Then

$$H(f) \circ \rho(q) = \frac{1}{f \circ \rho(q)} = \frac{1}{\rho \circ \eta_p^{\omega\omega}(q)} = \rho \circ g \circ \eta_p^{\omega\omega}(q) = \rho \circ \eta_{h(p)}^{\omega\omega}(q) \,,$$

and so $\eta_{h(p)}^{\omega\omega}$ is a (ρ, ρ)-realization of $H(f)$. Notice that a single computable function $h : \Sigma^\omega \rightarrow \Sigma^\omega$ determines the inverse of *all* partial continuous functions $f :\subseteq \mathbb{R} \rightarrow \mathbb{R}$. The operator H is computable w.r.t. the set-valued representation $[\rho^n \rightarrow \rho]_{\mathrm{set}}$ (Exercise 3.1.7 and Remark 3.3.17.3). Also the other statements from Theorem 6.2.1 have more general domain-independent versions (Exercise 6.2.5). $\qquad \square$

It is well known that for a continuous function $f : \mathbb{R} \rightarrow \mathbb{R}$, $f^{-1}[U]$ is open, if U is open, $f^{-1}[A]$ is closed, if A is closed, $f[K]$ is compact, if K is compact, and obviously, $\mathrm{cls}(f[A])$ is closed, if A is closed. Effective versions for a fixed computable function f are given in Examples 5.1.12.2 and 3, 5.1.19.2 and 5.2.11. We prove versions which are effective also in the function f.

Theorem 6.2.4. Consider the representations $\psi^n, \psi^n_<, \psi^n_>, \theta^n, \theta^n_<, \kappa^n, \kappa^n_>$ from Sects. 5.1 and 5.2, where n indicates the dimension.
1. The operation $(f, U) \mapsto f^{-1}[U]$ for $f \in C(\mathbb{R}^m, \mathbb{R}^n)$ and open $U \subseteq \mathbb{R}^n$ is $(\delta^{mn}_\rightarrow, \theta^n_<, \theta^m_<)$-computable.
2. The operation $(f, A) \mapsto f^{-1}[A]$ for $f \in C(\mathbb{R}^m, \mathbb{R}^n)$ and closed $A \subseteq \mathbb{R}^n$ is $(\delta^{mn}_\rightarrow, \psi^n_>, \psi^m_>)$-computable.
3. The operation $(f, A) \mapsto \text{cls}(f[A])$ for $f \in C(\mathbb{R}^m, \mathbb{R}^n)$ and closed $A \subseteq \mathbb{R}^m$ is $(\delta^{mn}_\rightarrow, \psi^m_<, \psi^n_<)$-computable.
4. The operation $(f, K) \mapsto f[K]$ for $f \in C(\mathbb{R}^m, \mathbb{R}^n)$ and compact $K \subseteq \mathbb{R}^m$ is $(\delta^{mn}_\rightarrow, \kappa^m_>, \kappa^n_>)$-computable and $(\delta^{mn}_\rightarrow, \kappa^m, \kappa^n)$-computable.

Proof: For simplifying the notations we formulate the proof for the case $m = n = 1$. The generalization is straightforward.

1. By Corollary 5.1.16 and Lemma 6.1.7 it suffices to show that the operation is $(\delta^\mathbb{R}_{oo}, \theta^{en}_<, \theta^{en}_<)$-computable. If $f = \delta^\mathbb{R}_{oo}(p)$, then p is a list of pairs $(J, L) \in (\text{Cb}^{(1)})^2$ such that $f[J] \subseteq L$ and for every x and n there is some pair (J, L) listed by p such that $x \in J$ and the diameter of L is less than 2^{-n}. If $U = \theta^{en}_<(q)$, then q is a list of cubes $L_i \in \text{Cb}^{(1)}$, $i \in \mathbb{N}$, such that $U = \bigcup_{i \in \mathbb{N}} L_i$. Therefore, $f(x) \in U$, iff there are a pair (J, L) listed by p and a cube L' listed by q such that $x \in J$ and $L \subseteq L'$. We conclude that $f^{-1}[U]$ is the union of all open cubes J for which there are cubes L, L' such that (J, L) is listed by p, L' is listed by q and $L \subseteq L'$. There is a Type-2 machine M which on input (p, q) ($p \in \text{dom}(\delta^\mathbb{R}_{oo})$, $q \in \text{dom}(\theta^{en}_<)$) lists all open cubes $J \in \text{Cb}^{(1)}$ for which there are cubes $L, L' \in \text{Cb}^{(1)}$ such that (J, L) is listed by p, L' is listed by q and $L \subseteq L'$. If $f = \delta^\mathbb{R}_{oo}(p)$ and $U = \theta^{en}_<(q)$, then $f^{-1}[U] = \theta^{en}_< \circ f_M(p, q)$. Therefore, $(f, U) \mapsto f^{-1}[U]$ is $(\delta^\mathbb{R}_{oo}, \theta^{en}_<, \theta^{en}_<)$-computable.

2. Let g be a realizing computable function from 1 above. For $f = \delta^\mathbb{R}_\rightarrow(p)$ we obtain

$$f^{-1}[\psi_>(q)] = \mathbb{R} \setminus f^{-1}[\mathbb{R} \setminus \psi_>(q)] = \mathbb{R} \setminus f^{-1}[\theta_<(q)]$$
$$= \mathbb{R} \setminus \theta_< \circ g(p, q) \qquad = \psi_> g(p, q) ,$$

and so $(f, A) \mapsto f^{-1}[A]$ is $(\delta^\mathbb{R}_\rightarrow, \theta_<, \theta_<)$-computable.

3. It suffices to show that the operation is $(\delta^\mathbb{R}_\rightarrow, \psi^{en}_<, \psi^{en}_<)$-computable (Lemma 5.1.10). If $D \subseteq A \subseteq \mathbb{R}$ is dense in A and $f : \mathbb{R} \to \mathbb{R}$ is continuous, then $f[D]$ is dense in $f[A]$ and also in $\text{cls}(f[A])$. By the utm-theorem and the smn-theorem for $\eta^{*\omega}$ and $\eta^{\omega\omega}$ there is a computable function $r : \Sigma^\omega \times \Sigma^\omega \to \Sigma^\omega$ such that $\eta^{\omega\omega}_q \circ \eta^{*\omega}_p(w) = \eta^{*\omega}_{r(q,p)}(w)$. There is a computable function $g :\subseteq \Sigma^\omega \times \Sigma^\omega \to \Sigma^\omega$ such that $g(q, 0^\omega) = 0^\omega$ and $g(q, 0^k 1p) = 0^k 1 r(q, p)$. Then the function g is a $(\delta^\mathbb{R}_\rightarrow, \psi^{en}_<, \psi^{en}_<)$-realization of $(f, A) \mapsto \text{cls}(f[A])$.

4. We modify the proof from Example 5.2.11. First we show that the operation is $(\delta_{oo}^{\mathbb{R}}, \kappa_c, \kappa_c)$-computable and so $(\delta_{\to}^{\mathbb{R}}, \kappa_>, \kappa_>)$-computable by Lemmas 5.2.5 and 6.1.7.

Suppose $f = \delta_{oo}^{\mathbb{R}}(q)$ and $K = \kappa_c(p)$. In order to leave the notation from Example 5.2.11 unchanged we may assume that q is a list of coded pairs $\iota(u, v)$ such that $f[I^1(u)] \subseteq I^1(v)$ plus the other properties from Definition 6.1.6. For any $w \in \Sigma^*$,

$$
\begin{aligned}
& f[K] \subseteq \bigcup \{I^1(v) \mid \iota(v) \lhd w\} \\
\Longleftrightarrow\ & (\forall x \in K)(\exists v)\Big(f(x) \in I^1(v), \iota(v) \lhd w\Big) \\
\Longleftrightarrow\ & (\forall x \in K)(\exists u, t, v)\Big(x \in I^1(u), \iota(u,t) \lhd q, I^1(t) \subseteq I^1(v), \iota(v) \lhd w\Big) .
\end{aligned}
$$

Since K is compact, finitely many open intervals $I^1(u)$ suffice to cover K, and so the last property is equivalent to

$$
(\exists w', \iota(w') \lhd p)(\forall u, \iota(u) \lhd w')(\exists t, v)\Big(\iota(u,t) \lhd q, I^1(t) \subseteq I^1(v), \iota(v) \lhd w\Big) .
$$

There is a Type-2 machine M which on input $(q, p) \in \text{dom}(\delta_{oo}^{\mathbb{R}}) \times \text{dom}(\kappa_c)$ prints a list of all words $\iota(w)$ such that $f[K] \subseteq \bigcup \{I^1(v) \mid \iota(v) \lhd w\}$. Therefore, f_M is a $(\delta_{oo}^{\mathbb{R}}, \kappa_c, \kappa_c)$-computable realization of the operation $(f, K) \mapsto f[K]$ for $f \in C(\mathbb{R})$.

We show that the operation is $(\delta_{\to}^{\mathbb{R}}, \kappa, \kappa)$-computable. Since the operation is $(\delta_{\to}^{\mathbb{R}}, \kappa_>, \kappa_>)$-computable, it is $(\delta_{\to}^{\mathbb{R}}, \kappa, \kappa_>)$-computable. By Theorem 6.2.4.3, the operation is also $(\delta_{\to}^{\mathbb{R}}, \psi_<, \psi_<)$-computable. Since $\kappa \leq \psi_<$ and $f[K]$ are compact, hence closed, the operation is $(\delta_{\to}^{\mathbb{R}}, \kappa, \psi_<)$-computable. Combining the two results, we obtain that the operation is $(\delta_{\to}^{\mathbb{R}}, \kappa, \kappa)$-computable, since $\kappa_< \wedge \psi_> \leq \kappa$ by Lemma 5.2.2. $\qquad \square$

Every continuous real-valued function f has a maximum value on every compact subset K of its domain. It can be computed from f and K:

Corollary 6.2.5 (maximum).   The maximum operation
$(f, K) \mapsto \max(f[K])$ for $f \in C(\mathbb{R}^n)$ and compact $K \subseteq \mathbb{R}^n$, $K \neq \emptyset$, is $(\delta_{\to}^{\mathbb{R}}, \kappa, \rho)$-computable.

Proof: Combine Theorem 6.2.4.4 and Lemma 5.2.6. $\qquad \square$

By Example 6.1.8.3 the multi-valued operator mapping a real function to its modulus of continuity at 0 is computable. We study the modulus of continuity in more generality.

Definition 6.2.6 (modulus of continuity). *Let $f :\subseteq \mathbb{R} \to \mathbb{R}$ be a real function.*

1. *A modulus of continuity of the function f at $x \in \mathrm{dom}(f)$ is any function $m : \mathbb{N} \to \mathbb{N}$ such that*

$$|x - y| \leq 2^{-m(n)} \implies |f(x) - f(y)| \leq 2^{-n}, \; \text{if } y \in \mathrm{dom}(f) \text{ and } n \in \mathbb{N} \, .$$

2. *A modulus of continuity of the function f on $A \subseteq \mathrm{dom}(f)$ is any function $m : \mathbb{N} \to \mathbb{N}$ such that*

$$|x - y| \leq 2^{-m(n)} \implies |f(x) - f(y)| \leq 2^{-n}, \; \text{if } x, y \in A \, .$$

By the (ε, δ)-definition of continuity, a real function is continuous, iff it has a modulus of continuity at each point x of its domain. Similarly, a real function is uniformly continuous on a subset A of its domain, iff it has a modulus of continuity on A. A continuous function is uniformly continuous on each compact subset of its domain. We show that a modulus of continuity can be determined from $f \in C(\mathbb{R})$ and $K \in \mathcal{K}$. Remember the standard representation $\delta_{\mathbb{B}}$ of the set $\mathbb{B} = \{m \mid m : \mathbb{N} \to \mathbb{N}\}$ (Definition 3.1.2).

Theorem 6.2.7 (modulus of continuity). The multi-valued function $F : C(\mathbb{R}) \times \mathcal{K} \rightrightarrows \mathbb{B}$ defined by its graph

$$R_F := \{(f, K, m) \mid m \text{ is a modulus of continuity of } f \text{ on } K\}$$

is $(\delta_{\to}^{\mathbb{R}}, \kappa_>, \delta_{\mathbb{B}})$-computable.

Proof: We show that F is $(\delta_{oo}^{\mathbb{R}}, \kappa_c, \delta_{\mathbb{B}})$-computable (Lemmas 5.2.5 and 6.1.7). For any open ball $B(x, r)$ define $3 \cdot B(x, r) := B(x, 3 \cdot r)$ and $\mathrm{diam}(B(x, r)) := 2 \cdot r$.

Assume $f = \delta_{oo}^{\mathbb{R}}(p)$, $K = \kappa_c(q)$ and $k \in \mathbb{N}$. We may assume that p is a list of coded pairs $\iota(u, v)$ such that $f[\mathrm{I}^1(u)] \subseteq \mathrm{I}^1(v)$ plus the other properties from Definition 6.1.6. Then

$$(\forall x \in K)(\exists u, v)\Big(x \in \mathrm{I}^1(u), \iota(u, v) \lhd p, \mathrm{diam}(\mathrm{I}^1(v)) \leq 2^{-k}\Big).$$

Since $\mathrm{I}^1(u)$ is open,
$$(\forall x \in K)(\exists t, u, v)\Big(x \in \mathrm{I}^1(t), 3 \cdot \mathrm{I}^1(t) \subseteq \mathrm{I}^1(u), \iota(u, v) \lhd p, \mathrm{diam}(\mathrm{I}^1(v)) \leq 2^{-k}\Big).$$

Since K is compact, finitely many cubes $\mathrm{I}^1(t)$ suffice to cover K. Since q is a list of all finite covers of K with rational cubes, there is some word w with $\iota(w) \lhd q$ such that

$$(\forall t, \iota(t) \lhd w)(\exists u, v)\Big(3 \cdot \mathrm{I}^1(t) \subseteq \mathrm{I}^1(u), \iota(u, v) \lhd p, \mathrm{diam}(\mathrm{I}^1(v)) \leq 2^{-k}\Big). \quad (6.1)$$

Let n_k be a natural number such that $2^{-n_k} < \min\{\mathrm{diam}(\mathrm{I}^1(t)) \mid \iota(t) \lhd w\}$ (where $\min(\emptyset) := \infty$). If $x, y \in K$ and $|x - y| \leq 2^{-n_k}$, then $x, y \in 3 \cdot \mathrm{I}^1(t)$ for some t with $\iota(t) \lhd w$ and so $|f(x) - f(y)| \leq 2^{-k}$.

There is a Type-2 machine M which on input (p, q), $p \in \mathrm{dom}(\delta_{\mathrm{oo}}^{\mathbb{R}})$ and $q \in \mathrm{dom}(\kappa_{\mathrm{c}})$, operates in stages $k = 0, 1, \ldots$ as follows.

Stage k: M searches for some word $\iota(w) \lhd p$ such that (6.1). Then it determines some number n_k such that $2^{-n_k} < \min\{\mathrm{diam}(\mathrm{I}^1(t)) \mid \iota(t) \lhd w\}$ and prints the word $0^{n_k}1$. The output $0^{n_0}10^{n_1}1\ldots \in \Sigma^\omega$ is the $\delta_{\mathbb{B}}$-name of a modulus of continuity of the function f on the compact set K. Therefore, f_M is a $(\delta_{\mathrm{oo}}^{\mathbb{R}}, \kappa_{\mathrm{c}}, \delta_{\mathbb{B}})$-realization of the function F. $\square$

The above multi-valued computable function has no $(\delta_{\rightarrow}^{\mathbb{R}}, \kappa_>, \delta_{\mathbb{B}})$-continuous choice function (Exercise 6.2.8, cf. Example 6.1.8.3).

Corollary 6.2.8 (modulus of continuity). The multi-valued function $G : \mathrm{C}(\mathbb{R}) \times \mathbb{R} \rightrightarrows \mathbb{B}$ defined by its graph

$$\mathrm{R}_G := \{(f, x, m) \mid m \text{ is a modulus of continuity of } f \text{ at } x\}$$

is $(\delta_{\rightarrow}^{\mathbb{R}}, \rho, \delta_{\mathbb{B}})$-computable.

Proof: The function $H : \mathbb{R} \rightarrow \mathcal{K}$, $x \mapsto [x - 2; x + 2]$, is $(\rho, \kappa_>)$-computable. Let $\overline{h}$ be a (ρ, κ)-realization of H and let $\overline{f}$ be a $(\delta_{\rightarrow}^{\mathbb{R}}, \kappa_>, \delta_{\mathbb{B}})$-realization of the function F from Theorem 6.2.7. Define $\overline{g}(p, q) := \overline{f}(p, \overline{h}(q))$. Then $\overline{g}$ is a $(\delta_{\rightarrow}^{\mathbb{R}}, \rho, \delta_{\mathbb{B}})$-realization of G. $\square$

In particular, a computable function has a computable modulus of continuity on each $\kappa_>$-computable subset K and at each computable point of its domain. Also the above multi-valued computable function G has no continuous choice function (Exercise 6.2.8). However, there is a $(\delta_{\rightarrow}^{\mathbb{R}}, \rho, \rho, \rho)$-computable function H such that

$$|f(x) - f(y)| < \varepsilon \quad \text{if} \quad |x - y| < H(f, x, \varepsilon)$$

(see [WZ97]). Theorem 6.2.7 and its corollary can be generalized from $\mathrm{C}(\mathbb{R})$ to $\mathrm{C}(\mathbb{R}^m, \mathbb{R}^n)$.

By Example 5.1.12.2, a closed set $A \subseteq \mathbb{R}^n$ is co-r.e. , iff $A = f^{-1}[\{0\}]$ for some computable function $f : \mathbb{R}^n \rightarrow \mathbb{R}$. We show that A can be computed from f and vice versa.

Theorem 6.2.9 (set of zeroes).
1. The function $f \mapsto f^{-1}[\{0\}]$ for $f \in \mathrm{C}(\mathbb{R}^n)$ is $(\delta_{\rightarrow}^{n1}, \psi_>)$-computable.
2. The multi-valued function $H : \mathcal{A} \rightrightarrows \mathrm{C}(\mathbb{R}^n)$ with graph

$$\mathrm{R}_H := \{(A, f) \mid A = f^{-1}[\{0\}]\}$$

is $(\psi_>, \delta_{\rightarrow}^{n1})$-computable.

Proof: 1. Since the set $\{0\}$ is $\psi^1_>$-computable, this follows from Theorem 6.2.4.2.

2. We show that the function H is $(\psi^{en}_>, \delta^{n1}_\rightarrow)$-computable (Lemma 5.1.10). Suppose $A = \psi^{en}_>(p)$. Then $\mathbb{R}^n \setminus A = \bigcup \{I^n(w) \mid \iota(w) \lhd p\}$. For each open cube $I^n(w) = B(a, r)$ define the "pyramid" function $f_w \in C(\mathbb{R}^n)$ by

$$f_w(x) := \max \left(0, \frac{r - d(a, x)}{r} \right).$$

Obviously, f_w is 0 outside $I^n(w)$ and > 0 inside $I^n(w)$. Fig. 6.4 shows a pyramid function on the real line.

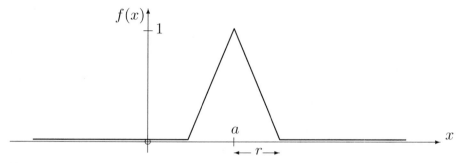

Fig. 6.4. The pyramid function over the interval $(a - r; a + r) \subseteq \mathbb{R}$

We define the function f as an appropriate linear combination of the pyramid functions f_w such that $\iota(w) \lhd p$: $f := \sum_{i \in \mathbb{N}} h_i$,

$$h_i(x) := \begin{cases} 0 & \text{if } p_{<i} \text{ has no suffix } \iota(w) \\ 2^{-i-1} \cdot f_w(x) & \text{if } \iota(w) \text{ is a suffix of } p_{<i}. \end{cases}$$

Then, $x \notin \psi^{en}_>(p) \iff (\exists w)(\iota(w) \lhd p \text{ and } x \in I^n(w)) \iff f(x) > 0$, and so $\psi^{en}_>(p) = f^{-1}[\{0\}]$. We show that $p \mapsto f$ is $(\text{id}_{\Sigma^\omega}, \delta^{n1}_\rightarrow)$-computable. The function h defined by $h(p, i, x) := h_i(x)$ is $(\text{id}_{\Sigma^\omega}, \nu_\mathbb{N}, \rho^n, \rho)$-computable. By Theorem 3.3.15, $(p, i) \mapsto h_i$ is $(\text{id}_{\Sigma^\omega}, \nu_\mathbb{N}, [\rho^n \rightarrow \rho]_{\mathbb{R}^n})$-computable, hence $(\text{id}_{\Sigma^\omega}, \nu_\mathbb{N}, \delta^{n1}_\rightarrow)$-computable, and $p \mapsto (h_i)_{i \in \mathbb{N}}$ is $(\text{id}_{\Sigma^\omega}, [\nu_\mathbb{N} \rightarrow \delta^{n1}_\rightarrow]_\mathbb{N})$-computable, hence $(\text{id}_{\Sigma^\omega}, [\delta^{n1}_\rightarrow]^\omega)$-computable by Lemma 3.3.16. Finally, by both parts of Theorem 6.2.2, $p \mapsto f = \sum_{i \in \mathbb{N}} h_i$ is $(\text{id}_{\Sigma^\omega}, \delta^{n1}_\rightarrow)$-computable. Any $(\text{id}_{\Sigma^\omega}, \delta^{n1}_\rightarrow)$-realization of this function is a $(\psi^{en}_<, \delta^{n1}_\rightarrow)$-realization of the multi-valued function $H : A \rightrightarrows C(\mathbb{R}^n)$ which, therefore, is $(\psi^{en}_<, \delta^{n1}_\rightarrow)$-computable. $\square$

The multi-valued function H mapping each closed set A to a continuous function f such that A is the set of zeroes of f, however, has no $(\psi^{en}_>, \delta^{n1}_\rightarrow)$-computable choice function (Exercise 6.2.9). We use Theorem 6.2.9 for proving a computable version of Urysohn's lemma ([MTY97, Eng89, Bra97, Sch98]). See Figure 6.5.

Theorem 6.2.10 (computable Urysohn lemma).
1. There is a $(\psi^n, \psi^n, \delta_{\rightarrow}^{n1})$-computable function $G :\subseteq \mathcal{A} \times \mathcal{A} \rightarrow C(\mathbb{R}^n)$ mapping every disjoint pair $A, B \subseteq \mathbb{R}^n$ of non-empty closed sets to a continuous function $f : \mathbb{R}^n \rightarrow \mathbb{R}$ such that $f(x) = 1$ for $x \in A$, $f(x) = 0$ for $x \in B$ and $0 < f(x) < 1$, otherwise.
2. The multi-valued function $F :\subseteq \mathcal{A} \times \mathcal{A} \rightrightarrows C(\mathbb{R}^n)$ mapping every disjoint pair $A, B \subseteq \mathbb{R}^n$ of closed sets to continuous functions $f : \mathbb{R}^n \rightarrow \mathbb{R}$ such that $f(x) = 1$ for $x \in A$, $f(x) = 0$ for $x \in B$ and $0 < f(x) < 1$, otherwise, is $(\psi_>^n, \psi_>^n, \delta_{\rightarrow}^{n1})$-computable.

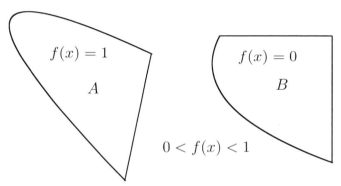

Fig. 6.5. An Urysohn function $f : \mathbb{R}^2 \rightarrow \mathbb{R}$ for the closed sets $A, B \subseteq \mathbb{R}^2$

Proof: 2. If $f_A, f_B : \mathbb{R}^n \rightarrow \mathbb{R}$ are continuous functions such that $A = f_A^{-1}[\{0\}]$ and $B = f_B^{-1}[\{0\}]$, then f with

$$f(x) := \frac{|f_B(x)|}{|f_A(x)| + |f_B(x)|}$$

has the desired property. By Theorem 6.2.9, $\delta_{\rightarrow}^{n1}$-names of suitable functions f_A and f_B from $\psi_>^n$-names of A and B, respectively, can be computed. From these names a $\delta_{\rightarrow}^{n1}$-name of the function f can be computed by Theorem 6.2.1.2 and 6.2.1.3.

1. As Exercise 6.2.10; use the distance functions d_A and d_B for f_A and f_B, respectively, in the above proof (see Definition 5.1.6 and Lemma 5.1.7). □

Many results of this section have more general versions with the representations $[\rho^m \rightarrow \rho^n]_{\text{set}}$, δ_1, δ_2 or δ_3 from Exercise 6.1.11 substituted for $\delta_{\rightarrow}^A$.

Exercises 6.2.

◇ 1. Complete the proof of Theorem 6.2.1.
2. Show that the operator

$$((f_i)_{i \in \mathbb{N}}, n) \mapsto \sum_{i=0}^{n} f_i$$

for $f_i \in C(\mathbb{R})$ is $([\delta_\rightarrow^\mathbb{R}]^\omega, \nu_\mathbb{N}, \delta_\rightarrow^\mathbb{R})$-computable.
3. Show that the operator

$$(f_i)_{i \in \mathbb{N}} \mapsto \sum_{i=0}^{\infty} \frac{f_i}{1 + f_i} \cdot 2^{-i}, \quad \text{where} \quad \frac{f_i}{1 + f_i}(x) := \frac{f_i(x)}{1 + f_i(x)}$$

for $f_i \in C(\mathbb{R})$ is $([\delta_\rightarrow^\mathbb{R}]^\omega, \delta_\rightarrow^\mathbb{R})$-computable.
4. Let $(f_i)_{i \in \mathbb{N}}$ be a $(\nu_\mathbb{N}, \delta_\rightarrow^{[0;1]})$-computable sequence of functions in $C[0; 1]$ such that $d(f_i, f_k) \leq 2^{-i}$ for $i < k$. Show that $\lim_{i \to \infty} f_i$ is computable.
5. For the set PR^{mn} of *all* partial continuous real functions $f :\subseteq \mathbb{R}^m \to \mathbb{R}^n$ consider the set-valued representation $[\rho^m \to \rho^n]_{\mathrm{set}}$ (Remark 3.3.17.3). Extend the operations from Theorem 6.2.1, Parts 2 - 5, to all partial continuous real functions with the given dimensions. Show that these extensions are computable w.r.t. the set-valued representations $[\rho^m \to \rho^n]_{\mathrm{set}}$.
6. Show that the function $(f, x) \mapsto \max\{f(x, y) \mid 0 \leq y \leq 1\}$ for $f \in C(\mathbb{R}^2)$ and $x \in \mathbb{R}$ is $(\delta_\rightarrow^{21}, \rho, \rho)$-computable.
7. Show that the function $(f, x) \mapsto \max\{f(y) \mid 0 \leq y \leq x\}$ for $f \in C(\mathbb{R})$ and $x \in \mathbb{R}$ (with value 0 for $x < 0$) is $(\delta_\rightarrow^{11}, \rho, \delta_\rightarrow^{11})$-computable.
8. Show that the multi-valued functions F and G from Theorem 6.2.7 and Corollary 6.2.8 have no continuous choice functions. (Apply Corollary 4.3.16, see Example 6.1.8.3)
9.◆ a) Show that there is no $(\psi_>^1, \delta_\rightarrow^{11})$-continuous single-valued function G such that A is the set of zeroes of $G(A)$ for each closed set $A \subseteq \mathbb{R}$.
   b) Show that there is a $(\psi^n, \delta_\rightarrow^{n1})$-computable single-valued function G such that A is the set of zeroes of $G(A)$ for each non-empty closed set $A \subseteq \mathbb{R}^n$.
   ◆ c) Show that the function $f \mapsto f^{-1}[\{0\}]$ is not $(\delta_\rightarrow^{11}, \psi_<^1)$-continuous.
10. Prove Theorem 6.2.10.1.
11. Prove generalizations of the statements from Theorem 6.2.4 for $[\rho^m \to \rho^n]_{\mathrm{set}}$ (Remark 3.3.17.3) substituted for $\delta_\rightarrow^{mn}$.
◆12. Show that there is a function $g : [0; 1] \to \mathbb{R}$ which has a computable modulus of continuity at every $x \in [0; 1]$ but has no computable modulus of continuity on $[0; 1]$. (Hint: consider the function f' from Example 6.4.9.) Of course, g cannot be computable.
◆13. Show that the Urysohn function F from Theorem 6.2.10.2 has no $(\psi_>^n, \psi_>^n, \delta_\rightarrow^{n1})$-continuous choice function.

6.3 Zero-Finding

A zero of a function $f :\subseteq X \to \mathbb{R}$ or $f :\subseteq X \to \mathbb{C}$ is an element $x \in X$ such that $f(x) = 0$. In this section we will study the problem of computing zeroes of continuous real functions $f :\subseteq \mathbb{R} \to \mathbb{R}$. It turns out that the simple classical existence statement "the function f has a zero" gives rise to a variety of questions about computability.

- Is every zero of a computable function a computable real number?
- Can we decide whether f has a zero?
- Can we compute the set of zeroes of f?
- Can we compute zeroes approximately, that is, can we compute for every continuous function f (which has a zero) and every number $n \in \mathbb{N}$ some real number x_n such that $|f(x_n)| \le 2^{-n}$?
- Under which conditions can we compute a zero of f by a multi-valued function?
- Under which conditions can we compute a zero of f by a single-valued function?

We will discuss these questions in detail applying concepts introduced in previous chapters. We start with an example.

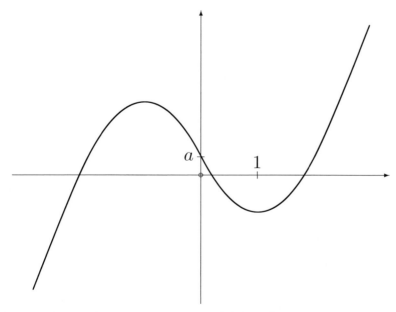

Fig. 6.6. Zeroes of the cubic polynomial $f_a(x) = x^3 - 3 \cdot x + a$

Example 6.3.1. Let $C := \{f_a \mid a \in \mathbb{R}\} \subseteq C(\mathbb{R})$ be the set of polynomials $f_a(x) := x^3 - 3 \cdot x + a$ of degree 3. Each of the functions f_a has one, two or three zeroes (Fig. 6.6). We will consider effectivity on C induced by the representation $\delta_\to^\mathbb{R} = [\rho \to \rho]_\mathbb{R}$ which has turned out to be very natural (Sects. 6.1 and 6.2, particularly Lemma 6.1.2). Which information about zeroes of a function $f \in C$ can be computed from a $\delta_\to^\mathbb{R}$-name of the function f ? Since the function $h : a \mapsto f_a$ is $(\rho, \delta_\to^\mathbb{R})$-computable and its inverse is $(\delta_\to^\mathbb{R}, \rho)$-computable (Exercise 6.1.10), it suffices to investigate which information about zeroes of a function $f_a \in C$ can be computed from $a \in \mathbb{R}$.

First we show that no continuous real function $g : \mathbb{R} \to \mathbb{R}$ maps every real number a to a zero of f_a. Suppose that such a function g exists. Observe that f_0 has its relative maximum at $(-1, 2)$ and its relative minimum at $(1, -2)$. Hence for $|a| > 2$, f_a has only a single zero (Fig. 6.6). Since $f_{-65/8}(5/2) = 0$ and $|-65/8| > 2$, $5/2$ is the only zero of $f_{-65/8}$, and so $g(-65/8) = 5/2$, hence $5/2 \in g[(-10; 2)]$. Since $f_2(1) = 0$, $1 \notin g[(-10; 2)]$. Since g is continuous, it must map an interval onto an interval, hence $g[(-10; 2)]$ is an interval containing $5/2$ but not 1, hence $0 \notin g[(-10; 2)]$. By symmetry, $g[(-2; 10)]$ is an interval containing $-5/2$ but not 0. Since $g[(-10; 10)] = g[(-10; 2)] \cup g[(-2; 10)]$, by continuity of g, $g[(-10; 10)]$ must be an interval containing $-5/2$ and $5/2$ but not 0. This is impossible. Therefore, no (ρ, ρ)-continuous function $g : \mathbb{R} \to \mathbb{R}$ maps every real number a to a zero of f_a.

However, the multi-valued function

$$F : \mathbb{R} \rightrightarrows \mathbb{R}, \quad R_F := \{(a, y) \mid f_a(y) = 0\} \,,$$

is (ρ, ρ)-computable. We will prove a more general result below.

Furthermore, the function $Z_{\min} : a \mapsto \min \circ f_a^{-1}[\{0\}]$ computing the minimal zero of f_a is $(\rho, \rho_<)$-computable. We sketch a proof. First of all there is a multi-valued $(\rho, \nu_\mathbb{Q})$-computable function which for any $a \in \mathbb{R}$ determines some rational lower bound $r \in \mathbb{Q}$ of $Z_{\min}(a)$ (Exercise 6.3.1). Let $r \in \mathbb{Q}$ be a number less than $Z_{\min}(a)$. Since $f_a(r) \neq 0$, for any $s \in \mathbb{Q}$,

$$s < Z_{\min}(a) \iff s < r \text{ or } 0 \notin f_a[[r; s]] \,.$$

By Theorem 6.2.4.4, for each $s \geq r$ and $a \in \mathbb{R}$ the compact interval $f_a[[r; s]]$ can be computed and $0 \notin f_a[[r; s]]$ can be assured, iff it is true. Therefore, from $a \in \mathbb{R}$ and some rational $r < Z_{\min}(a)$ a list of all $s \in \mathbb{Q}$ with $s < Z_{\min}(a)$ can computed and so $Z_{\min}$ is $(\rho, \rho_<)$-computable.

The above method can be used to determine a $\rho_<$-name and a $\rho_>$-name and hence a ρ-name of y from $a \in \mathbb{R}$ and $r, s \in \mathbb{Q}$, if y is the unique zero of f_a in the interval $(r; s)$.

By Theorem 6.2.4.2 the function $a \mapsto f_a^{-1}[\{0\}]$ is $(\rho, \psi_>)$-computable. But the function $a \mapsto \text{card}(f_a^{-1}[\{0\}])$ determining the number of zeroes of the function f_a cannot be $(\rho, \nu_\mathbb{N})$-computable by Lemma 4.3.15. $\quad\square$

In the following we will state and prove our theorems for the space C[0; 1]. More general cases can be treated similarly. By our first theorem there is no method for determining zeroes for arbitrary functions from C[0; 1].

Theorem 6.3.2. The multi-valued function

$$Z :\subseteq C[0;1] \rightrightarrows \mathbb{R} , \quad R_Z := \{(f,x) \mid f(x) = 0\} ,$$

is not $(\delta_{\rightarrow}^{[0;1]}, \rho)$-continuous.

Proof: For $a \in \mathbb{R}$ let $g_a \in C[0;1]$ be the polygon function (see Fig. 6.7) defined by the vertices

$$(0,-1), \ (1/3,a), \ (2/3,a), \ \text{and} \ (1,1) .$$

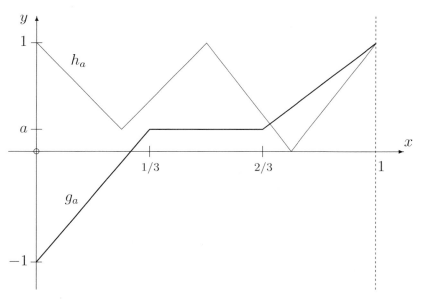

Fig. 6.7. The polygon functions g_a from the proof of Theorem 6.3.2 and h_a from Exercise 6.3.3

First, we show that the multi-valued function

$$H : \mathbb{R} \rightrightarrows \mathbb{R}, \quad R_H := \{(a,x) \mid g_a(x) = 0\} ,$$

is not (ρ, ρ)-continuous. Suppose, there is a continuous function $f :\subseteq \Sigma^\omega \to \Sigma^\omega$ mapping any ρ-name p of a real number a to a ρ-name $f(p)$ of a zero of g_a, that is, $\rho \circ f(p)$ is a zero of $g_{\rho(p)}$ for all $p \in \text{dom}(\rho)$. By Lemma 4.1.6 we may assume for the moment that ρ is the Cauchy representation from Definition 4.1.5. Let $w \in \mathrm{dom}(\nu_\mathbb{Q})$ such that $\nu_\mathbb{Q}(w) = 0$. Then $\rho(q) = 0$ for $q := \iota(w)\iota(w)\ldots$ and $y := \rho \circ f(q)$ is a zero of g_0, and so $1/3 \leq y \leq 2/3$.

Assume $y > 1/3$. Since ρ is admissible with final topology $\tau_\mathbb{R}$ (Lemma 4.1.4), it is $(\tau_C, \tau_\mathbb{R})$-continuous by Lemma 3.2.5, and so $\rho \circ f$ is $(\tau_C, \tau_\mathbb{R})$-continuous. Since $y > 1/3$, $y \in (1/3; 1)$. By continuity of $\rho \circ f$, $\rho \circ f[(\iota(w))^m \Sigma^\omega] \subseteq (1/3; 1)$ for some $m \in \mathbb{N}$. Let $w_m \in \mathrm{dom}(\nu_\mathbb{Q})$ such that $\nu_\mathbb{Q}(w_m) = 2^{-m}$ and $p_m := (\iota(w))^m \iota(w_m)\iota(w_m)\ldots$. Then $\rho(p_m) = 2^{-m} > 0$ and so $\rho \circ f(p_m) < 1/3$ by assumption on f. But on the other hand, $\rho \circ f(p_m) \in \rho \circ f[(\iota(w))^m \Sigma^\omega] \subseteq (1/3; 1)$ (contradiction).

The case $y < 2/3$ can be handled accordingly. Therefore, the function H is not (ρ, ρ)-continuous.

The function $h : a \mapsto g_a$ is $(\rho, \delta_{\to}^{[0;1]})$-computable (Exercise 6.3.2). Therefore, if Z is $(\delta_{\to}^{[0;1]}, \rho)$-continuous, then H is (ρ, ρ)-continuous, which has been disproved above.   □

Notice that we have proved the stronger result: The multi-valued function $Y :\subseteq \mathrm{C}[0;1] \rightrightarrows \mathbb{R}$, $\mathrm{R}_Y := \{(g_a, x) \mid a \in \mathbb{R},\ g_a(x) = 0\}$ is not $(\delta_{\to}^{[0;1]}, \rho)$-continuous.

Although there is no method for finding from $f \in \mathrm{C}[0;1]$ a (ρ-name of some) number y such that $f(y) = 0$, there is a method for finding numbers y such that $f(y)$ is as tight as we please to 0.

Theorem 6.3.3 (approximate zero). The multi-valued function

$$Z_{\mathrm{app}} :\subseteq \mathrm{C}[0;1] \times \mathbb{Q} \rightrightarrows \mathbb{Q}, \quad \mathrm{R}_{Z_{\mathrm{app}}} := \{(f, a, y) \mid |f(y)| < a\},$$

is $(\delta_{\to}^{[0;1]}, \nu_\mathbb{Q}, \nu_\mathbb{Q})$-computable.

Proof: By Lemma 6.1.7 it suffices to show that Z_{app} is $(\delta_{\mathrm{co}}^{[0;1]}, \nu_\mathbb{Q}, \nu_\mathbb{Q})$-computable. Suppose $|f(y)| < a$ for some $y \in [0;1]$. Then by continuity of f there is some open interval $K \in \mathrm{Cb}^{(1)}$ such that $K \cap [0;1] \neq \emptyset$ and $|f(y)| < a$ for all $y \in \overline{K} \cap [0;1]$. Therefore, $|f(y)| < a$ for some $y \in [0;1]$, iff there is some $K \in \mathrm{Cb}^{(1)}$ such that $K \cap [0;1] \neq \emptyset$ and $f[\overline{K}] \subseteq (-a; a)$. If $\delta_{\mathrm{co}}^{[0;1]}(p) = f$, then p is a (coded) list of all pairs (J, L) of open intervals $J, L \in \mathrm{Cb}^{(1)}$ such that $J \cap [0;1] \neq \emptyset$ and $f[\overline{J}] \subseteq L$ (Definition 6.1.3). Let M be a Type-2 machine which on inputs $p \in \mathrm{dom}(\delta_{\mathrm{co}})$ and $w \in \mathrm{dom}(\nu_\mathbb{Q})$ searches in p for a pair $(K, L) \in \mathrm{Cb}^{(1)} \times \mathrm{Cb}^{(1)}$ such that $L = (-a; a)$ ($a := \nu_\mathbb{Q}(w)$) and prints (a $\nu_\mathbb{Q}$-name of) the left endpoint of K. Obviously, the function f_M is a $(\delta_{\mathrm{co}}^{[0;1]}, \nu_\mathbb{Q}, \nu_\mathbb{Q})$-realization of Z_{app}.   □

The above function Z_{app} and its real-valued modification do not have continuous choice functions (Exercise 6.3.4). While a ρ-name of a zero of f cannot be computed in general, a $\rho_<$-name of the minimal zero can be computed.

Theorem 6.3.4. The function $Z_{\min} : f \mapsto \min \circ f^{-1}[\{0\}]$ is $(\delta_\to^{[0;1]}, \rho_<)$-computable.

Proof: We outline an algorithm for computing a $\rho_<$-name of $Z_{\min}(f)$ from a $\delta_\to^{[0;1]}$-name of $f \in C[0;1]$. Compute a $\delta_\to^\mathbb{R}$-name of the extension $f' \in C(\mathbb{R})$ of f defined by $f'(x) := f(0)$, if $x < 0$, and $f'(x) := f(1)$, if $x > 1$. Determine a $\psi_>$-name of $A := f'^{-1}[\{0\}]$ (Theorem 6.2.4.2). Determine a $\psi_>$-name of $A \cap [0;1] = f^{-1}[\{0\}]$ (Theorem 5.1.13). Determine a $\kappa_>$-name of $f^{-1}[\{0\}]$ (Definition 5.2.1). Determine a $\rho_<$-name of $\min \circ f^{-1}[\{0\}]$ (Lemma 5.2.6). For a more direct proof see Exercise 6.3.5. □

By symmetry, $f \mapsto \max \circ f^{-1}[\{0\}]$ is $(\delta_\to^{[0;1]}, \rho_>)$-computable. If a function f has a single zero, then it is the maximum and the minimum of $f^{-1}[\{0\}]$. Therefore:

Corollary 6.3.5. The function $Z_u :\subseteq C[0;1] \to \mathbb{R}$, defined by

$$Z_u(f) = y \quad :\Longleftrightarrow \quad y \text{ is the unique zero of } f,$$

is $(\delta_\to^{[0;1]}, \rho)$-computable.

In classical analysis the *intermediate value theorem* guarantees the existence of a zero.

Example 6.3.6 (intermediate value theorem). In its simplest form the intermediate value theorem can be stated as follows: "Every continuous function $f \in C[0;1]$ with $f(0) < 0 < f(1)$ has a zero." We give an elementary proof by *bisection*. Define sequences $(a_i)_{i\in\mathbb{N}}$ and $(b_i)_{i\in\mathbb{N}}$ of rational numbers inductively as follows:

$$a_0 := 0, \quad b_0 := 1,$$
$$a_{i+1} := m_i \text{ and } b_{i+1} := b_i, \quad \text{if } f(m_i) \le 0,$$
$$a_{i+1} := a_i \text{ and } b_{i+1} := m_i, \quad \text{if } f(m_i) > 0,$$

where $m_i := (a_i + b_i)/2$. Then for all $i \in \mathbb{N}$:

$$a_i \le a_{i+1} < b_{i+1} \le b_i, \quad b_i - a_i = 2^{-i}, \quad f(a_i) \le 0 \le f(b_i),$$

hence the sequences converge to a common limit y. Since the continuous function f commutes with lim, we obtain

$$0 \le \lim_{i\to\infty} f(b_i) = f(\lim_{i\to\infty} b_i) = f(y) = f(\lim_{i\to\infty} a_i) = \lim_{i\to\infty} f(a_i) \le 0$$

and so $f(y) = 0$. Although this proof looks constructive it cannot be realized directly w.r.t. $(\delta_\to^{[0;1]}, \rho)$ on a Type-2 machine, since tests "$f(m_i) \le 0$?" have to be performed repeatedly which are not decidable. For the subset of strictly increasing functions, however, there is a modification, the *trisection method*, which can be computed on a Type-2 machine (Exercise 6.3.6). □

The multi-valued function Z from Theorem 6.3.2 for determining a zero of a continuous function $f \in C[0; 1]$ is "discontinuous" at the function g_0 (see the proof of Theorem 6.3.2) which is constantly 0 on an open interval. The function Z is discontinuous also at non-negative functions with exactly two zeroes (Exercise 6.3.3). The restriction of Z to the functions which are not 0 on some open interval and which change sign is computable. Our proof is a general version of the trisection method [Abe80].

Theorem 6.3.7. Let C_0 be the set of all functions $f \in C[0; 1]$, such that
(C1)   For every open interval J intersecting $[0; 1]$ there is some $y \in J \cap [0; 1]$ such that $f(y) \neq 0$,
(C2) $f(x) \cdot f(z) < 0$ for some $x, z \in [0; 1]$.

Then the multi-valued function

$$Z_0 :\subseteq C[0; 1] \rightrightarrows \mathbb{R}, \quad (f, y) \in R_{Z_0} \iff (f \in C_0 \text{ and } f(y) = 0) ,$$

1. is $(\delta_\rightarrow^{[0;1]}, \rho)$-computable,
2. has no $(\delta_\rightarrow^{[0;1]}, \rho)$-continuous choice function.

Proof: 1. The following properties can be proved easily:

(P1)   If $f(x) > 0$ or $f(x) < 0$, then for some neighborhood J of x,
$(\forall y \in J) f(y) > 0$ or $(\forall y \in J) f(y) < 0$, respectively.
(P2)   The sets
$$\{(f, r) \in C[0; 1] \times \mathbb{Q} \mid f(r) > 0\} ,$$
$$\{(f, r) \in C[0; 1] \times \mathbb{Q} \mid f(r) < 0\}$$
are $(\delta_\rightarrow^{[0;1]}, \nu_\mathbb{Q})$-r.e. (Definition 3.1.3, apply Lemma 6.1.2.2).
(P3)   If $a < b$ and $f(a) < 0 < f(b)$, then there are rational numbers a', b' such that

$$a \leq a' \leq b' \leq b, \quad f(a') < 0 < f(b') \quad \text{and} \quad b' - a' \leq (b - a)/2 .$$

We outline a Type-2 machine M which on input $p \in \text{dom}(\delta_\rightarrow^{[0;1]})$ ($f := \delta_\rightarrow^{[0;1]}(p)$) computes a (coded) sequence $[a_0; b_0], [a_1; b_1], \ldots$ of nested intervals with rational endpoints.
First M determines rational numbers $a_0 < b_0$ such that $f(a_0) \cdot f(b_0) < 0$. (Such rational numbers exist by Conditions (C1) and (C2), they can be computed by Property (P2).) Then for $i = 1, 2, \ldots$ the machine determines a_i, b_i as follows:
Assume $f(a_0) < 0 < f(b_0)$. The machine determines rational numbers a_{i+1}, b_{i+1} such that

$$a_i \leq a_{i+1} \leq b_{i+1} \leq b_i, \quad f(a_{i+1}) < 0 < f(b_{i+1}) \quad \text{and} \quad b_{i+1} - a_{i+1} \leq (b_i - a_i)/2 .$$

(Such numbers exist by Property (P3), and by induction they can be computed by Property (P2).)

If $f(a_0) > 0 > f(b_0)$, the machine operates similarly.

As in Example 6.3.6, $y := \lim_{i \to \infty} a_i = \lim_{i \to \infty} b_i$ is a zero of f, and so the machine M maps any $\delta_{\to}^{[0;1]}$-name of any $f \in C_0$ to some ρ^b-name (Lemma 4.1.6) of a zero of f. Therefore, the function Z_0 is $(\delta_{\to}^{[0;1]}, \rho)$-computable.

2. The function $a \mapsto f_a$ where $f_a(x) := x^3 - 3 \cdot x + a$ is $(\rho, \delta_{\to}^{[0;1]})$-computable. Since the function F from Example 6.3.1 has no (ρ, ρ)-continuous choice function, Z_0 has no $(\delta_{\to}^{[0;1]}, \rho)$-continuous choice function. $\qquad\square$

After we have studied computability of operators which determine zeroes the next theorem considers zeroes of computable functions (see [Spe59] for Theorem 6.3.8.2).

Theorem 6.3.8 (computable zeroes).
1. Every computable function $f \in C[0; 1]$ such that $f(x) \cdot f(z) < 0$ for some $x, z \in [0; 1]$ has a computable zero.
2. There is a computable non-negative function $f \in C[0; 1]$ such that the set $f^{-1}[\{0\}]$ of zeroes has a Lebesgue measure $> 1/2$ but f has no computable zero.

Proof: 1. If f is constantly 0 on some open interval J intersecting $[0; 1]$, then $f(y) = 0$ for some rational number y, which is computable. Otherwise, $f \in \text{dom}(Z_0)$, where Z_0 is the multi-valued operator from Theorem 6.3.7 which has a computable $(\delta_{\to}^{[0;1]}, \rho)$-realization $h :\subseteq \Sigma^\omega \to \Sigma^\omega$. The function h maps any computable $\delta_{\to}^{[0;1]}$-name of f to a computable ρ-name of a zero of f, and so f has a computable zero.

2. Let U_2 be the set from Theorem 4.2.8 with Lebesgue measure $\leq 1/4$ containing all computable real numbers, which is $\theta_<^{\text{en}}$-computable. Its complement $\mathbb{R} \setminus U_2$ has a computable $\psi_>$-name $p \in \Sigma^\omega$. By Theorem 6.2.9 some computable function maps p to a computable $\delta_{\to}^{\mathbb{R}}$-name of a function $f' : \mathbb{R} \to \mathbb{R}$ such that $\mathbb{R} \setminus U_2 = (f')^{-1}[\{0\}]$. The restriction $f := f'|_{[0;1]}$ of f' to the interval $[0; 1]$ is non-negative, computable, has no computable zero and has a Lebesgue measure $> 1/2$. $\qquad\square$

We call a zero y of f isolated, iff it is the only zero of f in some open interval. From Corollary 6.3.5 we obtain:

Corollary 6.3.9 (isolated zeroes). Every isolated zero of a computable function $f \in C[0; 1]$ or $f \in C(\mathbb{R})$ is computable.

Proof: Consider $f \in C[0; 1]$. There are rational numbers $0 \leq a < b \leq 1$ such that y is the only zero of f in $[a; b]$. The function $g \in C[0; 1]$ defined by $g(x) := f(a)$, if $x < a$, $g(x) := f(x)$, if $a \leq x \leq b$ and $g(x) := f(b)$, if $b < x$, is computable and has y as its only zero. By Corollary 6.3.5, y is computable. For $f \in C(\mathbb{R})$, apply the transformation $x \mapsto a + (b - a) \cdot x$. Details are left as Exercise 6.3.9. $\qquad\square$

Since every zero of a polynomial function $f : \mathbb{R} \to \mathbb{R}$ is isolated, every zero of a polynomial function with computable coefficients is computable, therefore:

Corollary 6.3.10 (real closed field).   The computable real numbers form a real closed field.

As a further consequence of Corollary 6.3.5 we show that the inverse of any increasing computable function is computable.

Theorem 6.3.11 (inverse function).   If $f \in C[0;1]$ is increasing and computable, then its inverse f^{-1} is computable.

Proof: Define $A := [f(0); f(1)]$. We have

$$f^{-1}(y) = x \iff f(x) = y \iff f(x) - y = 0 \iff f_y(x) = 0 \,,$$

where $f_y(x) := f(x) - y$. Since $(y, x) \mapsto f(x) - y$ is (ρ, ρ, ρ)-computable, $H : y \mapsto f_y$ for $y \in A$ is $(\rho, \delta_{\to}^{[0;1]})$-computable by Theorem 3.3.15. Let Z_u be the function from Corollary 6.3.5. If x is the (only) zero of f_y for $y \in A$, the (ρ, ρ)-computable function $Z_u \circ H$ maps y to x. Therefore, $f^{-1} = Z_u \circ H$ is computable.   □

The theorem holds for decreasing functions and for other computable intervals accordingly (Exercise 6.3.7).

Finding the position of a *maximum value* is closely related to zero-finding. By Corollary 6.2.5 finding the maximum value of a function $f \in C[0;1]$ is $(\delta_{\to}^{[0;1]}, \rho)$-computable. The following two observations show that finding the position of a maximum, however, is as difficult as zero-finding. For any function $f \in C[0;1]$ and number $x \in [0;1]$,

1. x is a zero of f, iff x is a maximum point of g where $g(x) := -|f(x)|$,
2. x is a maximum point of f, iff x is a zero of h where
   $h(x) := f(x) - \max\{f(x) \mid 0 \leq x \leq 1\}$.

Notice that the above functions $f \mapsto g$ and $f \mapsto h$ are $(\delta_{\to}^{[0;1]}, \delta_{\to}^{[0;1]})$-computable.

Exercises 6.3.

1. For $a \in \mathbb{R}$ define $f_a : \mathbb{R} \to \mathbb{R}$ by $f_a(x) := x^3 - 3 \cdot x + a$.
   a) Show that the multi-valued function

   $$G : \mathbb{R} \rightrightarrows \mathbb{Q}, \quad R_G := \{(a, r) \mid r < \min \circ f_a^{-1}[\{0\}] \,,$$

   is $(\rho, \nu_{\mathbb{Q}})$-computable.
   b) Is there a single-valued $(\rho, \nu_{\mathbb{Q}})$-computable function $G : \mathbb{R} \to \mathbb{Q}$ such that $G(a) < \min \circ f_a^{-1}[\{0\}]$?

c) Is there a single-valued (ρ, ρ)-computable function $G : \mathbb{R} \to \mathbb{R}$ such that $G(a) < \min \circ f_a^{-1}[\{0\}]$?

2. Show that the function $h : a \mapsto g_a$, g_a from the proof of Theorem 6.3.2, is $(\rho, \delta_\to^{[0;1]})$-computable.

3. For any $a \in \mathbb{R}$ define a polygon function $h_a \in C[0;1]$ by the vertices $(0,1)$, $(1/4, a)$, $(1/2, 1)$, $(3/4, 0)$, $(1, 1)$, if $a \geq 0$ and by the vertices $(0,1)$, $(1/4, 0)$, $(1/2, 1)$, $(3/4, -a)$, $(1, 1)$, if $a < 0$ (Fig. 6.7). Show that the multi-valued function

$$Z :\subseteq C[0;1] \rightrightarrows \mathbb{R}, \quad R_Z := \{(f, x) \mid f(x) = 0\},$$

is not $(\delta_\to^{[0;1]}, \rho)$-continuous at h_0 (see the proof of Theorem 6.3.2). More generally, the function Z is discontinuous at any non-negative function $h \in C[0;1]$ which has exactly two zeroes.

♦ 4. Consider the function Z_{app} from Theorem 6.3.3 and define

$$Z' :\subseteq C[0;1] \times \mathbb{Q} \rightrightarrows \mathbb{R} \text{ by } R_{Z'} := \{(f, a, y) \mid |f(y)| < a\}.$$

a) Show that Z_{app} has no $(\delta_\to^{[0;1]}, \nu_\mathbb{Q}, \nu_\mathbb{Q})$-continuous choice function.
b) Show that Z' is $(\delta_\to^{[0;1]}, \nu_\mathbb{Q}, \rho)$-computable.
c) Show that Z' has no $(\delta_\to^{[0;1]}, \nu_\mathbb{Q}, \rho)$-continuous choice function f'. Hint: consider the family $(h_a)_{a \in \mathbb{R}}$ of functions from Exercise 6.3.3 and investigate the function $a \mapsto f'(h_a, 1/2)$.

5. Prove Theorem 6.3.4 directly by using the Cauchy representation $\delta_C^{[0;1]}$ (Definition 6.1.9, Lemma 6.1.10).

6. By Corollary 6.3.5 the function Z_s mapping each strictly increasing function $f \in C[0;1]$ with $f(0) < 0 < f(1)$ to its zero is $(\delta_\to^{[0;1]}, \rho)$-computable. Prove this fact using *trisection* instead of bisection (Example 6.3.6). Notice that $f(l_i) = f(r_i) = 0$ is impossible, if $a_i < l_i < r_i < b_i$.

7. Generalize Theorem 6.3.11 to functions from $C[a; b]$, $C(a; b)$, and $C(\mathbb{R})$ where $a < b$ are computable numbers.

8. Show that the function $f \mapsto f^{-1}$ for increasing $f : \mathbb{R} \to \mathbb{R}$ is $(\delta_\to^\mathbb{R}, \delta_\to^\mathbb{R})$-computable.

9. Complete the proof of Corollary 6.3.9.

♦10. Show that there is no $([\rho^2, \rho^2], [\rho^2, \rho^2])$-continuous function (Definition 3.3.3) $f : \mathbb{C}^2 \to \mathbb{C}^2$ such that $f(a_1, a_0)$ is a vector of the zeroes of the complex polynomial $z^2 + a_1 z + a_0$. Generalize the result to complex polynomials of degree $n > 2$.

♦11. (Computable version of the fundamental theorem of algebra) Consider $n \geq 1$. Show that the multi-valued function $Z : \mathbb{C}^n \rightrightarrows \mathbb{C}^n$ such that $((a_{n-1}, \ldots, a_0), (z_1, \ldots, z_n)) \in R_Z$, iff $\{z_1, \ldots z_n\}$ is the set of zeroes of the polynomial $P(z) := z^n + a_{n-1}z^{n-1} + \ldots + a_1 z + a_0$, is $([\rho^2]^n, [\rho^2]^n,)$-computable (Definition 3.3.3). Proceed as follows:
a) Find a $\theta_<^{\text{en}}$-name of $\{z \mid |P(z) > 0\}$ (Theorem 6.2.4.1).

b) Assume that $(f, J) \mapsto$ the number of zeroes of f in $J \in \mathrm{Cb}^{(2)}$
(f polynomial, f has no zero on the boundary of J) is computable
(Exercise 6.5.4).

c) For $k = 1, 2, \ldots$ compute a tuple $v_k = (w_1, n_1, \ldots, w_j, n_j)$ such that
the boundary of $\mathrm{I}^2(w_i)$ has no zero, the radius of $\mathrm{I}^2(w_i)$ is less than
2^{-k}, f has exactly $n_i > 0$ zeroes in $\mathrm{I}^2(w_i)$ and $n_1 + \ldots + n_j = n$.
(Hint: Search exhaustively, cover boundaries of the $\mathrm{I}^2(w_i)$ by finitely
many balls enumerated in 11a).

$\Diamond$12. Show that there is a (ρ, ρ)-computable function $f : \subseteq \mathbb{R} \to \mathbb{R}$ such that
$\mathrm{dom}(f) = [0; 1]$, $\max_{0 \le x \le 1} f(x) = 1$, $f(x) < 1$ for every computable real
number and $\{x \in [0; 1] \mid f(x) = 1\}$ has measure $\ge 1/2$.

6.4 Differentiation and Integration

For the sake of simplicity we will consider merely continuous functions from
$C(\mathbb{R})$ or $C[0; 1]$. We start with integration which is computable.

Theorem 6.4.1 (integration).

1. For $A = [0; 1]$ or $A = \mathbb{R}$ the function

$$F_A : f \mapsto \int_0^1 f(x) \, \mathrm{d}x \quad \text{for} \quad f \in C(A)$$

is $(\delta_\to^A, \rho)$-computable.

2. The function

$$G : (f, a, b) \mapsto \int_a^b f(x) \, \mathrm{d}x \quad \text{for} \quad f \in C(\mathbb{R}) \quad \text{and} \quad a, b \in \mathbb{R}, \quad a \le b,$$

is $(\delta_\to^\mathbb{R}, \rho, \rho, \rho)$-computable.

3. For any computable $a \in \mathbb{R}$ function

$$H : f \mapsto g, \quad \text{where} \quad g(y) := \int_a^y f(x) \, \mathrm{d}x, \quad f \in C(\mathbb{R}),$$

is $(\delta_\to^\mathbb{R}, \delta_\to^\mathbb{R})$-computable. ($H(f)$ is a primitive of f.)

Proof: 1. Consider the case $A = [0; 1]$. We use the Cauchy representations
(Lemma 4.1.6, Lemma 6.1.10) and show that the function $F_{[0;1]}$ is $(\delta_C^{[0;1]}, \rho_C''')$-
computable. There is a Type-2 machine M which on input $p := \iota(w_0)\iota(w_1) \ldots$
($w_i \in \mathrm{dom}(\nu_{\mathrm{Pg}})$, see Definition 6.1.9) prints a sequence $q := \iota(u_0)\iota(u_1) \ldots$
such that

$$\nu_\mathbb{Q}(u_i) = \int_0^1 \nu_{\mathrm{Pg}}(w_i)(x) \, \mathrm{d}x$$

for all $i \in \mathbb{N}$ (Fig. 6.3). Let $f := \delta_{\mathrm{C}}^{[0;1]}(p)$ and $f_i := \nu_{\mathrm{Pg}}(w_i)$. Since

$$\begin{aligned}
\left| F_{[0;1]}(f) - \nu_{\mathbb{Q}}(u_i) \right| &= \left| \int_0^1 f(x)\,\mathrm{d}x - \int_0^1 f_i(x)\,\mathrm{d}x \right| \\
&= \left| \int_0^1 (f(x) - f_i(x))\,\mathrm{d}x \right| \\
&\leq \int_0^1 |f(x) - f_i(x)|\,\mathrm{d}x \\
&\leq 2^{-i} ,
\end{aligned}$$

$f_M(p)$ is a ρ_{C}'''-name of $F_{[0;1]}(f)$. Therefore, $F_{[0;1]}$ is $(\delta_{\mathrm{C}}^{[0;1]}, \rho_{\mathrm{C}}''')$-computable. Consider the case $A = \mathbb{R}$. The function $F_1 : f \mapsto f\rfloor_{[0;1]}$ is $(\delta_{\rightarrow}^{\mathbb{R}}, \delta_{\rightarrow}^{[0;1]})$-computable (Exercise 6.1.6). Then $F_{\mathbb{R}} = F_{[0;1]} \circ F_1$ is computable.

2. We reduce this case to the first one. The function $H_1 : (f, a, b, x) \mapsto f(a + (b-a) \cdot x)$ is $(\delta_{\rightarrow}^{\mathbb{R}}, \rho, \rho, \rho, \rho)$-computable by Lemma 3.3.14. Define $H_1 : (f, a, b) \mapsto g \in \mathrm{C}(\mathbb{R})$ such that $g(x) := f(a + (b-a) \cdot x)$. Then by Theorem 3.3.15.2, H_1 is $(\delta_{\rightarrow}^{\mathbb{R}}, \rho, \rho, \delta_{\rightarrow}^{\mathbb{R}})$-computable and $\int_a^b f(x)\,\mathrm{d}x = (b-a) \cdot \int_0^1 g(x)\,\mathrm{d}x$. Since $G(f, a, b) = (b-a) \cdot F_{\mathbb{R}} \circ H_1(f, a, b)$, G is $(\delta_{\rightarrow}^{\mathbb{R}}, \rho, \rho, \rho)$-computable.

3. By Statement 2. above, the function

$$H_> : (f, y) \mapsto \int_a^y f(x)\,\mathrm{d}x, \quad (a \leq y)$$

is $(\delta_{\rightarrow}^{\mathbb{R}}, \rho, \rho)$-computable, since a is computable. Similarly, the function

$$H_< : (f, y) \mapsto \int_a^y f(x)\,\mathrm{d}x = - \int_y^a f(x)\,\mathrm{d}x, \quad (y \leq a)$$

is $(\delta_{\rightarrow}^{\mathbb{R}}, \rho, \rho)$-computable. Lemma 4.3.5 can be generalized easily to functions with a second variable $f \in \mathrm{C}(\mathbb{R})$ as a parameter. Therefore,

$$(f, y) \mapsto \int_a^y f(x)\,\mathrm{d}x, \quad (a \in \mathbb{R})$$

is $(\delta_{\rightarrow}^{\mathbb{R}}, \rho, \rho)$-computable. By Theorem 3.3.15.2, H is $(\delta_{\rightarrow}^{\mathbb{R}}, \delta_{\rightarrow}^{\mathbb{R}})$-computable. $\square$

In particular, every computable function $f \in \mathrm{C}(\mathbb{R})$ has a computable primitive $g \in \mathrm{C}(\mathbb{R})$, $g(y) := \int_0^y f(x)\,\mathrm{d}x$. While integration over intervals $[a; b]$ is computable in a, b, integration over compact sets is not even continuous (Exercise 6.4.4). As a corollary we show that two dimensional integration is computable. The corollary can be generalized to $n > 2$ dimensions.

Corollary 6.4.2 (two dimensional integral). The function

$$T : f \mapsto \int_0^1 \int_0^1 f(x, y)\,\mathrm{d}x\,\mathrm{d}y, \quad f \in \mathrm{C}(\mathbb{R}^2) ,$$

is $(\delta_{\rightarrow}^{\mathbb{R}^2}, \rho)$-computable.

Proof: Let $\delta := \delta_\rightarrow^{\mathbb{R}^2} = [\rho^2 \rightarrow \rho]_{\mathbb{R}^2}$. By Lemma 3.3.14, $(f, (x, y)) \mapsto f(x, y)$ for $f \in C(\mathbb{R}^2)$ and $x, y \in \mathbb{R}$ is (δ, ρ^2, ρ)-computable. By Lemmas 3.3.6 and 4.1.18, $((f, y), x) \mapsto f(x, y)$ is $([[\delta, \rho], \rho], \rho)$-computable, and by Theorem 3.3.15.2, $(f, y) \mapsto f_y$, $f_y(x) := f(x, y)$, is $([\delta, \rho], \delta_\rightarrow^{\mathbb{R}})$-computable. By Statement 1 above, $g \mapsto \int_0^1 g(x)\,dx$ for $g \in C(\mathbb{R})$ is $(\delta_\rightarrow^{\mathbb{R}}, \rho)$-computable. Therfore, $(f, y) \mapsto \int_0^1 f_y(x)\,dx$ is $([\delta, \rho], \rho)$-computable. By Theorem 3.3.15.1, $f \mapsto h$, $h(y) := \int_0^1 f_y(x)\,dx$, is $(\delta, \delta_\rightarrow^{\mathbb{R}})$-computable. Finally, by Statement 1 above, $f \mapsto \int_0^1 h(y)\,dy = T(f)$ is (δ, ρ)-computable. $\qquad\square$

In contrast to integration, differentiation is complicated. As usual, f' will denote the derivative of the function f. In the following we will consider the set

$$C^1[0; 1] := \{f \in C[0; 1] \mid f' \in C[0; 1]\}$$

of continuously differentiable functions $f :\subseteq \mathbb{R} \rightarrow \mathbb{R}$ with $\text{dom}(f) = [0; 1]$. While every (ρ, ρ)-computable function is continuous, there are computable functions which are not differentiable, for example the function $x \mapsto |x|$ (for a more sophisticated example see Exercise 6.4.10). By our first theorem differentiation is not computable w.r.t. names for continuous functions.

Theorem 6.4.3.
1. For every $a \in [0; 1]$ the function $f \mapsto f'(a)$, $f \in C^1[0; 1]$, is not $(\delta_\rightarrow^{[0;1]}, \rho)$-continuous.
2. The function $f \mapsto f'$, $f \in C^1[0; 1]$, is not $(\delta_\rightarrow^{[0;1]}, \delta_\rightarrow^{[0;1]})$-continuous.

Proof: 1. Suppose that the function $H : f \mapsto f'(a)$ is $(\delta_\rightarrow^{[0;1]}, \rho)$-continuous. Then it is $(\tau_{co}^{[0;1]}, \tau_{\mathbb{R}})$-continuous by Theorem 3.2.11, since $\delta_\rightarrow^{[0;1]} \equiv \delta_{co}^{[0;1]}$ which is admissible with final topology $\tau_{co}^{[0;1]}$ (Lemmas 6.1.5 and 6.1.7) and ρ is admissible with final topology $\tau_{\mathbb{R}}$ (Lemma 4.1.4). Let $f_0(x) := 0$ for all $x \in [0; 1]$. Then $H(f_0) = 0$, and by continuity of H there are open intervals $J_1, L_1, \ldots, J_n, L_n \in Cb^{(1)}$ such that $f_0 \in R(J_1, L_1) \cap \ldots \cap R(J_n, L_n)$ and $H[R(J_1, L_1) \cap \ldots \cap R(J_n, L_n)] \subseteq (-1; 1)$ (Definition 6.1.3). There is some $k \in \mathbb{N}$ such that $1/k \in L_1 \cap \ldots \cap L_n$. Define $g \in C^1[0; 1]$ by $g(x) := \sin(k(x - a))/k$. Then $g \in R(J_1, L_1) \cap \ldots \cap R(J_n, L_n)$ but $H(g) = 1 \notin (-1; 1)$ (contradiction).

2. This follows from Statement 1, since $g \mapsto g(1/2)$ is $(\delta_\rightarrow^{[0;1]}, \rho)$-continuous by Lemma 3.3.14. $\qquad\square$

Roughly speaking, the information available from $\delta_\rightarrow^{[0;1]}$-names of continuously differentiable functions is not sufficient to compute $\delta_\rightarrow^{[0;1]}$-names of the derivatives. By brute force we define a representation of $C^1[0; 1]$, such that $(f, x) \mapsto f(x)$ as well as $(f, x) \mapsto f'(x)$ become computable (Remark 3.3.18).

Definition 6.4.4 (standard representation of $C^1[0;1]$). *Define a representation $\delta^{(1)}$ of the set $C^1[0;1]$ by*

$$\delta^{(1)}\langle p,q \rangle = f \quad \Longleftrightarrow \quad \delta_\to^{[0;1]}(p) = f \quad and \quad \delta_\to^{[0;1]}(q) = f' \ .$$

In fact, $\delta^{(1)}$ is the poorest representation of $C^1[0;1]$ such that $(f,x) \mapsto f(x)$ and $(f,x) \mapsto f'(x)$ are computable (cf. Lemmas 3.3.14 and 6.1.2):

Lemma 6.4.5. For every representation δ of $C^1[0;1]$,
1. $(f,x) \mapsto f(x)$ and $(f,x) \mapsto f'(x)$ are (δ,ρ,ρ)-computable, iff $\delta \leq \delta^{(1)}$,
2. $(f,x) \mapsto f(x)$ and $(f,x) \mapsto f'(x)$ are (δ,ρ,ρ)-continuous, iff $\delta \leq_t \delta^{(1)}$.

The proof is left as Exercise 6.4.6. Remember that by Lemma 3.3.14, $(f,x) \mapsto f(x)$ is (δ,ρ,ρ)-computable, iff $f \mapsto f$ is $(\delta, \delta_\to^{[0;1]})$-computable, and $(f,x) \mapsto f'(x)$ is (δ,ρ,ρ)-computable, iff $f \mapsto f'$ is $(\delta, \delta_\to^{[0;1]})$-computable (correspondingly for continuous replacing computable).

By extending Definition 6.1.3 of the compact-open representation $\delta_{co}^{[0;1]}$ we obtain a computable topological space the canonical representation of which is equivalent to $\delta^{(1)}$ (Definitions 3.2.1 and 3.2.2).

Lemma 6.4.6. Define a computable topological space $\mathbf{S} = (C^1[0;1], \sigma^{(1)}, \nu^{(1)})$ by

$$\nu^{(1)}(0\iota(u)\iota(v)) := \{f \mid f[\bar{I}^1(u)] \subseteq I^1(v)\} \quad (I^1(u) \cap [0;1] \neq \emptyset) \ ,$$

$$\nu^{(1)}(1\iota(u)\iota(v)) := \{f \mid f'[\bar{I}^1(u)] \subseteq I^1(v)\} \quad (I^1(u) \cap [0;1] \neq \emptyset) \ .$$

Then
1. $\delta_\mathbf{S} \equiv \delta^{(1)}$,
2. the topology $\tau_\mathbf{S}$, the final topology of $\delta_\mathbf{S}$ and $\delta^{(1)}$, is generated by the metric $d^{(1)}$ on $C^1[0;1]$ defined by

$$d^{(1)}(f,g) := \max\left(d(f,g), d(f',g')\right) \ .$$

Proof: 1. Define a representation δ of $C^1[0;1]$ by $\delta\langle p,q \rangle = f$, iff $\delta_{co}^{[0;1]}(p) = f$ and $\delta_{co}^{[0;1]}(q) = f'$. Then $\delta \equiv \delta^{(1)}$ follows from Lemma 6.1.7 and $\delta \equiv \delta_\mathbf{S}$ can be proved easily.

2. Let $C := C[0;1]$, $C^{(1)} := C^{(1)}[0;1]$. For $I,J \in Cb^{(1)}$, $I \cap [0;1] \neq \emptyset$ let $R(I,J) := \{f \in C \mid f[\bar{I}] \subseteq J\}$ (Definition 6.1.3).

Since the representations $\delta_{co}^{[0;1]}$ and $\delta_C^{[0;1]}$ of C are equivalent, the compact-open topology τ_{co} on C (Definition 6.1.3) is generated by the maximum metric d on C (Definition 6.1.9) and has the set of all $R(I,J)$, $I,J \in Cb^{(1)}$ and $I \cap [0;1] \neq \emptyset$ as a subbase.

On the product space $C \times C$ the product topology $\tau_{co} \otimes \tau_{co}$ is generated by the metric d_2 defined by

$$d_2((f_1, f_2), (g_1, g_2)) := \max(\, d(f_1, g_1)\,,\, d(f_2, g_2)\,)$$

and has the set of all

$$R(I, J) \times R(K, L), \quad I, J, K, L \in \mathrm{Cb}^{(1)}, \ I \cap [0; 1] \neq \emptyset, \ I \cap [0; 1] \neq \emptyset$$

as a subbase. The restriction of this space to $E := \{(f, f') \mid f \in C^{(1)}\} \subseteq C \times C$ is generated by (the restriction of) d_2 and has the set of all $R(I, J) \times R(K, L) \cap E$ as a subbase. The bijection $H : E \to C^{(1)}$, $(f, f') \mapsto f$ induces a topology on $C^{(1)}$ which has the set of all $T(I, J, K, L) := H[R(I, J) \times R(K, L) \cap E]$ as a subbase and which is generated by the metric $d^{(1)}$, since $d_2((f, f'), (g, g')) = d^{(1)}(f, g) = d^{(1)}(H(f, f'), H(g, g'))$. Since $T(I, J, K, L) = \{f \in C^{(1)} \mid f[\overline{I}] \subseteq J \text{ and } f'[\overline{K}] \subseteq L\}$, the sets $T(I, J, K, L)$ are a subbase of $\tau_{\mathbf{S}}$. □

The final topology $\tau_{\mathbf{S}}$ of $\delta^{(1)}$ is the well-known standard topology on $C^1[0; 1]$. Two functions $f, g \in C^1[0; 1]$ are close w.r.t. the distance $d^{(1)}$, iff they are close w.r.t the distance d and their derivatives are close w.r.t. the distance d. A $\delta_{\mathbf{S}}$-name of a function $f \in C^1[0; 1]$ is a list of all compact-open boxes $R(I, J)$ (Definition 6.1.3) such that $f \in R(I, J)$ and all compact-open boxes $R(I, J)$ such that $f' \in R(I, J)$.

A $\delta^{(1)}$-name of a function $f \in C^1[0; 1]$ consists of a $\delta_{\to}^{[0;1]}$-name p of f and a $\delta_{\to}^{[0;1]}$-name q of its derivative f'. By the following theorem either q can be replaced by a name of a modulus of continuity of f' or p can be replaced by a ρ-name of $f(0)$. Recall that $\delta_{\mathbb{B}} \equiv [\nu_{\mathbb{N}} \to \nu_{\mathbb{N}}]_{\mathbb{N}}$ by Exercise 3.3.12.

Theorem 6.4.7 (further representations of $C^1[0; 1]$). Define representations $\delta_m^{(1)}$ and $\delta_i^{(1)}$ of $C^1[0; 1]$ as follows:

1. $\delta_m^{(1)}\langle p, q \rangle = f$, iff $\delta_{\to}^{[0;1]}(p) = f$ and $[\nu_{\mathbb{N}} \to \nu_{\mathbb{N}}]_{\mathbb{N}}(q)$ is a modulus of continuity of f' on $[0; 1]$,
2. $\delta_i^{(1)}\langle p, q \rangle = f$, iff $\rho(p) = f(0)$ and $\delta_{\to}^{[0;1]}(q) = f'$.

Then $\delta^{(1)} \equiv \delta_m^{(1)} \equiv \delta_i^{(1)}$.

Proof: $\delta_m^{(1)} \leq \delta^{(1)}$: The function

$$F_0 :\subseteq C^1[0; 1] \times \mathbb{N}^{\mathbb{N}} \times \mathbb{R} \times \mathbb{N} \to \mathbb{R}\,,$$

$$F_0(f, g, x, n) := 2^{g(n)+2} \cdot \left(f(x + 2^{-g(n)-2}) - f(x)\right)\,,$$

is $\left(\delta_{\to}^{[0;1]}, [\nu_{\mathbb{N}} \to \nu_{\mathbb{N}}]_{\mathbb{N}}, \rho, \nu_{\mathbb{N}}, \rho\right)$-computable.
Let g be a modulus of continuity of f' and consider $0 \leq x \leq 2/3$ and $n \in \mathbb{N}$. By the mean value theorem there is some y, $x \leq y \leq x + 2^{-g(n)-2}$, such that

$$2^{g(n)+2} \cdot \left(f(x + 2^{-g(n)-2}) - f(x)\right) = f'(y)\,.$$

Since $|x - y| \leq 2^{-g(n)}$, $|f'(x) - f'(y)| \leq 2^{-n}$, and so

$$|f'(x) - F_0(f, g, x, n)| \leq 2^{-n} \text{ and } f'(x) = \lim_{n \to \infty} F_0(f, g, x, n) \, .$$

By Theorem 3.3.15.2, the function $G_0 : (f, g, x) \mapsto \left(F_0(f, g, x, n + 2)\right)_{n \in \mathbb{N}}$ is $(\delta_\to^{[0;1]}, [\nu_\mathbb{N} \to \nu_\mathbb{N}]_\mathbb{N}, \rho, [\rho]^\omega)$-computable. By Theorem 4.3.7.4.1, the function $H_0 : (f, g, x) \mapsto f'(x)$ is $(\delta_\to^{[0;1]}, [\nu_\mathbb{N} \to \nu_\mathbb{N}]_\mathbb{N}, \rho, \rho)$-computable for $f \in C^1[0; 1]$, g a modulus of f' and $0 \leq x \leq 2/3$.
For similar reasons there is a function H_1 computing $f'(x)$ for $1/3 \leq x \leq 1$. Combining a machine computing H_0 and a machine computing H_1 we obtain a machine computing $H : (f, g, x) \mapsto f'(x)$ for $0 \leq x \leq 1$. By Theorem 3.3.15.2, $(f, g) \mapsto f'$ is $(\delta_\to^{[0;1]}, [\nu_\mathbb{N} \to \nu_\mathbb{N}]_\mathbb{N}, \delta_\to^{[0;1]})$-computable, and so $\delta_m^{(1)} \leq \delta^{(1)}$.

$\delta^{(1)} \leq \delta_m^{(1)}$: this follows from Theorem 6.2.7.
$\delta^{(1)} \leq \delta_i^{(1)}$: this is obvious.
$\delta_i^{(1)} \leq \delta^{(1)}$: this follows from Theorem 6.4.1.1.   □

Since a computable function maps computable points to computable points, we have:

Corollary 6.4.8 (computable derivatives). Let $f \in C[0; 1]$ be $\delta_\to^{[0;1]}$-computable.
1. The derivative f' is computable, if it has a computable modulus of continuity.
2. The derivative f' is computable, if f'' is continuous.
3. The n-th derivative $f^{(n)}$ is computable for all n, if $f^{(n)}$ exists for all n.

Proof: 1. This follows from $\delta_m^{(1)} \leq \delta^{(1)}$ (Theorem 6.4.7).
2. If f'' is continuous, then it has a maximum M on $[0; 1]$. By the mean value theorem, $|f'(x) - f'(y)| \leq M \cdot |x - y|$ for all $x, y \in [0; 1]$. For $2^m \geq M$ and $g(n) := m + n$, g is a computable modulus of continuity of f' on $[0; 1]$, and so f' is computable by Property 1.
3. This follows from Property 2.   □

While computable functions map computable points to computable ones, it is hardly surprising that the differentiation operator, which is not even continuous w.r.t. the standard topology on $C[0; 1]$, maps some computable function to a non-computable one. The following example is due to Myhill [Myh71]. We present a modification by Pour-El and Richards [PER89].

Example 6.4.9 (non-computable derivative).   There is a function $f \in C[0; 1]$ such that
1. f is computable,
2. f' is continuous but not computable,
3. f'' exists but is not continuous.

We define f' explicitly and obtain f by integration: $f(x) = \int_0^x f'(y)\,dy$. Let $g : \mathbb{R} \to \mathbb{R}$ be the canonical "pulse" function:

$$g(x) := \begin{cases} e^{-\frac{x^2}{1-x^2}} & \text{for } |x| < 1 , \\ 0 & \text{for } |x| \geq 1 . \end{cases}$$

Then g and all its derivatives are computable. Let $a : \mathbb{N} \to \mathbb{N}$ be an injective computable function such that $A := \text{range}(a)$ is not recursive and $0, 1 \notin A$. Define the k-th pulse $g_k \in C[0;1]$ by

$$g_k(x) := g\big(2^{k+a(k)+2} \cdot (x - 2^{-a(k)})\big) .$$

Then g_k is a pulse of height 1 such that $g_k(x) = 0$ for $x \notin J_{a(k)}$, where $J_n := \big(\frac{3}{4} \cdot 2^{-n} \;;\; \frac{3}{2} \cdot 2^{-n}\big)$. Notice that the left end-point of J_n is the right end-point of J_{n+1}. Define

$$f'(x) := \sum_{k=0}^{\infty} 4^{-a(k)} \cdot g_k(x) .$$

For all $n \in \mathbb{N}$ and $x \in \overline{J}_n$, if $n \notin A$, then $f'(x) = 0$, and if $n \in A$, then $f'(x) = 4^{-a(k)} \cdot g_k(x)$ for the single $k \in \mathbb{N}$ such that $a(k) = n$. Furthermore, $f'(0) = 0$.

We show that f'' exists. Since $g'(-1) = g'(1) = 0$, $f''(x)$ exists for all $0 < x \leq 1$. For determining $f''(0)$ we estimate $(f'(x) - f'(0))/(x - 0) = f'(x)/x$ for $0 < x \leq 1$. If $x \in \overline{J}_n$ and $n \notin A$, then $f'(x)/x = 0$. If $x \in \overline{J}_n$ and $n = a(k)$, then $f'(x)/x \leq 4^{-a(k)}/\big(\frac{3}{4} \cdot 2^{-a(k)}\big) \leq 2 \cdot 2^{-n}$. Therefore, $f''(0) = \lim_{x \to 0} f'(x)/x = 0$. Since f'' exists, f' is continuous.

We show that f' is not (ρ, ρ)-computable. We have

$$f'(2^{-n}) = \begin{cases} 4^{-n} & \text{for } n \in A \\ 0 & \text{otherwise} . \end{cases}$$

If f' is computable, we can decide $n \in A$ for $n \in \mathbb{N}$ as follows: determine an interval J of length $< 4^{-n}$ such that $f'(2^{-n}) \in J$. Then $n \in A$, iff $4^{-n} \in J$. Since A is not recursive, f' cannot be computable.

It remains to show that f is computable. The inverse function w of a, $w(n) := \max\{m \mid a(m) \leq n\}$, is non-decreasing and unbounded. If

$$f_i(x) := \sum_{k=0}^{i} 4^{-a(k)} \cdot g_k(x) ,$$

then for $i < j$, $|f_i(x) - f_j(x)| \leq \max\{4^{-a(k)} \mid i < k \leq j\}$, and so $|f_i(x) - f_j(x)| \leq 4^{-n}$ for $i \geq w(n)$. Therefore, the sequence $(f_i)_{i \in \mathbb{N}}$ converges uniformly to f', and for every x the sequence $i \mapsto \int_0^x f_i(y)\,dy$ converges to $\int_0^x f'(y)\,dy = f(x)$. Since the functions a and g are computable, the sequence $k \mapsto g_k$ is $(\nu_\mathbb{N}, \delta_{\to}^{[0;1]})$-computable and the sequence

$i \mapsto f_i$ is $(\nu_{\mathbb{N}}, \delta^{[0;1]}_{\rightarrow})$-computable by Theorem 6.2.2.1 (C[0; 1]-variant). Since $(h, x) \mapsto \int_0^x h(y) \, dy$ is $(\delta^{[0;1]}_{\rightarrow}, \rho, \rho)$-computable by Theorem 6.4.1.2 (C[0; 1]-variant), $(x, i) \mapsto \int_0^x f_i(y) \, dy$ is $(\rho, \nu_{\mathbb{N}}, \rho)$-computable, that is,

$$T : x \mapsto (y_i)_{i \in \mathbb{N}}, \quad y_i := \int_0^x f_i(y) \, dy, \text{ is } (\rho, [\rho]^\omega)\text{-computable.}$$

Since g_k is a very narrow pulse, $\int_0^1 g_k(y) \, dy$ is very small: $\int_0^1 g_k(y) \, dy = 2^{-k-a(k)-2} \cdot \int_{-1}^1 g(y) \, dy \leq 2^{k-1}$, and so the sequence $(y_i)_{i \in \mathbb{N}}$ has $e(n) := n$ as a modulus of convergence: for $n \leq i < j$,

$$|y_j - y_i| \leq \left| \int_0^x 4^{-a(i+1)} g_{i+1}(y) \, dy + \ldots + \int_0^x 4^{-a(j)} g_j(y) \, dy \right|$$
$$\leq \int_0^1 g_{i+1}(y) \, dy + \ldots + \int_0^1 g_j(y) \, dy$$
$$\leq 2^{-i-2} + \ldots + 2^{-j-1}$$
$$\leq 2^{-n} .$$

We obtain $f = L' \circ T$, L' from Theorem 6.2.2.2, and so f is (ρ, ρ)-computable.

As a consequence of Theorem 6.2.2.2 the computable sequence $i \mapsto f_i$ cannot have a computable modulus of convergence, since its limit f' is non-computable; it converges extremely slowly. Our modulus function w is very fast increasing (in fact faster than any computable function). However, the computable sequence $i \mapsto \int_0^x f_i(y) \, dy$ converges to $f(x)$ very fast. $\square$

By Corollary 6.4.8.3, the n-th derivative $f^{(n)}$ of f is computable for every n, if $f \in C^\infty([0; 1], \mathbb{R})$ is computable. However, the sequence $n \mapsto f^{(n)}$ is not $(\nu_{\mathbb{N}}, \delta^{[0;1]}_{\rightarrow})$-computable in general (Exercise 6.4.8). Furthermore, Corollaries 6.4.8.2 and 6.4.8.3 break down, if the domain [0; 1] is replaced by $\mathbb{R}$ (Exercise 6.4.9).

Exercises 6.4.

1. Show that $f \mapsto g$, $g(y) := \int_0^y f(x) \, dx$, $f \in C[0; 1]$, is $(\delta^{[0;1]}, \delta^{[0;1]})$-computable.
2. Show that the function

$$T_0 : f \mapsto \int_0^1 \int_0^1 f(x, y) \, dx \, dy, \quad f \in C([0; 1]^2) ,$$

is $(\delta^{[0;1]^2}_{\rightarrow}, \rho)$-computable
   a) by reduction to Corollary 6.4.2,
   b) by using the polynomial Cauchy representation δ_p of $C([0; 1]^2)$ (see the remarks after Lemma 6.1.10).

3. Let $\gamma :\subseteq Y \to M$ ($Y \in \{\Sigma^*, \Sigma^\omega\}$) be a naming system and let $f : M \times \mathbb{R} \to \mathbb{R}$ be (γ, ρ, ρ)-computable. Show that

$$z \mapsto \int_0^1 f(z, x)\, \mathrm{d}x \quad \text{is} \quad (\gamma, \rho)\text{-computable}.$$

♦ 4. Consider integration of the function $x \mapsto 1$, $x \in \mathbb{R}$, over subsets $A \subseteq \mathbb{R}$, that is, determining the Lebesgue measure $\mu(A)$ of a set A.
   Show that $A \mapsto \mu(A)$
   a) for open $A \subseteq \mathbb{R}$ is $(\theta_<, \overline{\rho}_<)$-computable (Exercise 4.1.21),
   b) for compact A is $(\kappa_>, \rho_>)$-computable,
   c) for compact A is not $(\kappa, \rho_<)$-continuous.

♦ 5. (Integration over compact sets)
   a) Show that the function

$$(f, A) \mapsto \int_A f(x)\, \mathrm{d}x\ , \quad f \in \mathrm{C}(\mathbb{R}^2)\ , \quad A \subseteq \mathbb{R}^2 \text{ compact,}$$

   is not $(\delta_\to^{\mathbb{R}^2}, \kappa, \rho_<)$-computable.
   b) Show that its restrictions to the convex subsets is $(\delta_\to^{\mathbb{R}^2}, \kappa, \rho)$-computable.

6. Prove Lemma 6.4.5.

7. Show that the function $(f, n, k) \mapsto f^{(n)}$ for $f \in C^\infty([0;1], \mathbb{R})$ and $\max\{|f^{(n+1)}(x)| \mid 0 \le x \le 1\} \le k$ is $(\delta_\to^{[0;1]}, \nu_\mathbb{N}, \nu_\mathbb{N}, \delta_\to^{[0;1]})$-computable.

♦ 8. Show that there is a computable function $f \in C^\infty([0;1], \mathbb{R})$ such that the sequence $n \mapsto f^{(n)}$ is not $(\nu_\mathbb{N}, \delta_\to^{[0;1]})$-computable. (See [PER89], Chap. 1, Theorem 3.)

♦ 9. Show that there is a computable function $f \in C^\infty(\mathbb{R}, \mathbb{R})$ such that f' is not computable. (See [PER89], p. 58.)

♦ 10. The function $f \in C[0;1]$ defined by

$$f(x) := \sum_{n=0}^\infty \left(\frac{3}{4}\right)^n \cdot g(4^n \cdot x)\ ,$$

where g is a sequence of peaks defined by $g(m+x) := g(m-x) := 1-x$ for all even $m \in \mathbb{N}$ and $0 \le x \le 1$, is continuous but nowhere differentiable [Rud64]. Show that f is computable.

6.5 Analytic Functions

In this section we deal with computability of analytic functions. A complex function $f :\subseteq \mathbb{C} \to \mathbb{C}$ is said to be *analytic* in the region $\Omega \subseteq \mathbb{C}$, iff $\Omega \subseteq \mathrm{dom}(f)$ and the derivative

$$f'(z) = \lim_{h \to 0} \frac{f(z+h) - f(z)}{h}$$

exists for all $z \in \Omega$, where a region is an open connected subset of $\mathbb{C}$ [Ahl66].

For every $z_0 \in \Omega$ there is a (unique) sequence $a_0, a_1, \ldots$ of complex numbers such that

$$f(z) = \sum_{n=0}^{\infty} a_n \cdot (z - z_0)^n \quad \text{for all} \ \ z \in B(z_0, r) \ ,$$

if the open ball $B(z_0, r)$ is contained in Ω. By Theorem 4.3.12 the function f is computable on the closed ball $\overline{B}(z_0, r)$, if z_0 and $n \mapsto a_n$ are computable and $r < R$ where $R := 1/\limsup_{n \to \infty} \sqrt[n]{|a_n|}$ is the radius of convergence of the power series. Below we will show that the sequence $n \mapsto a_n$ of coefficients can be computed from f, z_0 and a radius $r > 0$ such that $\overline{B}(z_0, r) \subseteq \Omega$.

In general it is impossible to compute the radius of convergence R of a power series $\sum_{i=0}^{\infty} a_i \cdot z^i$ from the sequence $(a_i)_{i \in \mathbb{N}}$ of coefficients. The function $(a_i)_{i \in \mathbb{N}} \mapsto R$ for real numbers a_i is not even ($[\rho]^\omega, \rho$)-continuous (Exercise 6.5.1). For a computable sequence $(a_i)_{i \in \mathbb{N}}$ of coefficients, the radius of convergence can not be even left- or right computable (use, for example, Exercise 4.3.16 and see [ZW99]. Although the power series $\sum_{i=0}^{\infty} a_i \cdot z^i$ converges for all $\{z \in \mathbb{C} \mid z < R\}$, for computable $(a_i)_{i \in \mathbb{N}}$ with even computable radius $R = 1/\limsup_{n \to \infty} \sqrt[n]{|a_n|}$ of convergence, the function $z \mapsto \sum_{i=0}^{\infty} a_i \cdot z^i$ is not computable on the full open disc $\{z \in \mathbb{C} \mid z < R\}$ in general:

Example 6.5.1 (analytic continuation [CPE75]). Let $A \subseteq \mathbb{N}$ be an r.e. set which is not recursive. There is an injective computable function $a : \mathbb{N} \to \mathbb{N}$ with $A = \text{range}(a)$. Define $w : \mathbb{N} \to \mathbb{N}$ by $w(n) := \max\{m \mid a(m) \leq n\}$. Then $n \in A \iff n \in \{a(0), \ldots, a(w(n))\}$. Assume that $w(n) \leq d(n)$ for some computable function $d : \mathbb{N} \to \mathbb{N}$. Then $n \in A \iff n \in \{a(0), \ldots, a(d(n))\}$, which gives a decision procedure for the set A, which is non-recursive by assumption. Therefore, w has no computable upper bound. Define

$$f(z) := \sum_{i=0}^{\infty} \frac{z^i}{a(i)^i} \ .$$

Since $\limsup_{i \to \infty}(1/a(i)) = 0$, the radius of convergence is ∞. We assume that f is computable on $\mathbb{C}$ (or on $\mathbb{R}$). Then the restriction $g : \mathbb{N} \to \mathbb{R}$ of f to $\mathbb{N}$ is ($\nu_\mathbb{N}, \rho$)-computable. From Example 4.1.10 we conclude that there is a computable function $h : \mathbb{N} \to \mathbb{N}$ with $g(n) \leq h(n)$ for all n. For any $n \in \mathbb{N}$ and $m := w(n)$ we obtain $a(m) \leq n$ and

$$h(2n) > \left(\frac{2n}{a(m)}\right)^m \geq \left(\frac{2n}{n}\right)^m = 2^m = 2^{w(n)} > w(n) \ .$$

Therefore, w has the computable upper bound $n \mapsto h(2n)$ (contradiction). Although by Theorem 4.3.12.1 the function f is computable on every disc $\{z \mid |z| \leq n\}$ it is not computable on the whole plane $\mathbb{C}$. □

We will say that $f :\subseteq \mathbb{C} \to \mathbb{C}$ is analytic on $A \subseteq \mathbb{C}$, iff $A \subseteq \Omega$ and f is analytic in Ω for some region Ω.

Theorem 6.5.2 (power series). Let $z_0 \in \mathbb{C}$ and $r \in \mathbb{R}$, $r > 0$, be computable and let $f :\subseteq \mathbb{C} \to \mathbb{C}$ be analytic on $A := \overline{B}(z_0, r)$. If $f\rfloor_A$ is (ρ^2, ρ^2)-computable, then the sequence $(a_i)_{i \in \mathbb{N}}$ such that $f(z) = \sum_{n=0}^{\infty} a_n \cdot (z - z_0)^n$ for $z \in A$ is $(\nu_{\mathbb{N}}, \rho^2)$-computable.

Proof: By Cauchy's integral formula [Ahl66]

$$a_n = \frac{1}{2\pi i} \int_C \frac{f(\zeta)}{(\zeta - z_0)^{n+1}} \, d\zeta$$

where C is the circle with center z_0 and radius r, and so

$$a_n = \int_0^{2\pi} \frac{r}{2\pi} \cdot \frac{f(z_0 + r \cdot e^{it}) \cdot e^{it}}{(r \cdot e^{it})^{n+1}} \, dt \ .$$

Since the constants and functions occurring in the integrand are computable, the integrand as well as its real part RE and its imaginary part IM are computable function of n and t:

$$a_n = \int_0^{2\pi} \text{RE}(n, t) \, dt + i \cdot \int_0^{2\pi} \text{IM}(n, t) \, dt \ .$$

By Theorem 3.3.15, the function $n \mapsto g_n$, where $g_n(t) := \text{RE}(n, t)$ is $(\nu_{\mathbb{N}}, \delta_{\to}^{[0;2\pi]})$-computable. By Theorem 6.4.1.1 (generalized to $A = [0; 2\pi]$), the function $n \mapsto \int_0^{2\pi} \text{RE}(n, t) \, dt$ is $(\nu_{\mathbb{N}}, \rho)$-computable. For the same reason the imaginary part of a_n is a computable function of n, and so $n \mapsto a_n$ is computable. $\qquad\square$

If z_0 and the sequence $n \mapsto a_n$ are computable, then by Theorem 4.3.12.1, $z \mapsto \sum_{n=0}^{\infty} a_n \cdot (z - z_0)^n$ is computable on every closed ball $\{z \in \mathbb{C} \mid |z - z_0| \le r\}$ such that r is less than the radius of convergence. Using analytic continuation we can prove the following generalization:

Theorem 6.5.3 (analytic continuation). Let $f :\subseteq \mathbb{C} \to \mathbb{C}$ be analytic in the region $\Omega \subseteq \mathbb{C}$. If f is computable on some non-empty open subset $U \subseteq \Omega$, then f is computable on every compact subset $K \subseteq \Omega$.

Proof: First we show how the domain of computability can be extended by a single ball.

Proposition 1: If f is computable on $B(c_1, r_1)$, $c_2 \in B(c_1, r_1)$ and $\overline{B}(c_2, r_2) \subseteq \Omega$ (c_1, c_2 rational), then f is computable on $B(c_2, r_2)$.

Proof 1: There is some radius $r \in \mathbb{Q}$ such that $\overline{B}(c_2, r) \subseteq B(c_1, r_1)$. By Theorem 6.5.2, the sequence $(a_n)_{n \in \mathbb{N}}$ such that $f(z) = \sum_{n=0}^{\infty} a_n \cdot (z - c_2)^n$ for

$z \in B(c_2, r)$ is computable. Since $\overline{B}(c_2, r_2) \subseteq \Omega$, $r_2 < R$ where R is the radius of convergence of the power series. By Theorem 4.3.12, the function f is computable on $B(c_2, r_2)$.

Proposition 2: For every point $z' \in \Omega$ there is a sequence $B(c_1, r_1), \ldots, B(c_n, r_n)$ of balls with rational centers and radii such that $c_1 \in U$, $\overline{B}(c_k, r_k) \subseteq \Omega$, $c_{k+1} \in B(c_k, r_k)$ and $z' \in B(c_n, r_n)$ (Fig. 6.8).

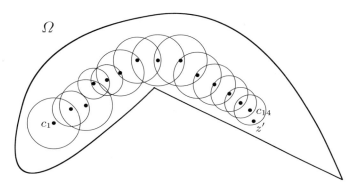

Fig. 6.8. A sequence $B(c_1, r_1), \ldots, B(c_{14}, r_{14})$ of balls from c_1 to z'

Proof 2: Let $V \subseteq \Omega$ be the set of all points z' with the above property. If $z' \in V$, then there is a sequence $B(c_1, r_1), \ldots, B(c_n, r_n)$ of balls with the above properties and $z' \in B(c_n, r_n)$. Then $B(z', \varepsilon) \subseteq B(c_n, r_n)$ and hence $B(z', \varepsilon) \subseteq V$ for some $\varepsilon > 0$. Therefore, every point in V has an open neighborhood in V, and so V is open.

On the other hand, suppose, $z'' \in \Omega \setminus V$. Then $B(z'', \varepsilon) \subseteq \Omega$ for some $\varepsilon > 0$. Suppose, $z' \in V$ for some $z' \in B(z'', \varepsilon/5)$. Then there is a sequence $B(c_1, r_1), \ldots, B(c_n, r_n)$ of balls with the above properties such that $z' \in B(c_n, r_n)$. There is some rational point $c_{n+1} \in B(c_n, r_n)$ such that $|c_{n+1} - z'| < \varepsilon/5$. Choose some rational number r_{n+1} such that $2\varepsilon/5 < r_{n+1} < \varepsilon/2$. Then $z'' \in B(c_{n+1}, r_{n+1})$ and $\overline{B}(c_{n+1}, r_{n+1}) \subseteq B(z'', \varepsilon) \subseteq \Omega$ and so $z'' \in V$ (contradiction). Therefore, the point z'' has an open neighborhood in $\Omega \setminus V$. Consequently, also $\Omega \setminus V$ is an open subset of Ω. Since Ω is connected, $V = \emptyset$ or $\Omega \setminus V = \emptyset$. Since $c_1 \in V$, $V = \Omega$. This proves Proposition 2.

Proposition 3: For every point $w \in \Omega$ there is an open ball $B(c_w, r_w) \subseteq \Omega$ with rational center and radius such that $w \in B(c_w, r_w)$ and the function f is computable on $B(c_w, r_w)$.

Proof 3: By Proposition 2 there is a sequence $B(c_1, r_1), \ldots, B(c_n, r_n)$ of balls with rational centers and radii such that $c_1 \in U$, $\overline{B}(c_k, r_k) \subseteq \Omega$, $c_{k+1} \in B(c_k, r_k)$ and $w \in B(c_n, r_n)$. Let $c_0 = c_1$ and $r_0 \in \mathbb{Q}$ such that $B(c_0, r_0) \subseteq U$. Applying Proposition 1 n-times we obtain that f is computable on $B(c_n, r_n)$. Define $c_w := c_n$ and $r_w := r_n$.

Finally we prove the statement of the theorem. By Proposition 3, for every $w \in K$ there is a rational ball $B(c_w, r_w) \subseteq \Omega$ such that $w \in B(c_w, r_w)$ and f is computable on $B(c_w, r_w)$. Therefore,

$$K \subseteq \bigcup \{ B(c_w, r_w) \mid w \in K \} .$$

Since K is compact, there are finitely many balls $B(c_1, r_1), \ldots, B(c_n, r_n)$ with rational centers and radii such that

$$K \subseteq B(c_1, r_1) \cup \ldots \cup B(c_n, r_n) \subseteq \Omega$$

and f is computable on $B(c_k, r_k)$ for $k = 1, \ldots, n$. Therefore, for every $k = 1, \ldots, n$ there is a Type-2 machine M_k computing the function f on $B(c_k, r_k)$. There is a Type-2 machine M_0 which for any $z \in K$ computes some index k such that $z \in B(c_k, r_k)$ (actually the machine M_0 computes a multi-valued function $F :\subseteq \mathbb{C} \rightrightarrows \mathbb{N}$). Combining M_0 with $M_1, \ldots, M_n$ we obtain a machine computing f on K. □

Interesting studies on computable analytic functions are [Mül95, Zho96, Her99a].

Exercises 6.5.

1. We use $[\rho]^\omega$ as our standard representation of sequences of real numbers (Definition 3.3.3). Show that the function $(b_i)_{i \in \mathbb{N}} \mapsto \limsup_{i \to \infty} b_i$ is not $([\rho]^\omega, \rho)$-continuous. (It is not even $([\nu_\mathbb{N}]^\omega, \rho)$-continuous for sequences of natural numbers.)

2. Show that the function $f(z) := \sum_{i=0}^\infty c_i \cdot z^i$ with $c_i := (1 + 2^{-a(i)})^i$ ($a(i)$ from Example 6.5.1) is computable on $\{ z \mid |z| \le r \}$ for every $r < 1$ but not on $\{ z \mid |z| < 1 \}$.

◆ 3. Construct a sequence $k \to f_k$ of analytic functions defined on $\mathbb{C}$ such that the sequence $k \mapsto f_k \rfloor_{A_1}$ is computable and the sequence $k \mapsto f_k \rfloor_{A_2}$ is not computable, where $A_1 := \overline{B}^e(0, 1)$ and $A_1 := \overline{B}^e(0, 2)$. (Hint: [PER89], Sect. 1.2)

4. Show that the operator H mapping every complex polynomial function $f : \mathbb{C} \to \mathbb{C}$ and every rectangle $Q \in \mathrm{Cb}^{(1)} \times \mathrm{Cb}^{(1)}$ such that f has no zero on the boundary of Q to the number of zeroes of f in Q is $([\rho^2 \to \rho^2], \mathrm{I}^1, \mathrm{I}^1, \nu_\mathbb{N})$-computable.
   (Hint: $H(f, Q) = \int_\gamma f'(z)/f(z) \, \mathrm{d}z/2\pi \mathrm{i}$ where the integral is along the boundary of Q [Ahl66].)

7. Computational Complexity

Conventional complexity theory refines computability theory. Total recursive word functions $f : \Sigma^* \to \Sigma^*$ or recursive subsets $X \subseteq \Sigma^*$ are classified with respect to the resource which machines need to compute or decide them, respectively. By means of notations complexity can be transferred to other sets. Complexity theory has grown to an extensive field with numerous important results.

In this chapter, we generalize traditional "discrete" computational complexity theory to TTE. In Sect. 7.1 we define time complexity of a Type-2 machine as the number of steps for producing an output of given length. We will consider merely *time complexity*. The basic concepts for *space complexity* are similar. However, seemingly no concrete results concerning space complexity in analysis are known. In addition to complexity we introduce *lookahead* of machines measuring the number of input symbols for producing an output of given length. In Sect. 7.2 we transfer complexity and lookahead from Cantor space to the real numbers, and in Sect. 7.3 we determine complexity bounds for some elementary functions. Finally, we suggest some definitions of the computational complexity of closed or compact subsets of $\mathbb{R}^n$.

7.1 Complexity of Type-2 Machine Computations

In this section we introduce time and lookahead for Type-2 machines and show, in particular, that complexity and lookahead can be bounded uniformly on compact subsets of the domain of a function.

Let M be a Type-2 machine such that $f_M :\subseteq (\Sigma^*)^m \to \Sigma^*$ (that is, an ordinary Turing machine with m input tapes). Then the computation time of M on input $(x_1, \ldots, x_m) \in (\Sigma^*)^m$ is defined by

$$\text{Time}_M(x_1, \ldots, x_m) := \begin{cases} \text{the number of steps which the machine M} \\ \text{on input } (x_1, \ldots, x_m) \text{ needs} \\ \text{until it reaches a HALT statement,} \end{cases}$$

and $\text{Time}_M(x_1, \ldots, x_m) := \infty$, if $f_M(x_1, \ldots, x_m)$ does not exist. This definition of Time_M can be extended straightforwardly to Type-2 machines M with finite or infinite inputs and finite output. We do not deal with this generalization.

However, for a machine M with infinite output the above definition is useless, since it does not halt on input $(x_1, \ldots, x_m)$, if $f_M(x_1, \ldots, x_m) \in \Sigma^\omega$ exists. In this case we introduce the "output precision" $k \in \mathbb{N}$ as a further parameter, and measure the time until the machine M has produced the k-th output symbol of its output $q \in \Sigma^\omega$. In addition to computation time the number of input symbols which the machine M uses for computing the first k output symbols is significant.

Definition 7.1.1 (time complexity and lookahead).

Let M be a Type-2 machine computing a function $f_M : \subseteq Y_1 \times \ldots \times Y_m \to \Sigma^\omega$, $(Y_1, \ldots, Y_m \in \{\Sigma^*, \Sigma^\omega\})$.
  1. For all $y \in Y_1 \times \ldots \times Y_m$ and for all $k \in \mathbb{N}$ define:

$$\mathrm{Time}_M(y)(k) := \begin{cases} \text{the number of steps which the machine } M \\ \text{on input } y \text{ needs} \\ \text{until it has printed the } k\text{-th output symbol,} \end{cases}$$

$$\mathrm{La}_M(y)(k) := \begin{cases} \text{the maximal number of symbols} \\ \text{which the machine } M \text{ on input } y \text{ reads} \\ \text{from some infinite input tape} \\ \text{until it has printed the } k\text{-th output symbol.} \end{cases}$$

  2. For all $A \subseteq \mathrm{dom}(f_M)$ and all $k \in \mathbb{N}$ define:

$$\mathrm{Time}_M^A(k) := \max_{y \in A} \ \mathrm{Time}_M(y)(k) \ ,$$
$$\mathrm{La}_M^A(k) := \max_{y \in A} \ \mathrm{La}_M(y)(k) \ .$$

Fig. 7.1 illustrates the definitions of time and lookahead.

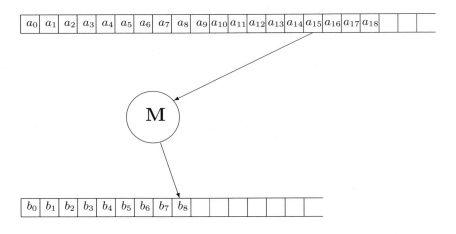

Fig. 7.1. In this situation, $\mathrm{Time}_M(y)(9) \geq 33$ (16 for moving the input head, 9 for writing the output symbols and 8 for moving the output head) and $\mathrm{La}_M(y)(9) = 16$

While for a Turing machine $\text{Time}_M(y)$ is a natural number, for a Type-2 machine with outputs in Σ^ω, $\text{Time}_M(y) :\subseteq \mathbb{N} \to \mathbb{N}$ and $\text{La}_M(y) :\subseteq \mathbb{N} \to \mathbb{N}$ are number functions (such that $\text{Time}_M(y)(0) = \text{La}_M(y)(0) = 0$ for all inputs y). Function values may exist for some but not for all $k \in \mathbb{N}$ (in this case $y \notin \text{dom}(f_M)$). Since $\text{La}_M(y)(k)$ is an input precision sufficient for M to determine $f_M(y)$ with precision k, the function $\text{La}_M(y)$ is a modulus of continuity of f_M at y.

As a fundamental fact, for every Turing machine M the property

$$\text{Time}_M(x) = t$$

is recursive in $x \in \Sigma^*$ and $t \in \mathbb{N}$. We generalize this theorem to Type-2 machines using Definition 2.4.1 and to lookahead on the input tapes.

Lemma 7.1.2. For every Type-2 machine M with $f_M :\subseteq Y \to \Sigma^\omega$
1. the function $(y, k) \mapsto \text{Time}_M(y)(k)$ is computable,
2. the function $(y, k) \mapsto \text{La}_M(y)(k)$ is computable,
3. the set $\{(y, k, t) \in Y \times \mathbb{N} \times \mathbb{N} \mid \text{Time}_M(y)(k) = t\}$ is decidable,
4. the set $\{(y, k, t) \in Y \times \mathbb{N} \times \mathbb{N} \mid \text{La}_M(y)(k) = t\}$ not decidable in general.

More concretely, there is a computable function $g :\subseteq Y \times \Sigma^* \to \Sigma^*$ such that $\text{Time}_M(y)(\nu_\mathbb{N}(u)) = \nu_\mathbb{N} \circ g(y, u)$, the set $\{(y, u, v) \mid \text{Time}_M(y)(\nu_\mathbb{N}(u)) = \nu_\mathbb{N}(v)\}$ is decidable etc..

Proof: 1. From M we can construct a Type-2 machine M'' which on input (y, u), $y \in Y$, $\nu_\mathbb{N}(u) = k$, simulates M on input y and counts the steps until the k-th output symbol has been written. Then it writes v, where $\nu_\mathbb{N}(v)$ is the number of steps.

2. (Similar to 1)

3. From M, we can construct a Type-2 machine M' which on input (y, u, v), $y \in Y$, $\nu_\mathbb{N}(u) = k$, $\nu_\mathbb{N}(v) = t$, simulates t steps of M on input y (if possible) and answers "yes", if the k-th output symbol has been written in the t-th step, and "no" otherwise.

4. (Exercise 7.1.5)   □

Lookahead is a lower bound of time, since reading a symbol from an input tape requires at least one step of computation.

Lemma 7.1.3. For every Type-2 machine M with $f_M :\subseteq Y \to \Sigma^\omega$, every $y \in Y$ and $A \subseteq Y$,

$$\text{La}_M(y)(k) \leq \text{Time}_M(y)(k) \quad \text{and} \quad \text{La}_M^A(k) \leq \text{Time}_M^A(k)$$

for all $k \in \mathbb{N}$.

A machine computing a function $f :\subseteq Y \to \Sigma^\omega$ can work economically with input information by reading for every output precision k only the input symbols which are necessary for fixing the first k output symbols. On the other hand, it can waste input information by reading many more symbols than necessary. Every computable function $f : \Sigma^\omega \to \Sigma^\omega$ can be computed by a machine not wasting input information [GS83]. Some functions have a trade-off between computation time and lookahead. Saving lookahead may increase the time complexity considerably [WK91](Exercise 7.1.7).

The composition $f \circ g$ of computable functions $f, g :\subseteq \Sigma^\omega \to \Sigma^\omega$ is computable. To prove this, consider Type-2 machines M_f and M_g computing f and g, respectively. The machine M for $f \circ g$ in the proof of Theorem 2.1.12 repeats the following statements using a common work tape for the output of M_g and the input of M_f:
- simulate M_g until it writes a symbol and interrupt,
- simulate one step of M_f.

This way, whenever M_f requires the next input symbol, it has already been computed by M_g. However, this method is not very efficient, since in general M_f produces symbols unnecessarily early. The following composition lemma can be obtained by a refined method, which determines an extension of the composition in general.

Lemma 7.1.4 (complexity of composition). Let $M, M_1, \ldots, M_n$ be Type-2 machines computing functions $f :\subseteq (\Sigma^\omega)^n \to \Sigma^\omega$, $g_1, \ldots, g_n :\subseteq (\Sigma^\omega)^m \to \Sigma^\omega$, respectively. Then there are a machine N computing an extension of $f \circ (g_1, \ldots, g_n)$ and a constant c such that

$$\mathrm{La}_N(p)(k) \leq \mathrm{La}'(p) \circ \mathrm{La}_M(g_1(p), \ldots, g_n(p))(k) \quad \text{and}$$
$$\mathrm{Time}_N(p)(k) \leq c \cdot \mathrm{Time}_M(g_1(p), \ldots, g_n(p))(k)$$
$$+ c \cdot \mathrm{Time}'(p) \circ \mathrm{La}_M(g_1(p), \ldots, g_n(p))(k) + c$$

for all $p \in (\Sigma^\omega)^m$ and $k \in \mathbb{N}$, where

$$\mathrm{La}'(p)(l) := \max\{\mathrm{La}_{M_1}(p)(l), \ldots, \mathrm{La}_{M_n}(p)(l)\} \quad \text{and}$$
$$\mathrm{Time}'(p)(l) := \max\{\mathrm{Time}_{M_1}(p)(l), \ldots, \mathrm{Time}_{M_n}(p)(l)\} .$$

Proof: First consider the special case $m = n = 1$. Let N be a Type-2 machine with a common work tape for the output of M_1 and the input of M which on input p iterates the following statements:
- simulate M until it requires the next input symbol and interrupt,
- simulate M_1 on input p until it writes the next output symbol and interrupt.

Consider $p \in \mathrm{dom}(f \circ g_1)$. Upon input p, N writes $g_1(p)$ on the common input/output tape and $f \circ g_1(p)$ on the output tape. For producing the first k output symbols, N reads the first $\mathrm{La}_M(g_1(p))(k)$ symbols of $g_1(p)$. These and only these symbols are written on the input/output tape by simulation of M_1 up to this moment. Therefore,

$$\mathrm{La}_N(p)(k) \le \mathrm{La}'(p) \circ \mathrm{La}_M(g_1(p))(k) \quad \text{and}$$

$$\mathrm{Time}_N(k) \le c \cdot \mathrm{Time}_M(g_1(p))(k) + c \cdot \mathrm{Time}'(p) \circ \mathrm{La}_M(g_1(p))(k) + c \ .$$

Now consider $m = 1$ and $n > 1$. In this case, N has n additional intermediate input/output tapes and simulates one of the machines $M_1, \ldots, M_n$, whenever the machine M needs a new input symbol. However, notice that now n machines need access to the single input tape containing $p \in \Sigma^\omega$. The time estimation remains true, if N has n independent read-only, one-way heads on its input tape. A straightforward simulation of such a multi-head tape by an ordinary one-way input tape and n work tapes as input tapes for $M_1, \ldots, M_n$, respectively, requires time $O(k^2)$ in the worst case for reading at most the first k symbols with each head. However, there is a simulation working in time $O(k)$ (Exercise 7.1.8).

The generalization to $m > 1$ is straightforward. □

In ordinary complexity theory computation time is usually measured as a function of the size (for example, $\mathrm{size}(x_1, \ldots, x_m) := \max_{j=1}^m |x_j|$ or $\mathrm{size}(x_1, \ldots, x_m) := |x_1| + \ldots + |x_m|$) of the input $(x_1, \ldots, x_m) \in (\Sigma^*)^m$ by taking the maximum over all inputs of size n:

$$\mathrm{time}_M(n) := \max\{\mathrm{Time}_M(x_1, \ldots, x_m) \mid \mathrm{size}(x_1, \ldots, x_m) = n\} \ . \quad (7.1)$$

The maximum exists, since there are only finitely many inputs of each size n. As an example consider multiplication of natural numbers in binary notation $\nu_\mathbb{N}$. Applying the school method for multiplication a Turing machine M can be constructed such that for some constant c,

$$\nu_\mathbb{N}(u) \cdot \nu_\mathbb{N}(v) \;=\; \nu_\mathbb{N} \circ f_M(u, v) \ ,$$

$$\mathrm{Time}_M(u, v) \;\le\; c \cdot n^2 \quad \text{where} \quad n = \max(|u|, |v|)$$

for all $u, v \in \mathrm{dom}(\nu_\mathbb{N})$. Therefore, there is a Turing machine which multiplies natural numbers in binary notation in time $\mathrm{time}_M \in O(n^2)$.

Since $|x|$ is not defined for $x \in \Sigma^\omega$, Definition (7.1) for uniform time complexity $\mathrm{time}_M(n)$ cannot be generalized to machines with infinite input tapes. In this more general case, we will consider maxima over *compact* subsets $A \subseteq Y_1 \times \ldots \times Y_m$ (Definition 2.2.1) instead of finite subsets. This is a generalization, since a subset of the space $(\Sigma^*)^m$ with discrete topology is compact, iff it is finite. For every computable time bound $t : \mathbb{N} \to \mathbb{N}$, the set of inputs for which a Type-2 machine works in time t is a closed and hence compact subset of $\mathrm{dom}(f_M)$ with r.e. complement (Exercise 7.1.6). The complexity of a Type-2 machine can be bounded by a computable function uniformly on every compact subset of its domain which has an r.e. complement (Definitions 2.4.1, 2.4.9), and this is true for lookahead replacing time accordingly.

Theorem 7.1.5 (uniform complexity bounds on compact sets).
Let M be a Type-2 machine with $f_M :\subseteq Y \to \Sigma^\omega$. For every compact set $A \subseteq \mathrm{dom}(f_M)$ with r.e. complement there are computable functions $h, h' : \mathbb{N} \to \mathbb{N}$ such that

$$\mathrm{Time}_M^A(k) \le h(k) \quad \text{and} \quad \mathrm{La}_M^A(k) \le h'(k)$$

for all $k \in \mathbb{N}$.

Proof: We prove only the first statement. The proof of the second one is almost identical. First we show that the set A has a very simple compact superset $B \supseteq A$. Since $A \subseteq \bigcup \{v \circ Y \mid v \in (\Sigma^*)^m\}$ (Definition 2.2.2), every set $v \circ Y$ is open and A is compact, there are words $v_1, \ldots, v_r$ such that $A \subseteq v_1 \circ Y \cup \ldots \cup v_r \circ Y$. Every set $v_i \circ Y$ is the product of sets $\{u\} \subseteq \Sigma^*$ or $u\Sigma^\omega \subseteq \Sigma^\omega$ which are not only open but also compact, and so $v_i \circ Y$ is compact. Therefore, also the set $B := v_1 \circ Y \cup \ldots \cup v_r \circ Y$ is compact.

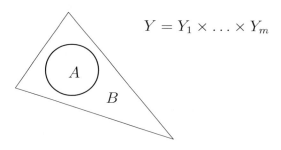

$$Y = Y_1 \times \ldots \times Y_m$$

Fig. 7.2. The compact set A and its simple compact superset B. The set $Y \setminus A$ is recursively enumerable

For a Type-2 machine L with $f_L :\subseteq Y \to \Sigma^\omega$, let T_L be the set of all $(k, u, t) \in \mathbb{N} \times (\Sigma^*)^m \times \mathbb{N}$ such that for any input $y = (y_1, \ldots, y_m) \in u \circ Y$, the machine L writes the k-th output symbol in t steps reading at most the prefix u_i of y_i for all i with $Y_i = \Sigma^\omega$ (where $u = (u_1, \ldots, u_m)$). The set T_L is recursive.

By Definition 2.4.1 there is a Type-2 machine N such that $\mathrm{dom}(f_N)$ is the set $Y \setminus A$ (Fig. 7.2). Therefore, for every $k \in \mathbb{N}$ and $y \in B$ we have some $u_y \in (\Sigma^*)^m$ and $t_y \in \mathbb{N}$ such that $(k, u_y, t_y) \in T_M$ or $(k, u_y, t_y) \in T_N$. Clearly, $B \subseteq \bigcup \{u_y \circ Y \mid y \in B\}$. By compactness of B, there is a finite cover, and so there are finitely many pairs $(u^1, t^1), \ldots, (u^s, t^s)$ such that $B \subseteq u^1 \circ Y \cup \ldots \cup u^s \circ Y$ and $(k, u^i, t^i) \in T_M$ or $(k, u^i, t^i) \in T_N$ for $i = 1, \ldots, s$.

Since $B \subseteq u^1 \circ Y \cup \ldots \cup u^s \circ Y$ is recursive in $\iota(u^1) \ldots \iota(u^s) \in \Sigma^*$ and T_M and T_N are recursive, there is a Turing machine L which on input $k \in \mathbb{N}$

- determines by exhaustive search pairs $(u^1, t^1), \ldots, (u^s, t^s)$ such that $B \subseteq u^1 \circ Y \cup \ldots \cup u^s \circ Y$ and $(k, u^i, t^i) \in T_M$ or $(k, u^i, t^i) \in T_N$ for $i = 1, \ldots, s$.
- determines $t := \max\{t^1, \ldots, t^s\}$ as its output.

Consider $y \in A$ and $k \in \mathbb{N}$. Then $y \notin \mathrm{dom}(f_N)$ and $y \in u^i \circ Y$ for some i. Therefore, $(k, u^i, t^i) \in T_M$ and so $\mathrm{Time}_M(y)(k) \leq t^i \leq f_L(k)$. $\qquad \square$

If the complement of the compact set A is not r.e., bounds h and h' still exist, which, however, might be non-computable (Exercise 7.1.2). If a compact set $A \subseteq Y$ has a recursive complement (Definition 2.4.9), then the functions Time_M^A and La_M^A are not only recursively bounded but are recursive functions themselves (Exercise 7.1.3). For both cases versions computable in A can be proved (Exercise 7.1.4).

The exact computation time functions $\mathrm{Time}_M(y) :\subseteq \mathbb{N} \to \mathbb{N}$ for $y \in Y$ or $\mathrm{Time}_M^A :\subseteq \mathbb{N} \to \mathbb{N}$ for $A \subseteq \mathrm{dom}(f_M)$ are very sensitive to modifications of the Turing machine model. Such a modification may change complexities by additive or multiplicative constants. Therefore, also in Type-2 theory commonly "robust" properties like $\mathrm{Time}_M(y) \in O(h)$, where

$$O(h) := \{g : \mathbb{N} \to \mathbb{N} \mid (\exists c)(\forall k) g(k) \leq c \cdot h(k) + c\},$$

are considered instead of the sensitive properties $(\forall k)\mathrm{Time}_M(y)(k) \leq h(k)$. On the other hand, lookahead is robust. Translating Turing machines from one model to the other usually can be arranged in such a way that the lookahead is preserved exactly. Therefore, it is reasonable to consider the functions $\mathrm{La}_M(y)$ and La_M^A exactly and not only their order of magnitude.

By Theorem 7.1.5, every machine is recursively time bounded on every compact subset of its domain with r.e. complement. On the other hand, by the next theorem for every computable time bound $t : \mathbb{N} \to \mathbb{N}$ there is a computable function not computable in time t. The next theorem is a way of expressing this.

Theorem 7.1.6 (complexity hierarchy). For every computable element $y \in Y$ and every computable $t : \mathbb{N} \to \mathbb{N}$ there is a sequence $q \in \Sigma^\omega$ such that the following holds.
If M is a Type-2 machine such that $f_M(y) = q$, then

$$\mathrm{Time}_M(y)(k) > t(k) \quad \text{for infinitely many} \quad k \in \mathbb{N}.$$

Proof: For the sake of simplicity consider $Y = \Sigma^\omega$. The general case can be proved similarly. Since y is computable, $y = f_L()$ for some Type-2 machine L by Definition 2.1.3. From Type-1 complexity theory [HU79] we know that there is some $q \in \Sigma^\omega$ such that there are no Type-2 machine K and no constant c such that $f_K() = q$ and

$$\text{Time}_K()(k) \leq c \cdot t(k) + c \cdot \text{Time}_L() \circ t(k) + c \quad \text{for all} \ \ k \in \mathbb{N} . \qquad (7.2)$$

Assume that there is some machine M such that

$$\text{Time}_M(y)(k) \leq t(k) \quad \text{for almost all} \ \ k \in \mathbb{N} . \qquad (7.3)$$

By the composition lemma 7.1.4 there are a machine N and a constant c such that $f_N = f_M \circ f_L$ (hence $f_N() = q$) and for all $k \in \mathbb{N}$,

$$\text{Time}_N()(k) \leq c \cdot \text{Time}_M(y)(k) + c \cdot \text{Time}_L() \circ \text{La}_M(y)(k) + c .$$

Since $\text{La}_M(y)(k) \leq \text{Time}_M(y)(k)$ (Lemma 7.1.3) and $\text{Time}_L()$ is increasing,

$$\text{Time}_N()(k) \leq c \cdot \text{Time}_M(y)(k) + c \cdot \text{Time}_L() \circ \text{Time}_M(y)(k) + c$$

for all k and so

$$\text{Time}_N()(k) \leq c \cdot t(k) + c \cdot \text{Time}_L() \circ t(k) + c$$

for almost all k. Therefore, there is a constant $c' \geq c$ such that

$$\text{Time}_N()(k) \leq c' \cdot t(k) + c' \cdot \text{Time}_L() \circ t(k) + c'$$

for all k, contradicting (7.2). $\square$

Notice that the sequence q is so complicated that no machine can compute it from y in time t. By Theorems 7.1.5 and 7.1.6 our definition of Time induces a non-trivial complexity hierarchy on the computable Σ^ω-valued functions.

Definition 7.1.7 (complexity class). *Let $s, t : \mathbb{N} \to \mathbb{N}$ be number functions and let $Y = Y_1 \times \ldots \times Y_m$ ($Y_1, \ldots, Y_m \in \{\Sigma^*, \Sigma^\omega\}$).*
1. *A point $p \in \Sigma^\omega$ is computable in time t, iff there is some Type-2 machine M such that $f_M() = p$ and $\text{Time}_M() \in O(t)$.*
2. *A function $f :\subseteq Y \to \Sigma^\omega$ is computable on $A \subseteq \text{dom}(f)$ in time t (and simultaneously with lookahead s), iff there is some Type-2 machine M such that $f_M(y) = f(y)$ for all $y \in A$ and $\text{Time}_M^A \in O(t)$ (and $\text{La}_M^A(k) \leq s(k)$ for all $k \in \mathbb{N}$).*

Property 1 refines computability from Definition 2.1.3. Notice that in Property 2 the least upper bound of all functions $\text{Time}_M(y)$ with $y \in A$ must be in $O(t)$. Notice that we do not use the weaker condition "$\text{Time}_M(y) \in O(t)$ for all $y \in A$" where for every $y \in A$ there is some constant c_y depending on y such that $\text{Time}_M(y)(k) \leq c_y \cdot t(k) + c_y$.

Definition 7.1.8 (M is finite on X). *Consider a Type-2 machine M, which for any input $y \in X$ ($X \subseteq \text{dom}(f_M)$) uses at most d_X tape cells on its work tapes (where d_X is a constant not depending on y). In this case we will say that M is finite on X.*

If M is finite on X, then there is a constant c such that $\mathrm{Time}_M(y)(k) \le c \cdot \mathrm{La}(y)(k) + c$ for all $y \in X$ and $k \in \mathbb{N}$.

Exercises 7.1.

$\Diamond$ 1. Show that for every Type-2 machine M, the property $\mathrm{Time}_M(y)(k) \le t$ is decidable in y, k and t.

2. Let M be a Type-2 machine with $f_M :\subseteq Y \to \Sigma^\omega$. Show that for every compact set $A \subseteq \mathrm{dom}(f_M)$ $\mathrm{Time}_M^A(k)$ and $\mathrm{La}_M^A(k)$ exist for all $k \in \mathbb{N}$.

$\blacklozenge$ 3. Let M be a Type-2 machine and let $A \subseteq \mathrm{dom}(f_M)$ be a compact set with recursive open complement. Show that $\mathrm{Time}_M^A : \mathbb{N} \to \mathbb{N}$ and $\mathrm{La}_M^A : \mathbb{N} \to \mathbb{N}$ are computable functions.

$\blacklozenge$ 4. For $Y := Y_1 \times \ldots \times Y_m$ $(Y_1, \ldots, Y_m \in \{\Sigma^*, \Sigma^\omega\})$ we define representations $\kappa_>^Y$ and κ^Y (which correspond to the representations $\kappa_>$ and κ of the compact subsets on $\mathbb{R}^n$, (see Sect. 5.2)) of the compact subsets of Y:

$$\kappa_>^Y \langle v, p \rangle = X : \Longleftrightarrow \begin{cases} v = \langle v_1, \ldots, v_s \rangle, \quad X \subseteq v_1 \circ Y \cup \ldots \cup v_s \circ Y \\ \text{and } Y \setminus X = \bigcup \{ w \circ Y \mid \iota(w) \lhd p \}, \end{cases}$$

$$\kappa^Y \langle v, p, q \rangle = X : \Longleftrightarrow \begin{cases} \kappa_>^Y \langle v, p \rangle = X \quad \text{and} \\ (\forall\, w)(w \circ Y \cap X \ne \emptyset \iff \iota(w) \lhd q) \end{cases}$$

(Definitions 2.2.2, 2.4.9). Let M be a Type-2 machine such that $f_M :\subseteq Y \to \Sigma^\omega$.

   a) Show that the multi-valued function mapping every compact set $X \subseteq \mathrm{dom}(f_M)$ to some upper bound $h : \mathbb{N} \to \mathbb{N}$ of Time_M^X is $(\kappa_>^Y, [\nu_\mathbb{N} \to \nu_\mathbb{N}]_\mathbb{N})$-computable.

   b) Show that the function $X \mapsto \mathrm{Time}_M^X$ for compact $X \subseteq \mathrm{dom}(f_M)$ is $(\kappa^Y, [\nu_\mathbb{N} \to \nu_\mathbb{N}]_\mathbb{N})$-computable.

   Show that the above properties hold for lookahead substituted for time accordingly.

5. Show that the relation $\mathrm{La}_M(y)(k) = t$ is not decidable in general.

$\blacklozenge$ 6. Let $\kappa_>^Y$ be the representation defined in Exercise 7.1.4. Show that for every Type-2 machine M the function

$$t \mapsto \{ y \in Y \mid (\forall k)\ \mathrm{Time}_M(y)(k) \le t(k) \}, \quad t : \mathbb{N} \to \mathbb{N},$$

is $([\nu_\mathbb{N} \to \nu_\mathbb{N}]_\mathbb{N}, \kappa_>^Y)$-computable. (Is this function $([\nu_\mathbb{N} \to \nu_\mathbb{N}]_\mathbb{N}, \kappa^Y)$-computable?) Show that the statement holds accordingly with lookahead replacing time.

$\blacklozenge$ 7. For a continuous function $f : \Sigma^\omega \to \Sigma^\omega$, define the minimal modulus of continuity or *dependence* in $p \in \Sigma^\omega$ by

$$\mathrm{Dep}_f(p)(k) := \min\{ n \in \mathbb{N} \mid f[p_{<n}\Sigma^\omega] \subseteq \left(f(p) \right)_{<k} \Sigma^\omega \}$$

for all $k \in \mathbb{N}$. Let M be a Type-2 machine computing $f : \Sigma^\omega \to \Sigma^\omega$. Construct a Type-2 machine N computing f such that $\mathrm{Dep}_f = \mathrm{La}_M$. Find an upper bound of Time_N in terms of Time_M. (There is a trade-off between lookahead and time complexity, see [GS83, WK91].)

8. Show that a Turing tape with k one-way read-only heads can be simulated in linear time by a tape with a single one-way read-only head and k ordinary Turing tapes. Hint: For the sake of simplicity consider $k = 2$ and assume that Head 1 is never to the left of Head 2. Let Tape a simulate Head 1 and Tape b simulate Head 2. The simulating machine copies a new symbol from the input tape to Tape a and reads it as soon as Head 1 in the original machine reads a new symbol. The symbols for Tape b are copied from Tape a: whenever the symbols on Tape b are exhausted all available further symbols are copied from Tape a.

7.2 Complexity Induced by the Signed Digit Representation

In Definition 3.1.3 we have transferred computability of points and functions from the "concrete" sets Σ^* and Σ^ω to "abstract" represented sets. In this section we discuss how *computational complexity* can be transferred from Σ^ω to the real numbers and real functions via representations. In Sect. 4.1 we selected the three representations $\rho_<$, $\rho_>$ and ρ of the real numbers which induce important computability concepts. A complexity concept for the $\rho_<$-computable or the $\rho_>$-computable real numbers induces a complexity concept on the set of all r.e. subsets on $\mathbb{N}$ (Exercise 7.2.1). However, recursion theory offers no reasonable definition of computational complexity separating the r.e. sets. Therefore, we start from the computability concept on $\mathbb{R}$ induced by our standard representation ρ (and the equivalent ones).

First we show that computational complexity of real numbers and real functions defined straightforwardly via the Cauchy representation is useless. Then we introduce a new representation ρ_{sd}, the *signed digit representation*, of the real numbers, discuss its properties and show that it induces a reasonable concept of computational complexity on the real numbers. Finally, we list some modified definitions of complexity.

By Definition 3.1.3, a real number is ρ_C-computable, iff it has a computable ρ_C-name $p \in \Sigma^\omega$, and a real function is (ρ_C, ρ_C)-computable, iff it has a computable (ρ_C, ρ_C)-realization $g :\subseteq \Sigma^\omega \to \Sigma^\omega$. In the following example we show that straightforward extension of this definition to computational complexity is useless.

Example 7.2.1 (a useless definition of complexity). Consider the following definition:

A real number x is computable in time $t : \mathbb{N} \to \mathbb{N}$, iff it has a ρ_C-name $p \in \Sigma^\omega$ computable in time t.

We show that according to this definition every computable real number is computable in time $\mathrm{id}_\mathbb{N} : k \mapsto k$ (that is, linear time).

Let x be ρ_C-computable. Then it has a computable ρ_C-name $p = \iota(w_0)\iota(w_1)\ldots$ such that $x = \lim_{i\to\infty} \nu_{\mathbb{Q}}(w_i)$. There are words u_i, v_i such that $w_i = u_i/v_i$ and $\iota(w_i) = 11\widetilde{u_i}0/\widetilde{v_i}011$, where $\widetilde{v} = 0a_10a_2\ldots0a_k$, if $v = a_1a_2\ldots a_k$.

There is a Type-2 machine M such that $f_M(\lambda) = p = \iota(w_0)\iota(w_1)\ldots$. We will outline a Type-2 machine N which works in linear time on input λ such that $f_N(\lambda) = q = \iota(w_0')\iota(w_1')\ldots$, where $w_i' = u_i0^{n_i}/v_i0^{n_i}$ for some $n_i \in \mathbb{N}$. Then by Definition 3.1.2.4, $\rho_C(p) = \rho_C(q)$.

Let N be a Type-2 machine which using M as a submachine on input λ first writes the words $\widetilde{u_0}, \widetilde{v_0}$ on work tapes without writing on the output tape and then operates in stages $i = 0, 1, \ldots$ as follows:

Stage i: Suppose, N has already stored the words $\widetilde{u_i}$ and $\widetilde{v_i}$ on work tapes.

1. N writes 11 and then transfers $\widetilde{u_i}$ to the output tape.
2. Simulating the machine M, the machine N computes u_{i+1} and v_{i+1}, and for every step of simulation it writes 00 on the output tape and also on some work tape C.
3. N writes 0/ on the output tape and transfers $\widetilde{v_i}$ to the output tape.
4. N transfers the content of Tape C to the output tape and writes 011 on the output tape.

In fact, on input λ the machine N writes a sequence $f_N(\lambda) = q = \iota(w_0')\iota(w_1')\ldots$ where $\iota(w_i') = 11\widetilde{u_i}0^{2n_i}0/\widetilde{v_i}0^{2n_i}011$, if $\iota(w_i) = 11\widetilde{u_i}0/\widetilde{v_i}011$, where n_i is a computation time in Stage i), and so $\rho_C(q) = x$.

To produce the words $\widetilde{u_0}$ and $\widetilde{v_0}$ without writing on the output tape the machine N needs some number c_1 of steps. In the following stages N writes continuously symbols on the output tape needing at most c_2 elementary Turing steps of computation for each output symbol (for some constant c_2). Therefore, Time$_M(\lambda) \in O(k)$, that is, N computes q in linear time.

In the proof $\nu_{\mathbb{Q}}$-names of rational numbers are padded in such a way that the sequence of symbols can be computed in linear time. Since every computable real number is computable in linear time, the above definition is useless. For the same reason the following definition of computational complexity of a computable real function is useless.

A real function $f : \mathbb{R} \to \mathbb{R}$ is computable in time $t : \mathbb{N} \to \mathbb{N}$, iff it has a (ρ_C, ρ_C)-realization computable in time t.

The "padding" method works similarly for the other representations equivalent to the Cauchy representation ρ_C as introduced in Sect. 4.1. □

Padding does not decrease complexity, if we measure the time not as a function of the number of printed symbols but as a function of precision on the real line.

Example 7.2.2 (complexity of $\sqrt{2}$). By simple trial and error one can determine successively digits $a_1, a_2, \ldots \in \{0, 1\}$ such that

$$(1{\bullet}a_1a_2\ldots a_k)_2 < \sqrt{2} < (1{\bullet}a_1a_2\ldots a_k)_2 + 2^{-k}$$

where $(u)_2$ denotes the value of the finite binary fraction u. Since integers in binary representation of length k can be multiplied in time $O(k^2)$, the digits $a_1, a_2, \ldots, a_k$ can be determined in time $O(1^2 + \ldots + k^2) = O(k^3)$. If we define $w_i := \text{“}1a_1 \ldots a_i/10^{i}\text{”}$ (Definition 3.1.2), then $\iota(w_0)\iota(w_1) \ldots$ is a ρ_C-name of $\sqrt{2}$ and the prefix $\iota(w_0)\iota(w_1) \ldots \iota(w_k)$ can be computed in time $O(k^3)$. Therefore, $\sqrt{2}$ can be computed with precision 2^{-k} in time $O(k^3)$. $\square$

Whereas for constants measuring the time as a function of precision seems to be reasonable, this concept is useless for functions.

Example 7.2.3 (a further useless definition). Consider the following definition:

> A real function $f :\subseteq \mathbb{R} \to \mathbb{R}$ is computable at $x \in \text{dom}(f)$ in time $t : \mathbb{N} \to \mathbb{N}$, iff there are a Type-2 machine M and a constant c_x such that f_M is a (ρ_C, ρ_C)-realization of f and for every ρ_C-name p of x as an input the machine M computes the first k rational numbers of the resultant Cauchy sequence in at most $c_x \cdot t(k) + c_x$ steps.

Notice that a realizing machine must work correctly on *all* ρ_C-names of x, in particular, on extremely padded ones. As a consequence, $f(x) = f(y)$ for all $y \in \text{dom}(f)$ with $|x - y| < 1$, if $f :\subseteq \mathbb{R} \to \mathbb{R}$ is computable at $x \in \text{dom}(f)$ in time $t : \mathbb{N} \to \mathbb{N}$ (Exercise 7.2.2). Since most functions of interest are not piecewise constant, the above definition is useless. The argument holds similarly for the other representations equivalent to the Cauchy representation ρ_C as introduced in Sect. 4.1. $\square$

In Example 7.2.1, complexity classes are too big, and in Example 7.2.3 complexity classes are too small. In both cases the problems are caused by the "padded" or "highly redundant" ρ_C-names of real numbers. By Theorem 4.1.15, there is no representation δ of the real numbers which is equivalent to ρ_C (and hence induces the same computability theory on the real numbers as ρ_C) such that every $x \in \mathbb{R}$ has only a single δ-name.

We solve the problems by introducing the *signed digit representation* ρ_{sd} as a new standard representation of the real numbers which is equivalent to ρ_C but does not allow padding of names. It extends the ordinary representation $\rho_{b,2}$ (by infinite base-2 fractions, Definition 4.1.12) by admitting not only the digits 0 and 1 but also the digit −1 which we abbreviate by $\overline{1}$. Avizienis [Avi61] applies the signed digit notation for digital computer arithmetic and Wiedmer [Wie80] introduces the decimal signed digit representation of real numbers for computing real functions concretely.

Definition 7.2.4 (signed digit notation and representation).   *Let $\overline{1} \in \Sigma$ abbreviate −1. Define the signed digit notation ν_{sd} of the binary rational numbers and the signed digit representation $\rho_{\text{sd}} :\subseteq \Sigma^\omega \to \mathbb{R}$ of the real numbers as follows.*

$$\text{dom}(\rho_{\text{sd}}) := \begin{cases} \text{all } a_n \ldots a_0 \bullet a_{-1} a_{-2} \ldots \in \Sigma^\omega \quad \text{such that } n \geq -1, \\ a_i \in \{\overline{1}, 0, 1\} \quad \text{for } i \leq n, \\ a_n \neq 0, \quad \text{if } n \geq 0 \quad \text{and } a_n a_{n-1} \notin \{1\overline{1}, \overline{1}1\}, \text{ if } n \geq 1, \end{cases}$$

$$\text{dom}(\nu_{\text{sd}}) := \{u \bullet v \mid u, v \in \Sigma^*, \ u \bullet v 0^\omega \in \text{dom}(\rho_{\text{sd}})\},$$

$$\rho_{\text{sd}}(a_n \ldots a_0 \bullet a_{-1} a_{-2} \ldots) := \sum_{i=n}^{-\infty} a_i \cdot 2^i,$$
$$\nu_{\text{sd}}(u \bullet v) := \rho_{\text{sd}}(u \bullet v 0^\omega).$$

For $p := a_n \ldots a_0 \bullet a_{-1} a_{-2} \ldots \in \text{dom}(\rho_{\text{sd}})$ *and* $k \in \mathbb{N}$ *define*

$$p[k] := a_n \ldots a_0 \bullet a_{-1} \ldots a_{-k} \in \Sigma^*.$$

For reducing the number of names we have excluded the prefix 0, which could be omitted, the prefix $1\overline{1}$, which could be replaced by 1, and the prefix $\overline{1}1$, which could be replaced by $\overline{1}$.

Notice that $\nu_{\text{b},2}$ can be translated easily to ν_{sd} and $\rho_{\text{b},2}$ can be translated easily to ρ_{sd} (see Definition 4.1.12). Notice also that by separating negative and non-negative digits every ρ_{sd}-name p can be decomposed easily into two $\rho_{\text{b},2}$-names (Definition 4.1.12) q and r such that $\rho_{\text{sd}}(p) = \rho_{\text{b},2}(q) - \rho_{\text{b},2}(r)$. Similarly, every ν_{sd}-name w can be decomposed into two $\nu_{\text{b},2}$-names (Definition 4.1.12) u and v such that $\nu_{\text{sd}}(p) = \nu_{\text{b},2}(u) - \nu_{\text{b},2}(v)$. Therefore, the notations ν_{sd} and $\nu_{\text{b},2}$ can be translated to each other very easily.

Every ρ_{sd}-name p of a real number x can be considered as a strongly normalized nested sequence of closed intervals converging to x (cf. Lemma 4.1.6). Let $p := a_n \ldots a_0 \bullet a_{-1} a_{-2} \ldots \in \text{dom}(\rho_{\text{sd}})$. Then $\rho_{\text{sd}}(p[k]0^\omega)$ is the center of the closed interval $\rho_{\text{sd}}[p[k]\Sigma^\omega]$ and $\rho_{\text{sd}}(p[k]0^\omega) = z \cdot 2^{-k}$ for some integer $z \in \mathbb{Z}$. For this number z,

$$\rho_{\text{sd}}[p[k]\Sigma^\omega] = [z-1; z+1] \cdot 2^{-k} \quad (= [2z-2; 2z+2] \cdot 2^{-(k+1)}),$$
$$\rho_{\text{sd}}[p[k]\overline{1}\Sigma^\omega] = [2z-2; 2z] \cdot 2^{-(k+1)},$$
$$\rho_{\text{sd}}[p[k]0\Sigma^\omega] = [2z-1; 2z+1] \cdot 2^{-(k+1)},$$
$$\rho_{\text{sd}}[p[k]1\Sigma^\omega] = [2z; 2z+2] \cdot 2^{-(k+1)}.$$

Therefore, the name p determines a sequence $(J_k)_{k \in \mathbb{N}}$, $J_k := \rho_{\text{sd}}[p[k]\Sigma^\omega]$, such that

$$J_k \text{ has the center } \rho_{\text{sd}}(p[k]0^\omega) \text{ and length } 2 \cdot 2^{-k},$$

$$J_{k+1} \text{ is the } \begin{cases} \text{left} \\ \text{middle} \\ \text{right} \end{cases} \text{half of } J_k \text{ if } a_{k+1} = \begin{cases} \overline{1} \\ 0 \\ 1 \end{cases},$$

$$\{\rho_{\text{sd}}(p)\} = \bigcap_{i \in \mathbb{N}} J_i.$$

Notice that an ordinary $\rho_{\text{b},2}$-name determines a sequence $(J_k)_{k \in \mathbb{N}}$ of closed intervals such that J_{k+1} is either the left or the right half of J_k. As an example consider $p = 11\overline{1} \bullet \overline{1}01\overline{1}0 \ldots$. The first intervals $J_0, \ldots, J_5$ are shown in

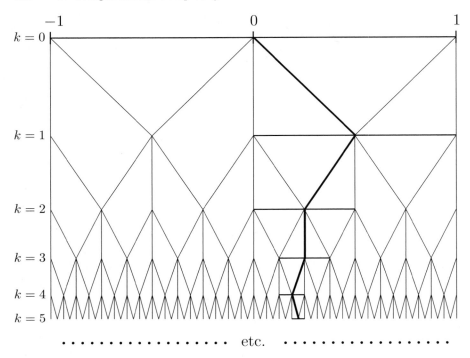

Fig. 7.3. The diagram of paths $k \mapsto \rho_{\mathrm{sd}}(p[k]0^\omega)$; the path for $p = {}_\bullet1\bar{1}0\bar{1}1\ldots$ and the closed intervals $\rho_{\mathrm{sd}}[p[k]\Sigma^\omega]$ in thick lines ($k = 0, \ldots, 5$)

Fig. 1.2. The sequences $k \mapsto \rho_{\mathrm{sd}}(p[k]0^\omega)$ can be visualized in a diagram. Fig. 7.3 shows the sequences for all $p \in \rho_{\mathrm{sd}}^{-1}[(-1;1)] \cup \{\bar{1}_\bullet0^\omega, 1_\bullet0^\omega\}$.

The condition $\big(a_n \neq 0$, if $n \geq 0$ and $a_n a_{n-1} \notin \{1\bar{1}, \bar{1}1\}$, if $n \geq 1\big)$ in Definition 7.2.4 implies

$$|\rho_{\mathrm{sd}}(u_\bullet p)| \leq 2^{|u|} \quad \text{and} \quad 2^{|u|-2} \leq |\rho_{\mathrm{sd}}(u_\bullet p)| \quad \text{if} \quad |u| \geq 2 \qquad (7.4)$$

and so for $|\rho_{\mathrm{sd}}(u_\bullet p)| \geq 4$,

$$2^{|u|-2} \leq |\rho_{\mathrm{sd}}(u_\bullet p)| \leq 2^{|u|} . \qquad (7.5)$$

Therefore, the length of the integer part of a name of x is roughly $\log_2 |x|$. If $q = u_\bullet a_{-1}a_{-2}\ldots$ and $x = \rho_{\mathrm{sd}}(q)$, then

$$|x - \rho_{\mathrm{sd}}(u_\bullet a_{-1}a_{-2}\ldots a_{-k}0^\omega)| \leq 2^{-k} .$$

If $|u| \geq 2$, then by (7.4),

$$\left| \frac{x - \rho_{\mathrm{sd}}(u_\bullet a_{-1}a_{-2}\ldots a_{-k}0^\omega)}{x} \right| \leq \frac{2^{-k}}{2^{|u|-2}} = 2^{-|u|-k+2} .$$

Therefore, the prefix $q[k] = u_\bullet a_{-1} \dots a_{-k}$ determines x up to an *absolute* error of 2^{-k}. And if $|u| \geq 2$ or $|x| \geq 4$, then $q[k]$ determines x up to a *relative* error which is $2^{-|u|-k+2}$. This is false for $|u| \leq 1$, since in this case x may be arbitrarily tight or even equal to zero.

While the set of ρ_C-names of a real number is "big", the set of ρ_{sd}-names is "small" where small means compact. More generally, $K := \rho_{sd}^{-1}[X]$ is compact for every compact subset X of $\mathbb{R}$. We prove a computable version using the representation $\kappa_>$ of the compact subsets of $\mathbb{R}^m$ (Definition 5.2.1) and the characterization of the r.e. open subsets of $(\Sigma^\omega)^m$ of Theorem 2.4.7.

Theorem 7.2.5 (signed digit representation).
1. $\rho_{sd} \equiv \rho$
2. For every $\kappa_>$-computable compact subset $X \subseteq \mathbb{R}^m$, the pre-image

$$\{(p_1, \dots, p_m) \mid (\rho_{sd}(p_1), \dots, \rho_{sd}(p_m)) \in X\} \subseteq (\Sigma^\omega)^m$$

is compact with r.e. complement.

Proof: 1. We prove $\rho_{sd} \leq \rho_C$: There is a Type-2 machine M which on input $p = a_m \dots a_0 a_{-1} a_{-2} \dots \in \mathrm{dom}(\rho_{sd})$ prints a sequence $q = \iota(w_0)\iota(w_1)\dots$ such that $\nu_{\mathbb{Q}}(w_i) = \rho_{sd}(p[i]0^\omega)$. Then $\rho_{sd}(p) = \rho_C(q)$, and so f_M translates ρ_{sd} to ρ_C.

We prove $\rho^a \leq \rho_{sd}$ (ρ^a from Lemma 4.1.6): there is a Type-2 machine M which on input $p = \iota(w_0)\iota(w_1)\dots \in \mathrm{dom}(\rho_C)$ determines digits $a_m, \dots, a_0 \in \{0, 1, \overline{1}\}$ such that

$$\mathrm{I}^1(u_0) \subseteq \rho_{sd}[a_m \dots a_0 \bullet \Sigma^\omega]$$

and prints "$a_m \dots a_0 \bullet$" and then successively for $k = 1, 2, \dots$ prints a symbol $a_{-k} \in \{0, 1, \overline{1}\}$ such that

$$\mathrm{I}^1(u_k) \subseteq \rho_{sd}[a_m \dots a_0 \bullet a_{-1} \dots a_{-k+1} a_{-k} \Sigma^\omega] .$$

Then the function f_M translates ρ^a to ρ_{sd}.

2. For the sake of simplicity we assume $m = 1$. The general case can be proved similarly. Let $X \subseteq \mathbb{R}$ be a $\kappa_>$-computable compact set. By Definition 5.2.1 there are a number $n \in \mathbb{N}$ and an r.e. set $B \in \Sigma^*$ such that

$$X \subseteq (-2^n; 2^n) \quad \text{and} \quad \mathbb{R} \setminus X = \bigcup_{w \in B} \mathrm{I}^1(w) .$$

Then the set

$$C := \{u \in \Sigma^* \mid (\exists w \in B)\, \rho_{sd}[u\Sigma^\omega] \subseteq \mathrm{I}^1(w) \ \text{ or } \ u\Sigma^\omega \cap \mathrm{dom}(\rho_{sd}) = \emptyset\}$$

is an r.e. subset of Σ^* by the projection theorem 2.4.11, and the sets

$$U_1 := \bigcup_{u \in C} u\Sigma^\omega \quad \text{and} \quad U_2 := \bigcup \{v\Sigma^\omega \mid v \in (\Sigma \setminus \{\bullet\})^*, \ |v| \geq n + 2\}$$

and $U := U_1 \cup U_2$ are r.e. We show $\rho_{\mathrm{sd}}^{-1}[X] = \Sigma^\omega \setminus U$.

Assume $p \in \rho_{\mathrm{sd}}^{-1}[X]$, that is, $\rho_{\mathrm{sd}}(p) \in X$.
Then $p \in v \boldsymbol{.} \Sigma^\omega$ for some $v \in \{0, 1, \overline{1}\}^*$ such that $|v| \leq n + 1$, since $X \subseteq (-2^n; 2^n)$, and so $p \notin U_2$. Suppose $p \in U_1$. Then there are some $u \in C$ and some $w \in B$ such that $p \in u\Sigma^\omega$ and $\rho_{\mathrm{sd}}[u\Sigma^\omega] \subseteq \mathrm{I}^1(w)$ or $u\Sigma^\omega \cap \mathrm{dom}(\rho_{\mathrm{sd}}) = \emptyset$. In both cases $p \notin \rho_{\mathrm{sd}}^{-1}[X]$ (contradiction).
From $p \notin U_1$ and $p \notin U_2$ we obtain $p \notin U$.

On the other hand, assume $p \notin \rho_{\mathrm{sd}}^{-1}[X]$.
First, consider $p \notin \mathrm{dom}(\rho_{\mathrm{sd}})$. If $p \in (\Sigma \setminus \{\boldsymbol{.}\})^\omega$, then $p \in U_2$. Assume $p \in v\{\boldsymbol{.}\}\Sigma^\omega$ for some $v \in (\Sigma \setminus \{\boldsymbol{.}\})^*$. Since $p \notin \mathrm{dom}(\rho_{\mathrm{sd}})$, there is some prefix u of p such that $u\Sigma^\omega \cap \mathrm{dom}(\rho_{\mathrm{sd}}) = \emptyset$, and so $u \in C$ and $p \in U_1$.
Finally, consider $p \in \mathrm{dom}(\rho_{\mathrm{sd}})$. Since $\rho_{\mathrm{sd}}(p) \notin X$, there are some prefix u of p and some $w \in B$ such that $\rho_{\mathrm{sd}}[u\Sigma^\omega] \subseteq \mathrm{I}^1(w)$, and so $p \in U_1$.
From $p \in U_1$ or $p \in U_2$ we obtain $p \in U$.

Therefore, U is the r.e. open complement of the set $\rho_{\mathrm{sd}}^{-1}[X]$ which is a closed and hence compact subset of Σ^ω. $\square$

More generally, the operator mapping every compact subset of $\mathbb{R}^m$ to its pre-image under ρ_{sd} is computable with respect to natural representations (Exercise 7.2.9).

Definition 7.2.6 (real complexity and lookahead).
1. *A Type-2 machine M computes a real function $f :\subseteq \mathbb{R}^m \to \mathbb{R}$ on $X \subseteq \mathrm{dom}(f)$ in time $t : \mathbb{N} \to \mathbb{N}$ with lookahead $s : \mathbb{N} \to \mathbb{N}$, iff*

$$f(\rho_{\mathrm{sd}}(p_1), \ldots, \rho_{\mathrm{sd}}(p_m)) = \rho_{\mathrm{sd}} \circ f_M(p_1, \ldots, p_m),$$

$$\mathrm{Time}_M^A \in \mathrm{O}(t) \quad and$$

$$\mathrm{La}_M^A(k) \leq s(k)$$

for all $(p_1, \ldots, p_m) \in A := (\rho_{\mathrm{sd}}, \ldots, \rho_{\mathrm{sd}})^{-1}[X]$ and all $k \in \mathbb{N}$.
2. *A real number a is computable in time $t : \mathbb{N} \to \mathbb{N}$, iff the constant function $() \mapsto a$ is computable in time t.*

In the first part the condition on computation time is stronger than the non-uniform condition

$$\mathrm{Time}_M(p_1, \ldots, p_m) \in \mathrm{O}(t), \quad \text{if } (\rho_{\mathrm{sd}}(p_1), \ldots, \rho_{\mathrm{sd}}(p_m)) \in X,$$

where the constant c with $\mathrm{Time}_M(p_1, \ldots, p_m)(k) \leq c \cdot t(k) + c$ may depend on the input $(p_1, \ldots, p_m)$. But it is sufficiently weak to make complexity robust to the usual modifications of the Turing machine model. As an immediate consequence of Theorems 7.1.5 and 7.2.5 we obtain:

> **Theorem 7.2.7.** If for some Type-2 machine M the function f_M is a $(\rho_{\mathrm{sd}}, \ldots, \rho_{\mathrm{sd}})$-realization of the real function $f :\subseteq \mathbb{R}^m \to \mathbb{R}$ and $X \subseteq \mathrm{dom}(f)$ is a compact $\kappa_>$-computable set, then there are computable functions $s, t : \mathbb{N} \to \mathbb{N}$ such that M computes f on X in time t with lookahead s.

Therefore, computable uniform bounds of complexity and lookahead can be guaranteed on every compact subset of the domain with r.e. open complement.

Example 7.2.8.

1. For every square $K_n := [-n; n] \times [-n; n] \subseteq \mathbb{R}^2$ $(n \in \mathbb{N})$ there are computable functions $s, t : \mathbb{N} \to \mathbb{N}$ such that addition (multiplication) is computable on K_n in time t with lookahead s. This follows from Theorem 7.2.7, since K_n is $\kappa_>$-computable and addition (multiplication) is computable by Theorem 4.3.2 and $\rho \equiv \rho_{\mathrm{sd}}$ by Theorem 7.2.5.

2. For every interval $[a; b] \in \mathbb{R}$ $(a, b \in \mathbb{Q}, \ 0 < a < b)$ there are computable functions $s, t : \mathbb{N} \to \mathbb{N}$ such that inversion $x \mapsto 1/x$ is computable on $[a; b]$ in time t with lookahead s. This follows from Theorem 7.2.7, since $[a; b]$ is $\kappa_>$-computable and inversion is computable by Theorem 4.3.2 and $\rho \equiv \rho_{\mathrm{sd}}$ by Theorem 7.2.5.

3. Consider inversion on the set $A := [-1; 1] \setminus \{0\}$. Assume that some machine M computes inversion on A in time $t : \mathbb{N} \to \mathbb{N}$. Then for every input $p \in \rho_{\mathrm{sd}}^{-1}[A]$, the machine M prints the leftmost output symbol in at most $t(0)$ steps. We obtain

$$f_M(0{\bullet}0^{t(0)}10^{\omega}) \in 1\Sigma^{\omega} \quad \text{and}$$

$$f_M(0{\bullet}0^{t(0)}\overline{1}0^{\omega}) \in \overline{1}\Sigma^{\omega} \ .$$

But this is impossible, since in $t(0)$ steps M can read at most the common prefix $0{\bullet}0^{t(0)}$of both inputs. Therefore, on the set A the complexity of inversion has no uniform bound. □

In the next section we will determine concrete complexity bounds for addition, multiplication and inversion. It remains to show that our definition of complexity classes admits non-trivial hierarchies.

Lemma 7.2.9. For every computable function $t : \mathbb{N} \to \mathbb{N}$ there is a computable real number $y \in \mathbb{R}$ which is not computable in time t.

Proof: For any set $A \subseteq \mathbb{N}$ define $p_A := {\bullet}a_1 0 a_2 0 a_3 0 \ldots$ with $a_i := 1$, if $i - 1 \in A$, and $a_i := \overline{1}$ otherwise, and $x_A := \rho_{\mathrm{sd}}(p_A)$. For $a \in \{1, \overline{1}\}$ define $\overline{a} := \overline{1}$, if $a = 1$, and $\overline{a} := 1$, if $a = \overline{1}$. Since

$$|\rho_{\mathrm{sd}}({\bullet}a_1 0 a_2 0 \ldots a_i 0 0^{\omega}) - x_A| < 2^{-2i} \quad \text{and} \tag{7.6}$$

$$|\rho_{\mathrm{sd}}({\bullet}a_1 0 a_2 0 \ldots a_i 0 0^{\omega}) - \rho_{\mathrm{sd}}({\bullet}a_1 0 a_2 0 \ldots \overline{a}_i 0 0^{\omega})| = 4 \cdot 2^{-2i} \ , \tag{7.7}$$

$$|\rho_{\mathrm{sd}}({\bullet}a_1 0 a_2 0 \ldots \overline{a}_i 0 0^{\omega}) - x_A| > 2^{-2i}$$

for $i \geq 1$, and so $(x_A = x_B \iff A = B)$.
Let $x_A = \rho_{\mathrm{sd}}(u{\bullet}d_1 d_2 \ldots)$. Since

$$|\rho_{\mathrm{sd}}(u{\bullet}d_1d_2\dots d_{2i}0^\omega) - x_A| \le 2^{-2i} , \tag{7.8}$$

$$|\rho_{\mathrm{sd}}({\bullet}a_10a_20\dots a_i00^\omega) - \rho_{\mathrm{sd}}(u{\bullet}d_1d_2\dots d_{2i}0^\omega)| < 2\cdot 2^{-2i} \tag{7.9}$$

by (7.6) and

$$|\rho_{\mathrm{sd}}({\bullet}a_10a_20\dots \overline{a}_i00^\omega) - \rho_{\mathrm{sd}}(u{\bullet}d_1d_2\dots d_{2i}0^\omega)| > 2\cdot 2^{-2i} \tag{7.10}$$

by (7.7).

Using Equations (7.9) and (7.10) we can construct a Type-2 machine L which for any set $A \subseteq \mathbb{N}$ on input $u{\bullet}d_1d_2\dots \in \Sigma^\omega$ with $x_A = \rho_{\mathrm{sd}}(u{\bullet}d_1d_2\dots)$ computes the sequence $p_A = {\bullet}a_10a_20a_30\dots$. By (7.9)

$$\rho_{\mathrm{sd}}({\bullet}a_10a_20\dots a_i00^\omega) - \rho_{\mathrm{sd}}(u{\bullet}d_1d_2\dots d_{2i}0^\omega) = b_i\cdot 2^{-2i} \tag{7.11}$$

for some $b_i \in \{1,0,\overline{1}\}$. The machine L first determines b_0 from the word u. Suppose, b_{i-1} has already been determined. Then L reads the symbols $d_{2i-1}d_{2i}$ and with b_{i-1} determines the symbol $a_i \in \{1,\overline{1}\}$ such that (7.9) holds (a_i is unique by (7.10)). Then L determines the symbol b_i for the difference (7.11) and prints a_i0 on the output tape. Since the intermediate computations are determined by finite decision tables, the machine L can be designed without any work tape (that is, the function f_L can be computed by a finite automaton).

From ordinary Type-1 complexity [HU79] we know that there is a recursive set $A \subseteq \mathbb{N}$ such that the set $\{0^i \mid i \in A\} \subseteq \Sigma^*$ is not decidable in time $t(2\cdot k+1)$, that is, there is no Turing machine M working in time $t(2\cdot k+1)$ (more precisely, $\mathrm{O}(t(2\cdot k+1))$) such that $f_M(0^i) = \lambda \iff i \in A$. Since A is recursive, the sequence $p_A := {\bullet}a_10a_20a_30\dots$ is computable and so the constant real function $f : () \mapsto x_A$ is computed with respect to ρ_{sd} by some Type-2 machine M. Combining M with L we get a machine N which on input () prints the sequence $p_A = {\bullet}a_10a_20a_30\dots$. Suppose the machine M computes f in time t. Then $\mathrm{Time}_N() \in \mathrm{O}(t)$ and $f_N() = p_A = {\bullet}a_10a_20\dots$. From N we can construct a Turing machine K which on input $0^i \in \Sigma^*$ determines the symbol a_i and operates in time $\mathrm{O}(t(2\cdot k+1))$. Therefore, A is decidable in time $\mathrm{O}(t(2\cdot k+1))$ (contradiction). $\square$

The result can be transferred to proper real functions as follows:

Corollary 7.2.10. Let $f :\subseteq \mathbb{R}^m \to \mathbb{R}$ be a real function and let $x \in \mathbb{R}$ be a real number not computable in time $t : \mathbb{N} \to \mathbb{N}$.
If $f(y) = x$ for some $y \in \mathbb{Q}^m$, then f is not computable in time t.

Proof: Suppose some Type-2 machine M computes f in time t. From M we can construct a machine N which on input () simulates the machine M on input $y \in \mathbb{Q}^m$. Then N computes the constant x in time t (contradiction). $\square$

Theorem 7.2.7 can be effectivized: there are computable functions which for every $\kappa_>$-name of a compact subset $X \subseteq \text{dom}(f)$ determine uniform bounds of time and lookahead, respectively, of M on X (Exercise 7.2.15). In particular, there are uniform bounds t and s even if X is compact but not $\kappa_>$-computable. However, these bounds may be non-computable (Exercise 7.2.16).

Since important functions like inversion $x \mapsto 1/x$ cannot be computed with uniformly bounded time on their (non-compact) domain (Example 7.2.8.3), the following definition of complexity classes is useful.

Definition 7.2.11 (time and lookahead real complexity class).
For $t, s : \mathbb{N} \to \mathbb{N}$ let $\text{TL}(t, s)$ be the set of all real functions $f :\subseteq \mathbb{R}^m \to \mathbb{R}$ ($m \in \mathbb{N}$), such that on every compact subset $X \subseteq \text{dom}(f)$, the function f can be computed in time t with lookahead s. For sets T, S of functions define $\text{TL}(T, S) := \bigcup \{\text{TL}(t, s) \mid t \in T, s \in S\}$.

From the composition lemma 7.1.4 for Σ^ω we can derive the following composition theorem for complexity classes of real functions.

Theorem 7.2.12 (composition in complexity classes). Let T, S be classes of non-decreasing number functions such that for all functions $s, s' \in S$ and $t, t' \in T$,

1. there are functions $s_+, s_M \in S$ such that
   $s' \circ s(k) \leq s_+(k)$ and $\max(s(k), s'(k)) \leq s_m(k)$, and
2. $t \circ s \in O(T)$ and $\max(t, t') \in O(T)$.

Then the class $\text{TL}(T, S)$ is closed under composition.

Proof: For the sake of simplicity we take $f, g :\subseteq \mathbb{R} \to \mathbb{R}$ leaving the general case $f :\subseteq \mathbb{R}^n \to \mathbb{R}$ and $g_1, \ldots, g_n :\subseteq \mathbb{R}^m \to \mathbb{R}$ to the reader. Let $f, g \in \text{TL}(T, S)$ and let $X \subseteq \text{dom}(f \circ g)$ be compact. Then $X \subseteq \text{dom}(g)$ and $g[X]$ is a compact subset of $\text{dom}(f)$. By Definitions 7.2.6 and 7.2.11 there are Type-2 machines M and M' and functions $s, s' \in S$ and $t, t' \in T$ such that M' computes g on X in time t' with lookahead s', and M computes f on $g[X]$ in time t with lookahead s. The function $f_M \circ f_{M'}$ realizes $f \circ g$. By Lemma 7.1.4, there is a Type-2 machine N computing $f_M \circ f_{M'}$ such that

$$\text{La}_N(p)(k) = \text{La}_{M'}(p) \circ \text{La}_M(f_{M'}(p))(k) \quad \text{and}$$
$$\text{Time}_N(p)(k) \leq c \cdot \text{Time}_M(f_{M'}(p))(k)$$
$$+ c \cdot \text{Time}_{M'}(p) \circ \text{La}_M(f_{M'}(p))(k) + c \ .$$

If $p \in \rho_{\text{sd}}^{-1}[X]$, then $\rho_{\text{sd}} \circ f_{M'}(p) \in g[X]$, and so

$$\text{La}_N(p)(k) \leq s' \circ s(k) \quad \text{and}$$
$$\text{Time}_N(p)(k) \leq c \cdot t(k) + c \cdot t' \circ s(k) + c$$

for some constant c not depending on p. Therefore, machine N computes $f \circ g$ on X in time t_1 with input lookahead s_1 for some $t_1 \in O(T)$ and $s_1 \in S$. The max-condition in the theorem is needed for the case $n \geq 2$. $\square$

Since lookahead is bounded by time (Lemma 7.1.3), $\mathrm{TL}(T, S) \subseteq \mathrm{TL}(T, \mathrm{O}(T))$ for all classes T and S. The composition theorem 7.2.12 applies to many interesting complexity classes.

Corollary 7.2.13 (complexity classes closed under composition).
For $S_0 := \{\mathrm{id}_\mathbb{N}\}$, $S_1 := \{k + c \mid c \in \mathbb{N}\}$ and $S_2 := \{c \cdot k + c \mid c \in \mathbb{N}\}$ the class $\mathrm{TL}(T, S)$ is closed under composition, if
1. $T = S_0$ and $S \in \{S_0, S_1, S_2\}$,
2. $T = \{k \cdot \log k\}$ and $S \in \{S_0, S_1, S_2\}$,
3. $T = \{k^\alpha\}$ $(\alpha > 1)$ and $S \in \{S_0, S_1, S_2\}$,
4. $T = \{k^l \mid l = 1, 2, \ldots\}$ and $S \in \{S_0, S_1, S_2, T\}$.

By Property 4, the class of real function computable in polynomial time is closed under composition. Also for real functions there is a trade-off between computation time and lookahead [Wei91b]

In Type-1 computability theory, computational complexity is defined explicitly for word functions $f : \Sigma^* \to \Sigma^*$ (for example time or space complexity of Turing machines) and then transferred to other sets X by means of notations $\nu :\subseteq \Sigma^* \to X$. Examples are the natural numbers denoted binary by $\nu_\mathbb{N}$, decimal or "unary" $(\nu(i) := 0^i)$, finite graphs denoted by adjacency lists or adjacency matrices, or regular languages denoted by right-linear grammars, deterministic finite automata or regular expressions [HU79]. The examples show that in practice, various notations are used for modelling computational complexity on a set X . Usually, these notations are computationally equivalent and induce the same complexity classes for big complexity bounds like exponential space. For small complexity bounds like time n^2, however, complexity classes induced by the various notations may differ considerably.

Similarly the definition of computational complexity of real numbers and functions can be modified in various ways. We give some examples.

Example 7.2.14 (other definitions of computational complexity).
1. In Definition 7.2.4 the digits $1, 0$ and $\bar{1}$ can be replaced by the signed decimal digits $\{9, 8, \ldots, 1, 0, \bar{1}, \ldots, \bar{8}, \bar{9}\}$. In this case the prefixes 0, $1\bar{9}$ and $\bar{1}9$ must be excluded. The subsequent definitions can be generalized straightforwardly.
2. In Definition 7.2.6.2, a ρ_{sd}-name of a real number must be computed in time t digit by digit from left to right. A somewhat weaker definition is as follows:
A real number x is computable in time $t : \mathbb{N} \to \mathbb{N}$, iff there is a function $h : \Sigma^* \to \Sigma^*$ computable in time t such that $|x - \rho_{\mathrm{sd}}(h(0^k)0^\omega)| \leq 2^{-k}$ for all $k \in \mathbb{N}$. Here $h(0^k)$ and $h(0^{k+1})$ may be completely different words.
3. If we allow arbitrary rational approximate numbers, we have the following definition: A real number x is computable in time $t : \mathbb{N} \to \mathbb{N}$, iff there is a function $h : \Sigma^* \to \Sigma^*$ computable in time t such that $|x - \nu_\mathbb{Q} \circ h(0^k)| \leq 2^{-k}$ for all $k \in \mathbb{N}$.

4. The following definition is similar to (3) above:
   A real number x can be computed in time $t : \mathbb{N} \to \mathbb{N}$, iff $x = \rho_C(p)$ for some $p = \iota(w_0)\iota(w_1)\dots$ and some Type-2 machine M computes p on input () such that the prefix $\iota(w_0)\iota(w_1)\dots\iota(w_k)$ of p is determined in at most $c \cdot t(k) + c$ steps (for some constant c).

5. If $p \in \mathrm{dom}(\rho_{\mathrm{sd}})$, then p has the form "$u.q$". The position of the dot can be encoded in binary notation at the beginning of the sequence $uq \in \{0, 1, \overline{1}\}$. We obtain the following representation ρ'_{sd} of $\mathbb{R}$:

$$\mathrm{dom}(\rho'_{\mathrm{sd}}) := \{u \# q \mid u \in \mathrm{dom}(\nu_{\mathbb{N}}), q \in \{1, 0, \overline{1}\}^{\omega},$$
$$q \notin \{0, 1\overline{1}, \overline{1}1\}\Sigma^{\omega} \text{ for } \nu_{\mathbb{N}}(u) > 0\},$$

$$\rho'_{\mathrm{sd}}(u \# a_1 a_2 \dots) := 2^{\nu_{\mathbb{N}}(u)} \cdot \sum_{i=1}^{\infty} a_i \cdot 2^{-i}.$$

   For every compact set $X \subseteq \mathbb{R}$ the restrictions of ρ_{sd} and ρ'_{sd} to X can be translated into each other very fast. Theorem 7.2.5 holds accordingly with ρ'_{sd} replacing ρ_{sd}.

6. By Definition 7.1.1, $\mathrm{Time}_M(p)(k)$ is the number of steps which the machine M on input $p \in \Sigma^{\omega}$ needs until it has printed the k-th output symbol. Since every input or output in $\mathrm{dom}(\rho_{\mathrm{sd}})$ has a dot, we can modify Definition 7.1.1 by counting symbols after the dot instead of from the beginning:

$$\mathrm{DotTime}_M(y)(k) := \begin{cases} \text{the number of steps which the machine M} \\ \text{on input } y \text{ needs until it has printed the } k\text{-th} \\ \text{symbol after the dot,} \end{cases}$$

$$\mathrm{DotLa}_M(y)(k) := \begin{cases} \text{the maximal number of symbols after the dot} \\ \text{which the machine M on input } y \text{ reads from some} \\ \text{infinite input tape until it has printed the } k\text{-th} \\ \text{symbol after the dot.} \end{cases}$$

   In this case the parameter k corresponds exactly with the precision 2^{-k} of the written output data or the read input data, respectively. For every compact subset $A \subseteq \mathrm{dom}(f_M)$ such that $f_M[A] \subseteq \mathrm{dom}(\rho_{\mathrm{sd}})$ there is some constant $c \in \mathbb{N}$ such that

$$\mathrm{Time}_M^A(k) \leq \mathrm{DotTime}_M^A(k) \text{ and } \mathrm{DotTime}_M^A(k) \leq \mathrm{Time}_M^A(k + c)$$

   for all $k \in \mathbb{N}$ (Exercise 7.2.13).

7. Definition 7.2.14.2 can be generalized to real functions as follows:
   A real function $f :\subseteq \mathbb{R} \to \mathbb{R}$ is computable on $X \subseteq \mathrm{dom}(f)$ in time $t : \mathbb{N} \to \mathbb{N}$, iff there is a Type-2 machine of type $f_M :\subseteq \Sigma^{\omega} \times \Sigma^* \to \Sigma^*$ such that for all $p \in \rho_{\mathrm{sd}}^{-1}[X]$ and all $k \in \mathbb{N}$ there are words $u \in \{1, 0, \overline{1}\}^*$ and $v \in \{1, 0, \overline{1}\}^k$ with

$$f_M(p, 0^k) = u \bullet v \ ,$$
$$|f \circ \rho_{sd}(p) - \rho_{sd}(u \bullet v 0^\omega)| \le 2^{-k} \quad \text{and}$$
$$M \ \text{halts on input} \ (p, 0^k) \ \text{in at most} \ t(k) \ \text{steps.}$$

This definition is similar to Definition 2.18 in Ko's book [Ko91, Ko98].

8. Define the set BI of binary intervals by

$$\text{BI} := \{2^z \cdot \rho_{sd}[\bullet v \Sigma^\omega] \mid z \in \mathbb{Z}, v \in \{0, 1, \overline{1}\}^*\} \ .$$

For each $K := 2^z \cdot \rho_{sd}[\bullet v \Sigma^\omega] \in \text{BI}$, define a representation ρ_K by

$$\rho_K(q) := 2^z \cdot \rho_{sd}(\bullet vq) \quad \text{for all} \ q \in \{0, 1, \overline{1}\}^\omega \ .$$

Suppose the input data are from a set $K = 2^z \cdot \rho_{sd}[\bullet v \Sigma^\omega] \in \text{BI}$. If the data z and v identifying the common *scale* K are omitted we get the representation ρ_K. If $f : \mathbb{R} \to \mathbb{R}$ is continuous and $K \subseteq \text{dom}(f)$, then $f[K]$ is compact and so $f[K] \subseteq L$ for some common scale $L \in \text{BI}$. If f is computable, then it is (ρ_K, ρ_L)-computable. Complexity and lookahead lookahead can be measured with respect to these simplified names.   □

For studying computational complexity we will usually consider the signed digit representation ρ_{sd} as our standard representation of the real numbers and use lengths of prefixes for defining complexity and lookahead.

Exercises 7.2.

1. Show that a definition of computational complexity for all $\rho_<$-computable real numbers induces a definition of computational complexity for all r.e. subsets of $\mathbb{N}$. Hint: Consider the real numbers $\sum\{4^{-i} \mid i \in A\}$ for r.e. subsets $A \subseteq \mathbb{N}$.
2. Prove $f(x) = f(y)$ for all $y \in \text{dom}(f)$ with $|x - y| < 1$, if $f :\subseteq \mathbb{R} \to \mathbb{R}$ is computable in $x \in \text{dom}(f)$ in time $t : \mathbb{N} \to \mathbb{N}$ according to Example 7.2.3.
3. Consider Definition 7.2.14.3. Show:
   a) For every computable real number x there is a computable function t such that x can be computed in time t.
   b) For every computable time bound $t : \mathbb{N} \to \mathbb{N}$ there is some computable real number not computable in time t.
4. Show that $\rho_{sd}^{-1}[\{a\}]$ is countable, iff $a = z/2^k$ for some $z \in \mathbb{Z}$ and $k \in \mathbb{N}$.
5. Show that there is some restriction ρ' of ρ_{sd} such that $\rho' \equiv \rho_{sd}$ and the real number 0 has the only ρ'-name $\bullet 000 \dots$.
6. Let δ be a representation obtained from ρ_{sd} by omitting one of the restrictions $a_n \ne 0$, $a_n a_{n-1} \ne 1\overline{1}$ or $a_n a_{n-1} \ne \overline{1}1$ in Definition 7.2.4. Show:
   a) $\delta \equiv \rho_{sd}$,
   b) $\delta^{-1}[K]$ is not compact for some compact $K \subseteq \mathbb{R}$ (cf. Theorem 7.2.5).

7. In his famous famous paper [Tur36] on "Turing machines", A.Turing used the binary representation $\rho_{b,2}$ (Definition 4.1.13) of the real numbers. Shortly after publication he recognized some difficulty arising from this representation. In his "Correction" [Tur37] he proposes the following representation $\delta :\subseteq \Sigma^\omega \to \mathbb{R}$ by overlapping intervals:
$\delta(p) = x$ iff $p = c_0 1^n 0 c_1 c_2 \ldots$ with $c_0, c_1, c_2, \ldots \in \{0, 1\}$ such that

$$x = (2c_0 - 1) \cdot n + \sum_{i=1}^{\infty} (2c_i - 1) \cdot \left(\frac{2}{3}\right)^i .$$

   Prove $\delta \equiv \rho$.

8. Show that $\mathrm{dom}(\rho_{sd})$ is
   a) not open,
   b) not closed,
   c) a countable union of compact sets,
   d) a countable intersection of open sets,
   e) the intersection of an open and a closed set.

9. Let $\kappa_>$ be the representation of the compact subsets of $\mathbb{R}^m$ (Definition 5.2.1) and let $\kappa_>^Y$ be the representation of the compact subsets of $(\Sigma^\omega)^m$ introduced in Exercise 7.1.4. Show that for every computable function $f :\subseteq \mathbb{R}^m \to \mathbb{R}$ the function

$$K \mapsto (\rho_{sd}, \rho_{sd}, \ldots, \rho_{sd})^{-1}[K]$$

   mapping every compact subset of $\mathrm{dom}(f)$ to the set of its names with respect to ρ_{sd} is $(\kappa_>, \kappa_>^Y)$-computable.

10. Show that there is no function $t : \mathbb{N} \to \mathbb{N}$ such that on the interval $(0; 1]$ inversion $x \mapsto 1/x$ is computable in time t.

11. Let $t : \mathbb{N} \to \mathbb{N}$ be a computable function. Construct a computable real function $f : \mathbb{R} \to \mathbb{R}$ such that $f[\mathbb{Q}] \subseteq \mathbb{Q}$ and on $[0; 1]$, f is not computable in time t (cf. Corollary 7.2.10).

12. Show that every representation δ of the real numbers which is equivalent to ρ has a restriction δ' such that $\delta' \equiv \rho$ and $\delta'^{-1}[K]$ is compact for every compact subset $K \subseteq \mathbb{R}$ (cf. Theorem 7.2.5).

13. Consider the definition of DotTime from Example 7.2.14.6. Let M be a Type-2 machine. Show that for every compact subset $A \subseteq \mathrm{dom}(f_M)$ such that $f_M[A] \subseteq \mathrm{dom}(\rho_{sd})$ there is some constant $c \in \mathbb{N}$ such that

$$\mathrm{Time}_M^A(k) \leq \mathrm{DotTime}_M^A(k) \quad \text{and} \quad \mathrm{DotTime}_M^A(k) \leq \mathrm{Time}_M^A(k + c)$$

   for all $k \in \mathbb{N}$.

14. Let $f :\subseteq \mathbb{R} \to \mathbb{R}$ be a real function with $\mathrm{dom}(f) = [0; 1]$. Show that f is computable in polynomial time (according to Definition 7.2.6), iff it is computable in polynomial time according to Definition 7.2.14.7.

15. Prove the following effective version of Theorem 7.2.7. Let M be a Type-2 machine realizing a real function $f : \mathbb{R}^m \to \mathbb{R}$. Then there are computable functions $f_T, f_I : \Sigma^\omega \to \Sigma^\omega$ such that for all $p \in \Sigma^\omega$ such that $\kappa_>(p) \subseteq \text{dom}(f)$ (Definition 5.2.1), M computes f on $\kappa_>(p)$ in time $[\nu_\mathbb{N} \to \nu_\mathbb{N}]_\mathbb{N} \circ f_T(p)$ with lookahead $[\nu_\mathbb{N} \to \nu_\mathbb{N}]_\mathbb{N} \circ f_I(p)$.

♦16. Show that there is a computable function $f :\subseteq \mathbb{R} \to \mathbb{R}$ such that every Type-2 machine M realizing f has the following property. For every function $s : \mathbb{N} \to \mathbb{N}$ there is some $x \in \text{dom}(f)$ such that $\text{La}_M(p)(k) \geq s(k)$, if $\rho_{\text{sd}}(p) = x$ and $k \in \mathbb{N}$ (notice that this implies $\text{Time}_M(p)(k) \geq s(k)$).
    Hint: Let $g :\subseteq \Sigma^\omega \to \Sigma^\omega$ be a computable translation from $\rho_{\text{sd}}]^{\mathbb{R} \backslash D}$ to $\rho_{\text{b},2}$, where $D := \{z/2^i \mid z \in \mathbb{Z}, i \in \mathbb{N}\}$ (cf. Theorem 4.1.13.1). Define $h :\subseteq \Sigma^\omega \to \Sigma^\omega$ by $h(0^{i_0}01a_00^{i_1}01a_1\ldots) := a_0a_1\ldots$ $(a_0, a_1, \ldots \in \Sigma)$. The function $h \circ g$ is a $(\rho_{\text{sd}}, \rho_{\text{sd}})$-realization of a real function f with $\text{dom}(f) = [0;1] \backslash D$. For $s : \mathbb{N} \to \mathbb{N}$ consider the real number $x := \rho_{\text{sd}}(\bullet 0^{s(0)}0110^{s(1)}011\ldots)$.

17. Show that the class $\text{TL}(T, S)$, $T := \{\text{id}_\mathbb{N}\}$ and $S := \{c \cdot \sqrt{k} + c \mid c \in \mathbb{N}\}$, is closed under composition.

7.3 The Complexity of Some Real Functions

As we know (Exercise 4.1.9) neither addition nor multiplication of real numbers can be computed on $\rho_{\text{b},2}$-names. Since the signed digit representation ρ_{sd} is equivalent to ρ, these functions can be computed on ρ_{sd}-names (reading and writing from left to right).

Real numbers can be added very fast. First we consider a machine operating correctly on the set $\bullet\{1, 0, \overline{1}\}^\omega$ of ρ_{sd}-names.

Theorem 7.3.1 (complexity of local addition).   There is a Type-2 machine M such that for all $p, q \in \bullet\{1, 0, \overline{1}\}^\omega$,
1. $\rho_{\text{sd}}(p) + \rho_{\text{sd}}(q) = \rho_{\text{sd}} \circ f_M(p, q)$,
2. M is finite on $\left(\bullet\{1, 0, \overline{1}\}^\omega\right)^2$,
3. $\text{La}_M(p, q)(k) \leq k + 2$ for all $k \in \mathbb{N}$.

See Definition 7.1.8. Notice that machine M operates on $\left(\bullet\{1, 0, \overline{1}\}^\omega\right)^2$ in linear time.

Proof: Let M be a Type-2 machine which on input (p, q), $p = \bullet a_1a_2\ldots \in \bullet\{1, 0, \overline{1}\}^\omega$ and $q = \bullet b_1b_2\ldots\bullet\{1, 0, \overline{1}\}^\omega$, works as follows. M determines a word $w \in \{1, \lambda, \overline{1}\}$ and a number $r_0 \in \{-2, -1, 0, 1, 2\}$ such that

$$\rho_{\text{sd}}(w\bullet 0^\omega) + r_0 \cdot 2^{-2} = \rho_{\text{sd}}(\bullet a_1a_20^\omega) + \rho_{\text{sd}}(\bullet b_1b_20^\omega) \tag{7.12}$$

and writes the word "$w\bullet$". (This is possible, since the distance of the sum on the right hand side to the next integer is at most $1/2$.) Then for $i = 1, 2, \ldots$

the machine M determines numbers $c_i \in \{1, 0, \bar{1}\}$ and $r_i \in \{-2, -1, 0, 1, 2\}$ such that

$$2r_{i-1} + a_{i+2} + b_{i+2} = 4c_i + r_i \qquad (7.13)$$

and writes c_i on the output tape. (Since, by induction, $r_{i-1} \in \{-2, -1, 0, 1, 2\}$, $|2r_{i-1} + a_{i+2} + b_{i+2}| \leq 6$ and so c_i and r_i can be determined easily.) Let $c_w := 1$ if $w = 1$, $c_w := 0$ if $w = \lambda$ and $c_w := -1$ if $w = \bar{1}$. Then

$$\sum_{i \leq n+2} a_i \cdot 2^{-i} + \sum_{i \leq n+2} b_i \cdot 2^{-i} = c_w + \sum_{i \leq n} c_i \cdot 2^{-i} + r_n \cdot 2^{-n-2} \qquad (7.14)$$

for all $n \geq 0$. This can be proved by induction from Equations (7.12) and (7.13). Taking $\lim_{n \to \infty}$ over Equation (7.14) we obtain $\rho_{sd}(p) + \rho_{sd}(q) = \rho_{sd}(r)$ where $r = w \bullet c_1 c_2 \dots$, and so f_M is a $(\rho_{sd}, \rho_{sd}, \rho_{sd})$-realization of addition. This proves the first statement. By Equation (7.13) for determining the output symbol c_i and the remainder r_i, the machine must merely read the symbols a_{i+2} and b_{i+2} and remember the remainder r_{i-1} from the previous stage. Therefore, M is finite on $\left(\bullet \{1, 0, \bar{1}\}^\omega \right)^2$ and $\mathrm{La}_M(p, q)(k) \leq k + 2$ for all $k \in \mathbb{N}$. $\qquad \square$

We extend the result to a machine operating on all ρ_{sd}-names.

Corollary 7.3.2 (complexity of global addition). There are a Type-2 machine M and a constant $c \in \mathbb{N}$ such that for all $n \in \mathbb{N}$ and $p, q \in \Sigma^{\leq n} \bullet \Sigma^\omega \cap \mathrm{dom}(\rho_{sd})$,
  1. f_M is a $(\rho_{sd}, \rho_{sd}, \rho_{sd})$-realization of addition,
  2. M is finite on every bounded set,
  3. $\mathrm{Time}_M(p, q)(k) \leq c \cdot (n + k) + c$ for all $k \in \mathbb{N}$,
  4. $\mathrm{La}_M(p, q)(k) \leq n + k + 2$ for all $k \in \mathbb{N}$.

Proof: Let M be a machine which on input (p, q), $p = u \bullet a_1 a_2 \dots$, $q = v \bullet b_1 b_2 \dots$, first reads the prefixes "$u \bullet$" and "$v \bullet$".
Suppose $|u| = |v| + j$ for some $j \geq 0$. Then M adds $\bullet u a_1 a_2 \dots$ and $\bullet 0^j v b_1 b_2 \dots$ using the machine from Theorem 7.3.1. The result $r := w \bullet c_1 c_2 \dots$ satisfies $\rho_{sd}(p) + \rho_{sd}(q) = 2^{|u|} \cdot \rho_{sd}(r)$. Therefore, a subsequent procedure in M must shift the dot $|u|$ positions to the right. Simultaneously every 0 at the beginning of the output must be cancelled, every prefix $1\bar{1}$ must be replaced by 1 and every prefix $\bar{1}1$ must be replaced by $\bar{1}$, before the first symbol is written in order to achieve the conditions for the prefix of a ρ_{sd}-name. If $|v| = |u| + j$ for some $j \geq 0$, M proceeds accordingly.

Obviously f_M realizes addition. Suppose, $\max(|u|, |v|) = n$. In its first phase, M reads at most the first $n + 1$ symbols from each input tape. In the worst case for time and lookahead, the result begins with "$\bullet$". For writing the final dot, the machine works at most $c_1 \cdot n + c_1$ steps and reads the first $n + 3$ symbols from one of the tapes. After writing the dot, M performs addition like the special machine from Theorem 7.3.1. Therfore, $\mathrm{La}(p,q)(k) \le n+k+2$ for all k and there is some c such that $\mathrm{Time}(p,q)(k) \le c \cdot (n+k) + c$ for all $k \in \mathbb{N}$.

Since X is bounded, $X \subseteq [-2^N; 2^N]^2$ for some $N \in \mathbb{N}$. If $u \bullet a_1 a_2 \ldots \in \mathrm{dom}(\rho_{\mathrm{sd}})$ and $|u| > N + 2$, then by Formula (7.4), $2^N < 2^{|u|-2} \le |\rho_{\mathrm{sd}}(u \bullet a_1 a_2 \ldots)|$. Therefore, $|u| \le N + 2$, if $|\rho_{\mathrm{sd}}(u \bullet a_1 a_2 \ldots)| \le 2^N$. For inputs (p,q), $p = u \bullet a_1 a_2 \ldots$, $q = v \bullet b_1 b_2 \ldots$, $|u|, |v| \le N + 2$, M needs only a finite number of tape cells which is of order n. Therefore, M is finite on X. □

Also the maximum and the minimum of two real numbers can be computed very fast.

Theorem 7.3.3 (complexity of local maximum/minimum).
There is a Type-2 machine M such that for all $p,q \in \bullet\{1,0,\overline{1}\}^\omega$,
  1. $\max(\rho_{\mathrm{sd}}(p), \rho_{\mathrm{sd}}(q)) = \rho_{\mathrm{sd}} \circ f_M(p,q)$,
  2. M is finite on $\left(\bullet\{1,0,\overline{1}\}^\omega\right)^2$
  3. $\mathrm{La}_M(p,q)(k) \le k$ for all $k \in \mathbb{N}$.

This is also true for a minimum replacing the maximum.

Proof: There is a Type-2 machine M which on input (p,q), $p = \bullet a_1 a_2 \ldots$, $q = \bullet b_1 b_2 \ldots$, prints a sequence $\bullet c_1 c_2 \ldots$ according to the following rules. Let $d_0 := 0$. Consider $i \ge 1$. Let

$$d_i := 2 \cdot d_{i-1} + a_i - b_i \ .$$

If $d_i \ge 2$ (then $\rho_{\mathrm{sd}}(p) \ge \rho_{\mathrm{sd}}(q)$), then $c_j := a_j$ for all $j \ge i$.
If $d_i \le -2$ (then $\rho_{\mathrm{sd}}(p) \le \rho_{\mathrm{sd}}(q)$), then $c_j := b_j$ for all $j \ge i$.
If $d_i \in \{1,0\}$, then $c_i := a_i$,
If $d_i = -1$, then $c_i := b_i$.
We leave the verification of the algorithm as Exercise 7.3.3 □

As for addition we can extend the result to a machine operating on all ρ_{sd}-names.

Corollary 7.3.4 (complexity of global maximum/minimum).
There are a Type-2 machine M and a constant $c \in \mathbb{N}$ such that for all $n \in \mathbb{N}$ and $p,q \in \Sigma^{\le n} \bullet \Sigma^\omega \cap \mathrm{dom}(\rho_{\mathrm{sd}})$,
  1. f_M is a $(\rho_{\mathrm{sd}}, \rho_{\mathrm{sd}}, \rho_{\mathrm{sd}})$-realization of the maximum function,
  2. M is finite on every bounded set,
  3. $\mathrm{Time}(p,q)(k) \le c \cdot (n+k) + c$ for all $k \in \mathbb{N}$,
  4. $\mathrm{La}(p,q)(k) \le n + k$ for all $k \in \mathbb{N}$.

Proof: Similar to the proof of Corollary 7.3.2 (see Exercise 7.3.4). □

By the following technical lemma any initial part of a ρ_{sd}-name of a real number can be extended, if some other better approximation for x is known. This can be done very fast.

Lemma 7.3.5 (improvement lemma).
Let $J := \rho_{sd}[u_\bullet a_1 a_2 \dots a_m \Sigma^\omega] \cap \rho_{sd}[v_\bullet b_1 b_2 \dots b_{m+k} \Sigma^\omega] \neq \emptyset$.
Then there are symbols $a_{m+1}, \dots, a_{m+k} \in \{1, 0, \overline{1}\}$ such that

$$J = \rho_{sd}[u_\bullet a_1 a_2 \dots a_{m+k} \Sigma^\omega] \cap \rho_{sd}[v_\bullet b_1 b_2 \dots b_{m+k} \Sigma^\omega].$$

There is a Turing machine which determines the word $a_{m+1} \dots a_{m+k}$ from $u_\bullet a_1 a_2 \dots a_m$ and $v_\bullet b_1 b_2 \dots b_{m+k}$ in time $n := |u| + m + k$.

Proof: Since $J \neq \emptyset$, for $i = 0, \dots, m$, there is a number $d_i \in \{-2, -1, 0, 1, 2\}$ such that

$$\rho_{sd}(u.a_1 \dots a_i 0^\omega) - \rho_{sd}(v.b_1 \dots b_i 0^\omega) = d_i \cdot 2^{-i}$$

(Fig. 7.3). The number d_0 can be determined from u and v in time $|u| + |v|$, since u and v can be written as the difference of two binary numbers, and addition and subtraction of binary numbers are computable in linear time. Since $d_i = 2 \cdot d_{i-1} + a_i - b_i$, d_i can be determined easily from d_{i-1} and so d_m can be determined in time $|u| + |v| + m$. In the following we abbreviate

$$K_i := \rho_{sd}[u_\bullet a_1 a_2 \dots a_{m+i} \Sigma^\omega], \quad L_i := \rho_{sd}[v_\bullet b_1 b_1 \dots b_{m+i} \Sigma^\omega],$$

$$d_{m+i} := 2^{m+i}\big(\rho_{sd}(u_\bullet a_1 a_2 \dots a_{m+i} 0^\omega) - \rho_{sd}(v_\bullet b_1 b_2 \dots b_{m+i} 0^\omega)\big).$$

By assumption, $K_0 \cap L_k = J \neq \emptyset$, and so $K_0 \cap L_i \neq \emptyset$, since $L_{i+1} \subseteq L_i$ $(i = 0, \dots, k)$.
Proposition: Suppose

$$K_0 \cap L_i = K_i \cap L_i \quad \text{and} \quad d_{m+i} \in \{-2, \dots, 2\}.$$

Then we can find $a_{m+i+1} \in \{1, 0, \overline{1}\}$ such that

$$K_0 \cap L_{i+1} = K_{i+1} \cap L_{i+1} \quad \text{and} \quad d_{m+i+1} \in \{-2, \dots, 2\}.$$

Consider various cases in the proof:
Case $d_{m+i} = 2$: Then $K_i \cap L_i$ consists of a single point and K_i must be a left end of K_0. Since $K_0 \cap L_{i+1} \neq \emptyset$, $b_{m+i+1} = 1$. If we define $a_{m+i+1} := \overline{1}$, then $K_0 \cap L_{i+1} = K_{i+1} \cap L_{i+1}$ and $d_{m+i+1} = 2$.
Case $d_{m+i} = 1$: Then $K_i \cap L_i$ is a proper interval K' and K_i must be a left end of K_0. Define $a_{m+i+1} := \overline{1}$ (and so $K_{i+1} = K'$). Then $K_0 \cap L_{i+1} = K_{i+1} \cap L_{i+1}$ and $d_{m+i+1} = 2$, if $b_{m+i+1} = \overline{1}$, $d_{m+i+1} = 1$, if $b_{m+i+1} = 0$ and $d_{m+i+1} = 0$, if $b_{m+i+1} = 1$.

Case $d_{m+i} = 0$: In this case $K_i = L_i$. If we define $a_{m+i+1} := b_{m+i+1}$, then $K_0 \cap L_{i+1} = K_{i+1} \cap L_{i+1}$ and $d_{m+i+1} = 0$.

The cases $d_{m+i} = -1$ and $d_{m+i} = -2$ are symmetrical to the first two cases. Thus, the proposition is proved. The first statement of the lemma follows from the proposition by induction on i. Since the symbols a_{m+i} and d_{m+i} can be computed easily, the sequence $a_{m+1} \ldots a_{m+k}$ can be computed in time $|u| + |v| + m + k$.

By $J \neq \emptyset$ and Formula (7.4) after Definition 7.2.4, v and u have almost the same length. Therefore, we obtain the time bound $|u| + m + k$. $\square$

We will apply this lemma to find a fast algorithm for real multiplication. By the school method, natural numbers in binary notation $\nu_\mathbb{N}$ can be multiplied in time $O(n^2)$. There are faster methods, for example a divide-and-conquer algorithm working in time $O(n^{\log_2 3})$ [AHU83] and the Schönhage/Strassen algorithm with upper bound $O(n \cdot \log n \cdot \log \log n)$ [SGV94]. These methods apply also to rational numbers notated by finite prefixes of ρ_{sd}-names. The above time bounds are *regular* [FS74, Mül86b].

Definition 7.3.6 (regular time bounds). *A function* $t : \mathbb{N} \to \mathbb{N}$ *is regular, iff* f *is non-decreasing and there are numbers* $N, c \in \mathbb{N}$ *such that* $t(N) > 0$ *and*

$$2 \cdot t(n) \leq t(2n) \leq c \cdot t(n) \quad \text{for all} \ \ n \geq N \ .$$

We will use the following properties of regular functions.

Lemma 7.3.7.   For every regular function $t : \mathbb{N} \to \mathbb{N}$,
1. $n \in O(t)$,
2. $t \in O(n^k)$ for some $k \in \mathbb{N}$,
3. $t(cn + c) \in O(t)$ for every $c \in \mathbb{N}$,
4. $\sum \{ t(2^i) \mid 2^i < 2n \} \in O(t)$.

Proof: See Exercise 7.3.7 [Mül86b]. $\square$

Convention 7.3.8.   In the following let $t_m : \mathbb{N} \to \mathbb{N}$ be a regular time bound for multiplication of natural numbers in binary notation $\nu_\mathbb{N}$.

First we consider multiplication on ρ_{sd}-names of real numbers beginning with "$\bullet$". The algorithm applies the "doubling method". For $j = 0, 1, \ldots$ we multiply the finite prefixes of (approximately) length 2^j (using a fast algorithm) for obtaining better and better approximations of the exact product. The improvement lemma is used for extending the output of Stage j to that of Stage $j + 1$.

Theorem 7.3.9 (complexity of local multiplication).   There are a Type-2 machine M and a constant c such that for all $p, q \in \bullet\{1, 0, \overline{1}\}^\omega$,
1. $\rho_{\mathrm{sd}}(p) \cdot \rho_{\mathrm{sd}}(q) = \rho_{\mathrm{sd}} \circ f_M(p, q)$,
2. $\mathrm{La}_M(p, q)(k) \leq \max(4, 2k - 1)$,
3. $\mathrm{Time}_M(p, q)(k) \leq c \cdot t_m(k) + c$.

Proof: There is a Type-2 machine M which on input (p, q), $p = \bullet a_1 a_2 \ldots$ and $q = \bullet b_1 b_2 \ldots$, produces a sequence $\bullet c_1 c_2 \ldots$ as follows:

First it prints "$\bullet$" and then operates in stages $j = 0, 1, 2, \ldots$.

Stage j: Let $\bullet c_1 \ldots c_k$ be the output computed so far and let $n := 2^j$. Then $k \leq n$. M determines digits $e_i \in \{1, 0, \overline{1}\}$, $1 \leq i \leq n$, such that

$$|\rho_{\mathrm{sd}}(\bullet e_1 \ldots e_k \ldots e_n 0^\omega) - \rho_{\mathrm{sd}}(\bullet a_1 \ldots a_{n+2} 0^\omega) \cdot \rho_{\mathrm{sd}}(\bullet b_1 \ldots b_{n+2} 0^\omega)| \leq 2^{-n-1} .$$

From $\bullet c_1 \ldots c_k$ and $\bullet\bullet e_1 \ldots e_k \ldots e_n$, M determines an extension $\bullet c_1 \ldots c_k c_{k+1} \ldots c_n$ according to the improvement lemma 7.3.5.

We show that f_M realizes multiplication. Define $x := \rho_{\mathrm{sd}}(\bullet a_1 a_2 \ldots)$, $y := \rho_{\mathrm{sd}}(\bullet b_1 b_2 \ldots)$, $x_i := \rho_{\mathrm{sd}}(\bullet a_1 a_2 \ldots a_i 0^\omega)$ and $y_i := \rho_{\mathrm{sd}}(\bullet b_1 b_2 \ldots b_i 0^\omega)$. For every $j \in \mathbb{N}$ and $n = 2^j$, we have $|y| \leq 1$, $|x_{n+2}| \leq 1$, $|x - x_{n+2}| \leq 2^{-n-2}$, $|y - y_{n+2}| \leq 2^{-n-2}$ and $|x_{n+2} y_{n+2} - \rho_{\mathrm{sd}}(\bullet e_1 \ldots e_n)| \leq 2^{-n-1}$ by rounding. We obtain

$$\begin{aligned}
|xy - \rho_{\mathrm{sd}}(\bullet e_1 \ldots e_n 0^\omega)| &\leq |xy - x_{n+2} y_{n+2}| + |x_{n+2} y_{n+2} - \rho_{\mathrm{sd}}(\bullet e_1 \ldots e_n 0^\omega)| \\
&\leq |xy - x_{n+2} y| + |x_{n+2} y - x_{n+2} y_{n+2}| + 2^{-n-1} \\
&\leq |x - x_{n+2}| \cdot |y| + |x_{n+2}| \cdot |y - y_{n+2}| + 2^{-n-1} \\
&\leq 2^{-n-2} + 2^{-n-2} + 2^{-n-1} \\
&\leq 2^{-n}
\end{aligned}$$

and so

$$xy \in \rho_{\mathrm{sd}}[\bullet e_1 \ldots e_{2^j} \Sigma^\omega] .$$

We show $xy \in \rho_{\mathrm{sd}}[\bullet c_1 \ldots c_{2^j} \Sigma^\omega]$ for all j by induction. In Stage 0, the output "$\bullet$" is extended by $c_1 := e_1$. Since $xy \in \rho_{\mathrm{sd}}[\bullet e_{2^0} \Sigma^\omega]$, $xy \in \rho_{\mathrm{sd}}[\bullet c_{2^0} \Sigma^\omega]$. Suppose $xy \in \rho_{\mathrm{sd}}[\bullet c_1 \ldots c_{2^{j-1}} \Sigma^\omega]$. Since $xy \in \rho_{\mathrm{sd}}[\bullet e_1 \ldots e_{2^j} \Sigma^\omega]$, in Stage j we obtain $xy \in \rho_{\mathrm{sd}}[\bullet c_1 \ldots c_{2^j} \Sigma^\omega]$ by the improvement lemma 7.3.5. We obtain $xy = \rho_{\mathrm{sd}}(\bullet c_1 c_2 \ldots)$ and so f_M realizes multiplication.

We estimate the lookahead. In Stage 0, M determines c_1 from $a_1 a_2 a_3$ and $b_1 b_2 b_3$, and so $\mathrm{La}_M(p, q)(1) = 0$ and $\mathrm{La}_M(p, q)(2) = 4$. In Stage j $(j \geq 1)$, M determines the symbols c_i for $2^{j-1} + 1 \leq i \leq 2^j$ from the symbols a_i, b_i for $i \leq 2^j + 2$. Hence, $\mathrm{La}_M(p, q)(i) = 2^j + 3 \leq 2i - 1$ for $2^{j-1} + 2 \leq i \leq 2^j + 1$. Therefore, $\mathrm{La}_M(p, q)(i) \leq 2i - 1$ for all $i \geq 3$.

We estimate $\mathrm{Time}_M(p, q)(i)$ for $i \geq 2$. The time for Stage k is bounded by

$$d_1 \cdot t_m(2^k + 2) + d_2 \cdot 2^k + d_3$$

for multiplication of $(2^k + 2)$-bit numbers and the remaining manipulations by Lemma 7.3.5 where d_1, d_2, d_3 are constants not depending on k. By Lemma 7.3.7 there are constants d_5, d_6 such that

$$d_1 \cdot t_m(2^k + 2) + d_2 \cdot 2^k + d_3 \le d_5 \cdot t_m(2^k) + d_6 .$$

The symbol c_i ($i \ge 2$) is determined in Stage j where j is uniquely determined by $2^{j-1} + 1 \le i \le 2^j$. Therefore, $\text{Time}_M(p, q)(1) \le d_4$ and for $i \ge 2$,

$$\text{Time}_M(p, q)(i) \le \sum_{k=0}^{j} (d_5 \cdot t_m(2^k) + d_6) + d_4$$

$$\le d_5 \cdot \sum_{k=0}^{j} t_m(2^k) + (j + 1) \cdot d_6 + d_4 .$$

Since $2^{j-1} + 1 \le i$ and so $2^j < 2i$, $\sum_{k=0}^{j} t_m(2^k) \le d_7 \cdot t_m(i) + d_8$ and $j + 1 \le d_9 \cdot t_m(i) + d_9$ by Lemma 7.3.7. Therefore, $\text{Time}_M(p, q) \in O(t_m)$. $\square$

The above algorithm is fast but it has a bad lookahead. The first $2k$ digits are used from each input tape for determining the first k digits of the output, although $k + 2$ digits are sufficient. We obtain a machine with lookahead $\text{La}_M(p, q)(k) = k + 2$, if in Stage j we determine c_j from $c_1 \ldots c_{j-1}$, $a_1 \ldots a_{j+2}$ and $b_1 \ldots b_{j+2}$. But this machine works only in time $k \cdot t_m(k)$. It can be modified to a machine working with the same lookahead in time k^2 (Exercise 7.3.1). Schröder [Sch97] has shown that there is a machine multiplying real numbers in time $t_m(k) \cdot \log(k)$ with dot lookahead $k + 3$ (if applied to inputs from $\bullet\{1, 0, \overline{1}\}^\omega$). In [Sch99] the results are extended to differentiable functions.

A machine multiplying arbitrary real numbers fast can be constructed from the machine from Theorem 7.3.9.

Corollary 7.3.10 (complexity of global multiplication). There is a Type-2 machine M such that
1. f_M is a $(\rho_{\text{sd}}, \rho_{\text{sd}}, \rho_{\text{sd}})$-realization of multiplication on $\mathbb{R}$,
2. For every bounded set $X \subseteq \mathbb{R}^2$ there is a constant c_X such that M operates on X in time t_m with lookahead $2k + c_X$.

Proof: Let M be a machine which on input $(p, q) \in (\rho_{\text{sd}}, \rho_{\text{sd}})^{-1}[X]$, $p = u_\bullet a_1 a_2 \ldots$ and $q = v_\bullet b_1 b_2 \ldots$,

- multiplies the inputs using the method from Theorem 7.3.9 ignoring the dots on the inputs,
- simultaneously counts the number of symbols of u and v and prints the dot on the output at the correct position,
- simultaneously avoids printing the prefixes 0, $1\overline{1}$ and $\overline{1}1$ by cancelling 0 and replacing $1\overline{1}$ by 1 and $\overline{1}1$ by $\overline{1}$.

Since X is bounded, there is a constant c such that $|u| + |v| \le c$, if $(\rho_{\text{sd}}(u_\bullet a_1 a_2 \ldots), \rho_{\text{sd}}(v_\bullet b_1 b_2 \ldots)) \in X$, and so in the third procedure at most c symbols can be deleted. Therefore, our machine works in time $t_m(k + c) \in O(t_m(k))$ with lookahead $2k + c_X$ for some constant c_X. $\square$

In addition, inversion of real numbers can be reduced to multiplication of integers.

We apply Newton's method for zero-finding.

Newton's method for zero finding: Let $J \subseteq \mathbb{R}$ be an open interval and let $f : J \to \mathbb{R}$ be a function such that the first derivative f' is continuous on $\overline{J}$ and the second derivative f'' exists on J. Then by Taylor's Theorem for all $x, y \in \overline{J}$ there is some $\xi \in J$ such that

$$f(y) = f(x) + f'(x)(y - x) + \frac{1}{2}f''(\xi)(y - x)^2 . \tag{7.15}$$

Assume $f(a) = 0$ for some $a \in J$. Under suitable conditions for the function f and for a suitable initial value $x_0 \in J$ the sequence $x_0, x_1, x_2, \ldots$ defined by

$$x_{n+1} := x_n - \frac{f(x_n)}{f'(x_n)} . \tag{7.16}$$

converges very fast to the zero a of f (Fig. 7.4).

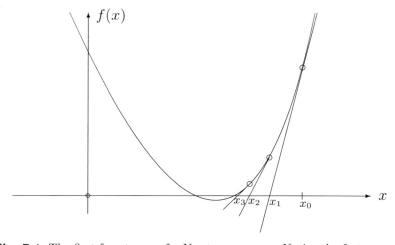

Fig. 7.4. The first four terms of a Newton sequence. Notice the fast convergence!

Assume, for example, that there are positive constants c, d such that
- $[a - d; a + d] \subseteq J$,
- $f'(x) \neq 0$ and $|f''(y)/f'(x)| \leq c$ for all $x, y \in J$,
- $|x_0 - a| \leq \min(d, 1/c)$.

Then for all n,

$$x_n \in J \text{ and } |x_n - a| \leq \frac{2}{c} \cdot 2^{-2^n} . \tag{7.17}$$

This follows by induction from (7.15) and (7.16) (Exercise 7.3.6).

In the proof of Theorem 7.3.11 below we reduce fast inversion to fast multiplication via Newton iteration. For $a \in \mathbb{R}$, $a > 0$, define $f_a :\subseteq \mathbb{R} \to \mathbb{R}$ by

$$f_a(x) := 1/x - a \ .$$

Then $1/a$ is the zero of f_a, the Newton iteration formula is

$$x_{n+1} := x_n - \frac{f(x_n)}{f'(x_n)} = x_n - \frac{\frac{1}{x_n} - a}{-\frac{1}{x_n^2}} = x_n + x_n - ax_n^2 = x_n \cdot (2 - ax_n)$$

and in this special case

$$\left| x_{n+1} - \frac{1}{a} \right| = \left| 2x_n - ax_n^2 - \frac{1}{a} \right| = a \cdot \left| x_n - \frac{1}{a} \right|^2 . \tag{7.18}$$

First, we consider a realization of inversion working on ρ_{sd}-names $p \in 1{\bullet}a_1a_2a_3\Sigma^\omega$ with $a_1, a_2, a_3 \in \{0, 1\}$.

Theorem 7.3.11 (complexity of local inversion). There are a Type-2 machine M and a constant c such that for all $p \in 1{\bullet}\{1, 0\}^3\{0, 1, \bar{1}\}^\omega$,
  1. $1/\rho_{\mathrm{sd}}(p) = \rho_{\mathrm{sd}} \circ f_M(p)$,
  2. $\mathrm{La}_M(p)(k) \leq \max(k + 2, 2k - 4)$,
  3. $\mathrm{Time}_M(p)(k) \leq c \cdot t_m(k) + c$.

Proof: Let M be a Type-2 machine which on input $p \in 1{\bullet}\{1, 0\}^3\{0, 1, \bar{1}\}^\omega$, $p = 1{\bullet}a_1a_2 \ldots$, produces a sequence $1{\bullet}c_1c_2 \ldots \in \mathrm{dom}(\rho_{\mathrm{sd}})$ in stages $n = 0, 1, \ldots$ as follows.
Stage 0: From $a_1 \ldots a_6$ the machine M determines symbols $c_1, \ldots, c_4$ such that $1/x \in \rho_{\mathrm{sd}}[1{\bullet}c_1 \ldots c_4\Sigma^\omega]$, if $x \in \rho_{\mathrm{sd}}[1{\bullet}a_1 \ldots a_6\Sigma^\omega]$ and prints $1{\bullet}c_1 \ldots c_4$. Define $z_0 := \rho_{\mathrm{sd}}(1{\bullet}c_1 \ldots c_40^\omega)$.
Stage n $(n \geq 1)$: The machine M determines

$$y_n := z_{n-1} \cdot (2 - r_nz_{n-1}) ,$$

where $r_n := \rho_{\mathrm{sd}}(1{\bullet}a_1 \ldots a_{k_n+3}0^\omega)$ and $k_n := 2^n + 3$, and determines a word $w_n \in \{1, 0, \bar{1}\}^{k_n}$ such that

$$|\rho_{\mathrm{sd}}(1{\bullet}w_n0^\omega) - y_n| \leq 2^{-k_n-1} . \tag{7.19}$$

Let $z_n := \rho_{\mathrm{sd}}(1{\bullet}w_n0^\omega)$. Then from $1{\bullet}c_1 \ldots c_{k_{n-1}}$ and $1{\bullet}w_n$, M determines an improvement $1{\bullet}c_1 \ldots c_{k_n}$ according to Lemma 7.3.5.

We have to prove correctness and to estimate time and lookahead. Let $p \in 1{\bullet}\{1, 0\}^3\{0, 1, \bar{1}\}^\omega$ and $a = \rho_{\mathrm{sd}}(p)$. Then $7/8 \leq \rho_{\mathrm{sd}}(p) \leq 2$. In Stage 0, $\rho_{\mathrm{sd}}[1{\bullet}a_1 \ldots a_6\Sigma^\omega]$ is a subinterval of $[7/8; 2]$ of length 2^{-5}. By the mean value theorem,

$$\mathrm{length}(f[\rho_{\mathrm{sd}}[1{\bullet}a_1 \ldots a_6\Sigma^\omega]]) = \mathrm{length}(\rho_{\mathrm{sd}}[1{\bullet}a_1 \ldots a_6\Sigma^\omega]) \cdot |f'(\xi)| \tag{7.20}$$

for some $\xi \in \rho_{sd}[1\bullet a_1 \ldots a_6 \Sigma^\omega] \subseteq [7/8;2]$, where $f(x) := 1/x$. Since $f'(\xi) = -1/\xi^2$ and $\max\{1/\xi^2 \mid \xi \in [7/8;2]\} = (8/7)^2 < 2$, the interval $f[\rho_{sd}[1\bullet a_1 \ldots a_6 \Sigma^\omega]]$ is shorter than $2 \cdot 2^{-5} = 2^{-4}$. Therefore, there are symbols $c_1, \ldots, c_4$ in Stage 0. In particular, $|z_0 - 1/a| \leq 2^{-4} = 2^{-k_0}$ where $k_0 := 2^0 + 3$.

Now consider $n \geq 1$ and assume that $z_{n-1} = \rho_{sd}(1\bullet w_{n-1}0^\omega)$ has been determined such that $|z_{n-1} - 1/a| \leq 2^{-k_{n-1}}$. If $x_n := z_{n-1}(2 - a z_{n-1})$, then $|x_n - 1/a| \leq a \cdot (2^{-k_{n-1}})^2$ by Equation (7.18), and so

$$|x_n - 1/a| \leq 2^{-k_n - 2} .$$

Since $|r_n - a| \leq 2^{-k_n - 3}$ and $z_{n-1} \leq 5/4$ (since $|z_{n-1} - 1/a| \leq 2^{-k_{n-1}} \leq 2^{-4}$ and $1/a \leq 8/7$),

$$\begin{aligned} |x_n - y_n| &= |z_{n-1} \cdot (2 - a z_{n-1}) - z_{n-1} \cdot (2 - r_n z_{n-1})| \\ &= z_{n-1}^2 |a - r_n| \\ &\leq 2^{-k_n - 2} . \end{aligned}$$

Since $|y_n - z_n| \leq 2^{-k_n - 1}$ by (7.19),

$$\begin{aligned} |z_n - 1/a| &\leq |z_n - y_n| + |y_n - x_n| + |x_n - 1/a| \\ &\leq 2^{-k_n - 1} + 2^{-k_n - 2} + 2^{-k_n - 2} \\ &\leq 2^{-k_n} . \end{aligned}$$

By induction, $1/a \in \rho_{sd}[1\bullet w_n \Sigma^\omega]$ for all n. By Lemma 7.3.5, $1/a \in \rho_{sd}[1\bullet c_1 \ldots c_{k_n} \Sigma^\omega]$ for all n, and so $1/a = \rho_{sd}(1\bullet c_1 c_2 \ldots)$. Therefore, machine M realizes inversion.

We estimate the lookahead. Simple numerical calculations show that c_1 can already be determined from $a_1 a_2$, $c_1 c_2$ from $a_1 \ldots a_4$ and $c_1 c_2 c_3$ from $a_1 \ldots a_5$. The machine M determines $c_1 \ldots c_4$ from $a_1 \ldots a_6$ and for $n \geq 1$, the symbol c_i ($k_{n-1} + 1 \leq i \leq k_n$) from $a_1 \ldots a_{k_n + 3}$. It follows that $\text{La}_M(p)(i)$ is bounded by $i + 2$ for $i \leq 6$ and by $2i - 4$ for $i \geq 7$, and so $\text{La}_M(p)(i) \leq \max(i + 2, 2i - 4)$.

Since in Stage n, $2^n + 3$-digit integers must be multiplied, the time for Stage n can be bounded by $d_1 \cdot t_m(2^n + 3) + d_2 \cdot 2^n + d_3$. Since t_m is regular, we obtain $\text{Time}_M(p)_M(k) \leq c \cdot t_m(k) + c$ for some c as in the proof of Theorem 7.3.9. □

Inversion of arbitrary real numbers $x \in \mathbb{R} \setminus \{0\}$ can be reduced to the special case of Theorem 7.3.11.

Theorem 7.3.12 (complexity of global inversion). There is a Type-2 machine M such that
1. $1/\rho_{sd}(p) = \rho_{sd} \circ f_M(p)$ for $p \in \text{dom}(\rho_{sd})$,
2. for every compact set $X \subseteq \mathbb{R} \setminus \{0\}$ there is a constant c_X such that M operates in time t_m with lookahead $2k + c_X$.

Proof: We will explain how the machine M inverts positive real numbers. For negative inputs it works similarly. We want to reduce inversion of x to the special case of Theorem 7.3.11.

Since X is compact, there is a constant $m \in \mathbb{N}$ such that $2^{-m} \leq |x| \leq 2^m$ for all $x \in X$. Let $x \in X$ and $p = u_{\bullet}q \in \mathrm{dom}(\rho_{\mathrm{sd}})$ such that $\rho_{\mathrm{sd}}(p) = x > 0$. Then $x = 2^{|u|} \cdot \rho_{\mathrm{sd}}({}_{\bullet}uq)$. The first non-zero digit of ${}_{\bullet}uq$ must be 1, since $x > 0$. Suppose $x = 2^{|u|} \cdot \rho_{\mathrm{sd}}({}_{\bullet}p')$ and the first four significant digits of p' are $1a_1a_2a_3$ ($a_j \in \{0, 1, \overline{1}\}$). If the digit $\overline{1}$ occurs in this word, its first occurrence can be deleted by one of the subword substitutions

$$1\overline{1} \mapsto 01, \quad 10\overline{1} \mapsto 011, \quad 100\overline{1} \mapsto 0111 \qquad (7.21)$$

which produce an equivalent ρ_{sd}-name. (The substitutions can be visualized very well in Fig. 7.3.) By repeated application of these subword substitutions to the first four significant digits the name ${}_{\bullet}uq$ can be transformed step by step to an equivalent name ${}_{\bullet}0^i1b_1b_2b_3(uq)_{>i+4}$ of x such that $i \in \mathbb{N}$ and $b_1, b_2, b_3 \in \{0, 1\}$ (where $(a_1a_2\ldots)_{>j} := a_{j+1}a_{j+2}\ldots$). The substitutions need to be applied only finitely often: If $x = 2^{|u|}\rho_{\mathrm{sd}}({}_{\bullet}0^j1s)$, then

$$2^{-m} \leq x \leq 2^{|u|}\rho_{\mathrm{sd}}({}_{\bullet}0^j1s) \leq 2^{|u|} \cdot 2^{-j} ,$$

and so $j \leq m + |u|$. We obtain

$$x = 2^{|u|-i-1}\rho_{\mathrm{sd}}(1_{\bullet}b_1b_2b_3(uq)_{>i+4}) . \qquad (7.22)$$

A machine M for inverting positive real numbers can work as follows on input $p = u_{\bullet}q \in \mathrm{dom}(\rho_{\mathrm{sd}})$. M determines $|u|$, $i \in \mathbb{N}$ and $b_1, b_2, b_3 \in \{0, 1\}$ such that $\rho_{\mathrm{sd}}(u_{\bullet}q) = 2^{|u|-i-1}\rho_{\mathrm{sd}}(1_{\bullet}b_1b_2b_3(uq)_{>i+4})$. Then it inverts $1_{\bullet}b_1b_2b_3(uq)_{>i+4}$ using the machine from Theorem 7.3.11 and simultaneously inserts the dot at the correct position.

We estimate time and lookahead on the compact set X. Since $2^{-m} \leq |x| \leq 2^m$, $|u| \leq m + 2$ by Formula (7.4) and $i \leq |u| + m$ (see above). Therefore, $\mathrm{Time}_M(u_{\bullet}q)(k) \leq c \cdot t(k + c) + c$ where $t \in \mathrm{O}(t_m)$ is the time from Theorem 7.3.11 and c is a constant. Therefore, $\mathrm{Time}_M(u.q) \in \mathrm{O}(t_m)$ by Lemma 7.3.7. Since inversion from Theorem 7.3.11 has lookahead $\leq 2k + c$, this machine M has lookahead $\mathrm{La}_M(u_{\bullet}q)(k) \leq 2k + c_X$ where c_X is a constant depending on X. $\qquad \square$

A function $f :\subseteq \mathbb{R}^m \to \mathbb{R}$ is *rational*, iff $f(x) = \sum_{i \in I} \alpha_i x^i / \sum_{i \in J} \beta_i x^i$ where I, J are finite subsets of $\mathbb{N}^m$, $\alpha_i, \beta_i \in \mathbb{R}$ and $x^{(i_1,\ldots,i_m)} := x_1^{i_1} \cdot \ldots \cdot x_m^{i_m}$. Since real addition, multiplication and inversion are computable on every compact subset of their domains in time t_m with lookahead $s \in \{c \cdot k + c \mid c \in \mathbb{N}\}$, we obtain immediately from Theorem 7.2.12:

Corollary 7.3.13 (complexity of rational functions). Every rational function $f :\subseteq \mathbb{R}^m \to \mathbb{R}$ with coefficients computable in time t_m is computable in time t_m with input lookahead $c \cdot k + c$ for some c.

In the proof of Theorem 7.3.11 we have reduced fast inversion to fast multiplication via Newton iteration. In further steps many other problems can be reduced to fast multiplication and inversion by means of Newton iteration or other methods.

We now merely list some interesting results. By Corollary 6.4.8, a computable real function on $[0; 1]$ has a computable derivative, if its second derivative is continuous. This property can be refined as follows.

Theorem 7.3.14 (complexity of derivative). [Mül86b] Let $t : \mathbb{N} \to \mathbb{N}$ be a regular function with $t_m \in O(t)$ and let $f :\subseteq \mathbb{R} \to \mathbb{R}$, $\mathrm{dom}(f) = [0; 1]$, be a real function computable in time t with a continuous second derivative. Then the derivative f' is computable in time t.

Lemma 7.3.15 (complexity of a zero). [Mül86b] Let $t : \mathbb{N} \to \mathbb{N}$ be a regular function such that $t_m \in O(t)$. Let $f :\subseteq \mathbb{R} \to \mathbb{R}$, $\mathrm{dom}(f) = J$ for some open interval J, be a two times differentiable function computable in time t. If $a \in J$ is a simple zero of f, then a is a real number computable in time t.

A real closed field F is a subset of the real numbers such that for every polynomial with coefficients from F, every real root is in F.

Corollary 7.3.16 (real closed field). [Mül86b] For every regular time bound t such that $t_m \in O(t)$, the class of real numbers computable in time t is a real closed field.

Consequently, the set of numbers computable in polynomial time is a real closed field [Ko91], and the set of all computable real numbers is a real closed field (Corollary 6.3.10).

A function $f :\subseteq \mathbb{R} \to \mathbb{R}$ is called real analytic on $J \subseteq \mathrm{dom}(f)$ for some open interval J, iff it can be computed by a real power series (cf. Sect. 6.5).

Theorem 7.3.17 (complexity of integral). [Mül87] Let $f :\subseteq \mathbb{R} \to \mathbb{R}$ be real analytic on $J \subseteq \mathrm{dom}(f)$. Let f be computable in time t, where t is regular and $k^2 \cdot t_m(k \log k) \in O(t)$. Then for every rational number $a \in J$, the integral function $x \mapsto \int_a^x f(\zeta) \, d\zeta$ is computable in time $k^2 t(k)$.

The proofs and further results can be found in [Alt85, Mül86a, Mül87, Sch90, Ko91, Ko98].

Theorem 7.3.18 (complexity of transcendental functions). [Bre76] On compact subsets of their domains the real functions exp, sin, cos, tan and their inverses can be computed in time $t_m(k) \cdot \log k$.

Exercises 7.3.

1. Let δ be the restriction of ρ_{sd} to $\bullet\{1, 0, \overline{1}\}^\omega$. Show that real numbers can be multiplied with respect to δ in time k^2 with lookahead $k + 2$.
2. Let a be a rational number. Show that the real function $x \mapsto a \cdot x$ can be computed by a machine which is finite on $\mathrm{dom}(\rho_{\mathrm{sd}})$ (Definition 7.1.8).
3. Verify the algorithm given in the proof of Theorem 7.3.3 for computing the maximum.
4. Prove Corollary 7.3.4.
5. Show that for a regular function $t : \mathbb{N} \to \mathbb{N}$ real numbers on $[-1; 1]$ can be multiplied in time t with respect to the representation ρ_{sd}, only if natural numbers in binary notation can be multiplied in time t.
$\Diamond$ 6. Let f, J, a, d, c satisfy the conditions for the Newton iteration (7.16) formulated above. Show by induction that $|x_n - a| \le d$ and $|x_n - a| \le \frac{2}{c} \cdot 2^{-2^n}$ for all n, if $|x_0 - a| \le \min(d, \frac{1}{c})$.
7. Prove Lemma 7.3.7.
8. Define $f(x) := x^2 - 2$. Then $\sqrt{2}$ is a zero of f which can be determined by Newton iteration. Show that $\sqrt{2}$ can be computed in time t_m (where t_m is a regular time bound for integer multiplication).

7.4 Complexity on Compact Sets

In this section we introduce various complexity definitions applicable to recursive compact subsets of Euclidean space.

In Sect. 5.2 we studied three computability concepts on the set of compact subsets or $\mathbb{R}^n$ via the representations $\kappa_<$, $\kappa_>$ and κ. For $A \subseteq \mathbb{N}$ the set $K_A := \{0\} \cup \{2^{-k} \mid k \in A\}$ is $\kappa_<$-computable, iff A is r.e. For r.e. subsets of numbers there is no reasonable definition of computational complexity, merely recursive sets have a computational complexity. Therefore, it is very unlikely that computational complexity can be defined reasonably for the $\kappa_<$-computable compact sets. For a similar reason the representation $\kappa_>$ is not suitable for introducing computational complexity. Thus, we try to define the computational complexity of the κ-computable compact sets.

For the sake of simplicity we will consider merely the set

$$\mathcal{K}^0 := \{K \subseteq [0; 1] \times [0; 1] \mid K \neq \emptyset, \ K \text{ compact}\}$$

of the non-empty compact subsets of the unit square $[0; 1] \times [0; 1]$. The concepts can be generalized straightforwardly to other squares and dimensions. We want to refine the computability theory induced by the representation κ on $\mathcal{K}^0$. Remember that restricted to $\mathcal{K}^0$, κ is equivalent to the representation κ_{mc} by minimal covers (Lemma 5.2.5), to the distance representation κ^{dist} (Remark 5.2.3) and the Cauchy representation (Theorem 5.2.9). As in the case of real numbers, for each of these representations the set of names of a single point (here a compact set) is too "big" such that no reasonable concept of complexity can be deduced. Furthermore, there is no injective representation equivalent to κ (Exercise 7.4.4). For the real numbers, the signed digit representation ρ_{sd} can be considered as a restriction of the Cauchy representation by admitting for every precision only finitely many rational numbers. Similarly we restrict the Cauchy representation κ_{H}.

If $D_n := \{m \cdot 2^{-n} \mid m = 0, 1, \ldots, 2^n\}$, then for every compact subset $K \in \mathcal{K}^0$ there is some subset B of the grid $D_n \times D_n$ such that $d_{\text{H}}(K, B) < 2^{-n}$ where d_{H} is the Hausdorff distance (Sect. 5.2). There may be several sets $B \subseteq D_n \times D_n$ with this property but only finitely many, since $D_n \times D_n$ has merely finitely many subsets. Fig. 7.5 shows a compact set $K \in [0; 1]^2$ and a set $B \subseteq D_5 \times D_5$ of points such that $d_{\text{H}}(A, B) \leq 2^{-5}$.

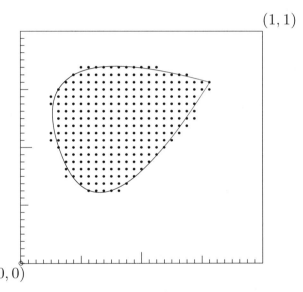

$(1, 1)$

$(0, 0)$

Fig. 7.5. A compact set $K \subseteq [0; 1]^2$ approximated in a 33×33 grid

Definition 7.4.1 (grid representation of $\mathcal{K}^0$).
1. *For $n \in \mathbb{N}$ define $D_n := \{m \cdot 2^{-n} \mid m = 0, 1, \ldots, 2^n\}$.*
2. *Define a notation $\nu_G :\subseteq \Sigma^* \to 2^{D_n \times D_n}$ such that $\operatorname{dom}(\nu_G) = \bigcup_n \Sigma^{(2^n+1)^2}$ and for $a_i \in \Sigma$,*

$$\nu_G(w) := \{(\frac{m}{2^n}, \frac{m'}{2^n}) \in D_n \times D_n \mid 0 \leq m, m' \leq 2^n, \ a_{m \cdot (2^n+1)+m'+1} = 1\},$$

if $w = a_1 a_2 \ldots a_{(2^n+1)^2}$.

3. *Define the grid representation κ_G of $\mathcal{K}^0$ by*

$$\kappa_G(p) = K \iff p = \iota(w_0)\iota(w_1)\ldots \quad \text{such that } |w_i| = (2^i + 1)^2$$
$$\text{and } d_{\mathrm{H}}(K, \nu_G(w_i)) \leq 2^{-i} \text{ for all } i \in \mathbb{N}.$$

Roughly speaking, every subset of $D_n \times D_n$ can be encoded by a boolean matrix of dimension $2^n + 1$ and ν_G is the "line by line" notation of the boolean matrices of dimension $2^n + 1$, $n \in \mathbb{N}$. A κ_G-name of a compact set K is a strongly normed Cauchy sequence of grids converging to K with respect to the Hausdorff metric.

First we state that this new representation induces the same computability concept on $\mathcal{K}^0$ an κ:

Lemma 7.4.2. On the non-empty compact subsets of $[0; 1]^2$ the representations κ and κ_G are equivalent.

Proof: See Exercise 7.4.2.   □

Although the representation κ_G is not injective, the set of names of a compact set K is "small", since at every level i of precision there are only finitely many choices for w_i. Theorem 7.2.5 holds accordingly (Exercise 7.4.3). In the following we suggest several definitions of computational complexity of compact subsets of the unit square which can be derived from the grid representation κ_G.

Definition 7.4.3 (complexity of compact sets). *Define the complexity of a compact set $K \in \mathcal{K}^0$ alternatively as follows: K has complexity $t : \mathbb{N} \to \mathbb{N}$, iff there is a sequence $w_0, w_1, \ldots$ of words with $K = \kappa_G(\iota(w_0)\iota(w_1)\ldots)$ such that*

1. *the sequence $p := \iota(w_0)\iota(w_1)\ldots \in \Sigma^\omega$ is computable in time t (Definition 7.1.7),*
2. *there is a word function $g : \Sigma^* \to \Sigma^*$ computable in time t such that $g(0^i) = w_i$ for all i.*
3. *the set $\{w_0, w_1, \ldots\} \subseteq \Sigma^*$ is decidable in time t,*
4. *the set of words*

   $$L := \{0^i \iota(u)\iota(v) \mid i \in \mathbb{N}, u, v \in \Sigma^*, (\nu_{\mathbb{N}}(u) \cdot 2^{-i}, \nu_{\mathbb{N}}(v) \cdot 2^{-i}) \in \nu_G(w_i)\}$$

   is decidable in time t.
5. *There is a function $f : (\Sigma^*)^3 \to \Sigma^*$ computable in time t such that*

   $$f(0^i, u, v) = \begin{cases} 1 & \text{if } (\nu_{\mathbb{N}}(u) \cdot 2^{-i}, \nu_{\mathbb{N}}(v) \cdot 2^{-i}) \in \nu_G(w_i) \\ 0 & \text{otherwise.} \end{cases}$$

The above definitions measure the complexity of a "picture" by different "scales". Definition 2 considers the time to construct the whole picture with (metric) precision 2^{-i}. Complexity from Definition 1 can be obtained by summing up complexities from Definition 2. Definition 5 is the pointwise version of Definition 2. Definition 3 and Definition 4 (cf. Exercise 5.1.11) reduce the complexity of a compact set to the complexity of a set of words. They allow definition of, for example, NP-computable compact sets. Definition 5, which is equivalent to Definition 4, can be considered as a way to generalize the concept of "characteristic functions" to subsets of $\mathbb{R}^2$. The above scales are recursively related (Exercise 7.4.5).

Notice that the definitions begin with "there is a sequence $w_0, w_1, \ldots$ of words", and so the sets and functions the complexity of which determine the complexity of the set K are not uniquely defined by K.

All the definitions can be applied, in particular, to singleton sets $\{x\}$ and so induce various ways to define the computational complexity of a point. The definitions can be generalized straightforwardly to the compact subsets of $[0; 1]^n$ for $n \in \mathbb{N}$. A compact set not in $[0; 1]^n$ can be mapped into $[0; 1]^n$ by an affine function $x \mapsto a \cdot x + b$ with simple binary rational numbers $a \in \mathbb{Q}$ and $b \in \mathbb{Q}^n$.

Example 7.4.4 (complexity of some sets).
1. The set $K_1 := \{(x, y) \in [0; 1]^2 \mid x \leq y\}$ has complexity $\mathrm{id}_\mathbb{N}$ with respect to Definition 7.4.3.5. Define

$$f(0^i, u, v) := \begin{cases} 1 & \text{if } 0 \leq \nu_\mathbb{N}(u) \leq \nu_\mathbb{N}(v) \leq 2^i \\ 0 & \text{otherwise.} \end{cases}$$

Obviously, the function f is computable in linear time. If

$$\nu_G(w_i) = \{(\nu_\mathbb{N}(u) \cdot 2^{-i}, \nu_\mathbb{N}(v) \cdot 2^{-i}) \mid f(0^i, u, v) = 1\},$$

then $K_1 = \kappa_G(\iota(w_0)\iota(w_1)\ldots)$.
2. The Euclidean closed Ball $K_2 := \overline{B}^e((1/2, 1/2), 1/2)$ has complexity t_m with respect to Definition 7.4.3.5. For precision 2^{-i} choose the grid

$$\left\{ \left(\frac{a}{2^i}, \frac{b}{2^i} \right) \mid a, b \in \mathbb{N}, 0 \leq a, b \leq 2^i, \ \left(\frac{a}{2^i} - \frac{1}{2} \right)^2 + \left(\frac{b}{2^i} - \frac{1}{2} \right)^2 \leq \frac{1}{4} \right\}.$$

The function f can be computed in time t_m (Convention 7.3.8). This is a worst case bound, for the majority of inputs linear time is sufficient.

A non-empty compact subset $K \subseteq [0; 1]^2$ is κ-computable, iff it is κ^{dist}-computable (Remark 5.2.3). Therefore, K is κ-computable, iff its distance function $d_K : x \mapsto d(x, K)$ is computable (see Fig. 7.6. If d_K is computable, then by Theorem 7.2.7 it can be computed by some Type-2 machine on $[0; 1]^2$ in time t for some computable function $t : \mathbb{N} \to \mathbb{N}$. Such a function t can be considered as a complexity of the set K.

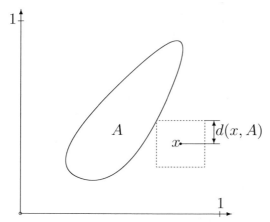

Fig. 7.6. The ball with center x and radius $d(x, A)$ touches the set A

Definition 7.4.5 (distance complexity of a compact set). *A non-empty compact subset $K \subseteq [0; 1]^2$ has distance complexity $t : \mathbb{N} \to \mathbb{N}$, iff its distance function $d_K : \mathbb{R}^2 \to \mathbb{R}$, $d_K(x) = d(x, K)$, is computable on $[0; 1]^2$ in time t.*

Distance complexity is recursively related to the complexities from Definition 7.4.3.5. As we have already mentioned in Sect. 1.3.4, the distance function $d_K : \mathbb{R}^2 \to \mathbb{R}$ can be considered as a generalization of the discrete characteristic function $cf_B : \mathbb{N} \to \mathbb{N}$ for $B \subseteq \mathbb{N}$. In comparison with the (non-computable) characteristic function $cf_K : \mathbb{R}^2 \to \mathbb{N}$ the distance function provides more information for points $x \notin K$, namely the distance to K and not only the answer "no". According to this aspect the function f from Definition 7.4.3.5 is more similar to the characteristic function than the distance function. Notice that this function f is not determined uniquely by K.

Since the representation κ_G is formally similar to the representation ρ_{sd} of the real numbers, it can be used to define the computational complexity of computable functions $T : \mathcal{K}^0 \to \mathcal{K}^0$ transforming black and white pictures.

Exercises 7.4.

1. Define a representation κ' of $\mathcal{K}^0$ by $\kappa'(p) := \kappa(p) \cap [0; 1]^2$ (κ from Definition 5.2.1). Show $\kappa' \not\leq \kappa$. Hint: Exercises 5.1.15 or 5.2.11.
2. Show that on the non-empty compact subsets of $[0; 1]^2$ the representations κ and κ_G are equivalent.
3. Let $\mathcal{L}$ be a subset of $\mathcal{K}^0$ which is compact with respect to the Hausdorff metric. Show that $\kappa_G^{-1}[\mathcal{L}]$ is compact.
4. Show that there are no injective representations equivalent to κ or to κ_G. Hint: Exercise 5.2.3 and Theorem 4.1.15.

5. For every pair (i, j) of complexity definitions from 7.4.3, find a simple operator H such that K has complexity $H(t)$ according to Definition j, if it has complexity t according to Definition i.
6. Compare the complexity of a real number $x \in [0; 1]$ according to Definition 7.2.6 with the complexity of the set $\{x\}$ according to Definition 7.4.3.5.
7. Call a closed subset $X \subseteq \Sigma^\omega$ recursive, iff

$$A_X := \{y \in \Sigma^* \mid y\Sigma^\omega \cap X \neq \emptyset\}$$

is recursive (Definition 2.4.9) and define: X has complexity $t : \mathbb{N} \to \mathbb{N}$, iff A_X is decidable in time t.
   a) Introduce the Hausdorff metric on the closed subsets of Σ^ω and compare this definition with Definition 7.4.3.4.
   b) Find a complexity bound for the set $(\{0, 1\}0)^\omega \subseteq \Sigma^\omega$.

8. Some Extensions

8.1 Computable Metric Spaces

In this book, we study computability concepts which are induced by notations and representations. Although every representation $\delta :\subseteq \Sigma^\omega \to M$ of a set induces a computability concept, only very few of them are useful. As an important class we have studied representations constructed from *computable topological spaces* (Sect. 3.2). Remember that every T_0-space with countable base can be extended to a computable topological space by an injective notation of a subbase. Many important T_0-spaces with countable base can be generated from separable metric spaces (Definition 2.2.1). Therefore, it is useful to study computability on metric spaces separately. The reader who is not familiar with the mathematical concepts used in this section is referred to any standard textbook on real analysis, for example, Rudin [Rud64].

Continuity of functions on metric spaces can be defined via the induced topologies (Definition 2.2.1). There are two other well-known useful definitions of continuity. Let (M, d) and (M', d') be metric spaces with induced topologies τ and τ', respectively, and let $f :\subseteq M \to M'$ be a function and $x \in \mathrm{dom}(f)$. Then the following properties are equivalent:

- **topological continuity:**
  for all $V \in \tau'$ with $f(x) \in V$ there is $U \in \tau$ with $x \in U$ and $f[U] \subseteq V$.
- **(ε, δ)-continuity:**
  $(\forall \varepsilon > 0)(\exists \delta > 0)(\forall y \in \mathrm{dom}(f))(d(x, y) < \delta \implies d'(f(x), f(y)) < \varepsilon)$.
- **sequential continuity:**
  if $(\forall i)x_i \in \mathrm{dom}(f)$ and $\lim_{i \to \infty} x_i = x$, then $\lim_{i \to \infty} f(x_i) = f(x)$.

We will consider only representations $\delta :\subseteq \Sigma^\omega \to M$ of metric spaces (M, d) such that the distance $d : M \times M \to \mathbb{R}$ is (δ, δ, ρ)-continuous. Already a properly weaker condition has far-reaching consequences.

Lemma 8.1.1. Let $\delta :\subseteq \Sigma^\omega \to M$ be a representation of a metric space (M, d) such that the distance $d : M \times M \to \mathbb{R}$ is $(\delta, \delta, \rho_>)$-continuous. Then the representation δ is continuous and the space is separable.

Proof: By assumption there is a continuous function $f :\subseteq \Sigma^\omega \times \Sigma^\omega \to \Sigma^\omega$ such that $d(\delta(p), \delta(q)) = \rho_> \circ f(p, q)$ for all $p, q \in \text{dom}(\delta)$.

First, we show that δ is $(\tau_C, \tau_\mathbb{R})$-continuous. Since by Lemma 2.2.5 the Cantor topology on Σ^ω is generated by a metric d_C, it suffices to show that δ is sequentially continuous. Let $p_0, p_1, \ldots$ be a sequence in $\text{dom}(\delta)$ converging to $p \in \text{dom}(\delta)$. Since f is sequentially continuous on Σ^ω, the sequence $q_0, q_1 \ldots$, $q_i := f(p, p_i)$, converges to $q := f(p, p)$. We obtain

$$\rho_>(q) = \rho_> \circ f(p, p) = d(\delta(p), \delta(p)) = 0 .$$

Consider $a \in \mathbb{R}$, $a > 0$. Since $\rho_>(q) = 0$, there is some prefix w of q such that $\rho_>(p') < a$ for all $p' \in w\Sigma^\omega \cap \text{dom}(\rho_>)$. Since $\lim q_i = q$, there is some $i_0 \in \mathbb{N}$ such that $q_i \in w\Sigma^\omega$ for all $i > i_0$ and so $\rho_>(q_i) < a$ for all $i > i_0$. We obtain $d(\delta(p), \delta(p_i)) = \rho_> \circ f(p, p_i) = \rho_>(q_i) < a$ for all $i > i_0$. Hence, the sequence $\delta(p_0), \delta(p_1), \ldots$ converges to $\delta(p)$. Therefore, the representation δ is continuous.

Next, we construct a countable dense subset of M. Let

$$V := \{v \in \Sigma^* \mid (\exists\, p \in \Sigma^\omega)\ vp \in \text{dom}(\delta)\} .$$

There is a function $h :\subseteq \Sigma^* \to \Sigma^\omega$ such that $vh(v) \in \text{dom}(\delta)$ for all $v \in V$. Let $X := \{vh(v) \mid v \in V\}$ and $Y := \{\delta(q) \mid q \in X\}$. We show that X is dense in $\text{dom}(\delta)$. Let $w\Sigma^\omega \cap \text{dom}(\delta) \neq \emptyset$. Then $w \in V$ and $wh(w) \in X \cap \text{dom}(\delta)$. Therefore, every non-empty open subset of $\text{dom}(\delta)$ has an element from X, and so X is dense in $\text{dom}(\delta)$. Since X is dense in $\text{dom}(\delta)$ and f is continuous, $Y = f[X]$ is dense in $M = \text{range}(\delta)$ (Exercise 8.1.1). $\qquad \square$

We define *effective metric spaces* and the associated *Cauchy representations*.

Definition 8.1.2 (effective metric space, Cauchy representation).

1. An *effective metric space* is a tuple $\mathbf{M} = (M, d, A, \alpha)$ such that (M, d) is a metric space and $\alpha :\subseteq \Sigma^* \to A$ is a notation of a dense subset $A \subseteq M$.
2. The *Cauchy representation* $\delta_\mathbf{M} :\subseteq \Sigma^\omega \to M$ associated with the effective metric space $\mathbf{M} = (M, d, A, \alpha)$ is defined by

$$\delta_\mathbf{M}(p) = x : \Longleftrightarrow \begin{cases} \text{there are words } w_0, w_1, \ldots \in \text{dom}(\alpha) \\ \text{such that } p = \iota(w_0)\iota(w_1)\ldots , \\ d(\alpha(w_i), \alpha(w_k)) \leq 2^{-i} \text{ for } i < k \\ \text{and } x = \lim_{i \to \infty} \alpha(w_i) . \end{cases} \tag{8.1}$$

3. A *computable metric space* is an effective metric space $\mathbf{M} = (M, d, A, \alpha)$ such that

$$\{(t, u, v, w) \in (\Sigma^*)^4 \mid \nu_\mathbb{Q}(t) < d(\alpha(u), \alpha(v)) < \nu_\mathbb{Q}(w)\} \text{ is r.e.} \tag{8.2}$$

The dense set A "spreads" the full metric space M and the notation α induces computability on A. Roughly speaking, a Cauchy name of a point

$x \in M$ is a sequence (of names) of elements from the dense set A which converges to x rapidly. By Property (8.2), for any two points $a, b \in A$, the distance can be approximated from below and from above arbitrarily closely. Condition (8.2) is equivalent to:

$$\text{dom}(\alpha) \text{ is r.e. and } d|_{A \times A} \text{ is } (\alpha, \alpha, \rho)\text{-computable.} \tag{8.3}$$

In [Wei93] an effective metric space is called computable, if

$$\{(u, v, w) \in (\Sigma^*)^3 \mid d(\alpha(u), \alpha(v)) < \nu_{\mathbb{Q}}(w)\} \text{ is r.e. ,} \tag{8.4}$$

or equivalently,

$$\text{dom}(\alpha) \text{ is r.e. and } d|_{A \times A} \text{ is } (\alpha, \alpha, \rho_>)\text{-computable.} \tag{8.5}$$

We suggest to call such metric spaces *semi-computable*. Definition 8.1.2 generalizes several earlier ones:

Example 8.1.3 (some computable metric spaces).
1. $\mathbf{M} := (\mathbb{R}, d, \mathbb{Q}, \nu_{\mathbb{Q}})$ where $d(x, y) := |x - y|$ is a computable metric space. In this case, the Cauchy representation $\delta_{\mathbf{M}}$ is exactly ρ_C from Definition 4.1.5. The generalization to $\mathbb{R}^n$ is obvious (Lemma 4.1.18).
2. Consider Definition 5.2.8. Then $\mathbf{M} := (\mathcal{K}^*, d_H, E^*(\mathbb{Q}^n), E\mathbb{Q}^n)$ is a computable metric space and κ_H is the associated Cauchy representation of the non-empty compact subsets of $\mathbb{R}^n$.
3. Consider Definition 6.1.9. Then $\mathbf{M} := (C(A), d, \text{Pg}, \nu_{\text{Pg}})$ where d is the maximum distance of continuous functions on the recursive interval A is a computable metric space and δ_C^A is the associated Cauchy representation.
4. Consider Definition 7.4.1. Then $\mathbf{M} := (\mathcal{K}^0, d_H, \text{range}(\nu_G), \nu_G)$ is a computable metric space. However, because of the condition "$|w_i| = (2^i + 1)^2$", κ_G is not the Cauchy representation associated with $\mathbf{M}$, but it is equivalent to it. □

For every effective metric space there is a canonical effective topological space. Often the derived representation (Definition 3.2.2) is equivalent to the Cauchy representation. This generalizes the property $\rho \equiv \rho_C$ from Theorem 4.1.6.

Theorem 8.1.4. Let $\mathbf{M} = (M, d, A, \alpha)$ be an effective metric space. Define an effective topological space $\mathbf{S} = (M, \sigma, \nu)$ by

$$\sigma := \{B(a, r) \mid a \in A, r \in \mathbb{Q}, r > 0\} \text{ and}$$

$$\nu\langle u, v \rangle := B(\alpha(u), \nu_{\mathbb{Q}}(v)) .$$

Then for $\delta_{\mathbf{M}}$, the Cauchy representation associated with $\mathbf{M}$, and $\delta_{\mathbf{S}}$, the representation derived from the computable topological space $\mathbf{S}$,
1. $\delta_{\mathbf{M}} \equiv_t \delta_{\mathbf{S}}$, and
2. $\delta_{\mathbf{M}} \equiv \delta_{\mathbf{S}}$, if $\mathbf{M}$ is computable and S is a computable topological space.

Proof: See Exercise 8.1.4 □

Let (M, d) be a pseudometric space. Then there may be elements $x, y \in M$ such that $x \neq y$ and $d(x, y) = 0$. A metric space (M', d') can be constructed canonically from (M, d) by identifying elements x, y with $d(x, y) = 0$. Formally, an equivalence relation is intrduced on M by $x \sim y \iff d(x, y) = 0$ and define $M' := \{x/_\sim \mid x \in M\}$ and $d'(x/_\sim, y/_\sim) := d(x, y)$, where $x/_\sim := \{y \in M \mid d(x, y) = 0\}$.

A metric space (M, d) is complete, iff every Cauchy sequence has a limit in M. Every metric space has a minimal complete extension (its *completion*) which is unique up to renaming. The completion can be constructed as follows. Let S be the set of all Cauchy sequences on M. Define a distance $d_S : S \times S \to \mathbb{R}$ by

$$d_S((x_0, x_1, \ldots), (y_0, y_1, \ldots)) := \lim_{i \to \infty} d(x_i, y_i) . \tag{8.6}$$

Then (S, d_S) is a pseudometric space. The completion (M', d') of (M, d) is defined as the canonical factorization of (S, d_S). The function $h : M \to M'$, $h(x) := (x, x, \ldots)/_\sim$, embeds M in M' such that $d(x, y) = d'(h(x), h(y))$. Up to renaming, every metric space is a subspace of its completion and every complete metric space is isometric to the completion of its restriction to any dense subset.

In the above completion, the set S of *all* Cauchy sequences can be replaced by the subset S_0 of all *rapidly converging* Cauchy sequences. Furthermore, (M, d) can be a pseudometric space. In this case, (M', d') is the completion of the canonical factorization of (M, d). From a notation of (M, d) a representation of its completion can be derived canonically:

Definition 8.1.5 (constructive completion).   *A noted pseudometric space is a triple $\mathbf{A}_0 := (A_0, d_0, \alpha_0)$ such that (A_0, d_0) is a pseudometric space and $\alpha_0 :\subseteq \Sigma^* \to A_0$ is a notation of A_0. Call $\mathbf{A}_0$ computable, iff*

$$\{(t, u, v, w) \in (\Sigma^*)^4 \mid \nu_{\mathbb{Q}}(t) < d_0(\alpha_0(u), \alpha_0(v)) < \nu_{\mathbb{Q}}(w)\} \tag{8.7}$$

is r.e.

The constructive completion $\mathbf{M} := (M, d, A, \alpha)$ of $\mathbf{A}_0$ is defined as follows.
1. Define S and $d_S : S \times S \to \mathbb{R}$ by

$$S := \{(a_0, a_1, \ldots) \mid a_i \in A_0 \ \text{ and } \ d_0(a_i, a_k) \leq 2^{-i} \ \text{ for all } \ i < k\} ,$$

$$d_S((a_0, a_1, \ldots), (b_0, b_1, \ldots)) := \lim_{i \to \infty} d_0(a_i, b_i) .$$

2. Let (M, d) be the factorization of the pseudometric space (S, d_S) with respect to $\sim$ where

$$(a_0, a_1, \ldots) \sim (b_0, b_1, \ldots) \iff \lim_{i \to \infty} d_0(a_i, b_i) = 0 .$$

3. Define A and $\alpha :\subseteq \Sigma^* \to A$ by

$$\text{dom}(\alpha) := \text{dom}(\alpha_0),$$
$$\alpha(w) := (\alpha_0(w), \alpha_0(w), \ldots)/\!\sim \ \ \text{for all} \ \ w \in \text{dom}(\alpha),$$
$$A := \text{range}(\alpha) \, .$$

Lemma 8.1.6. The constructive completion $\mathbf{M} = (M, d, A, \alpha)$ of a noted pseudometric space $\mathbf{A_0} = (A_0, d_0, \alpha_0)$ is a complete effective metric space, such that

$$\delta_{\mathbf{M}}(\iota(w_0)\iota(w_1)\ldots) = (\alpha_0(w_0), \alpha_0(w_1), \ldots)/\!\sim \, .$$

(where $\delta_{\mathbf{M}}$ is the Cauchy representation associated with $\mathbf{M}$).
If $\mathbf{A_0}$ is computable, then $\mathbf{M}$ is computable.

Proof: See Exercise 8.1.5.   □

Constructive completion extends the computability concept from a (countable) noted pseudometric space to its complete extension. If, in particular, $\mathbf{A_0}$ is a *metric* space, then the function $h : a \mapsto (a, a, \ldots)/\!\sim$ embeds $\mathbf{A_0}$ in $\mathbf{M}$ isometrically $(d_0(a, a') = d(h(a), h(a')))$, and so (A_0, d_0) can be identified with $(A, d|_{A \times A})$. Every effective metric space $\mathbf{M} = (M, d, A, \alpha)$ is (isometric to) a restriction of the constructive completion of the noted metric space $(A, d|_{A \times A}, \alpha)$. See Exercise 8.1.6 and Example 8.1.3. Constructive completion can be applied to define a natural representation of (up to isometric mappings) all separable complete metric spaces (Exercise 8.1.13, [Wei93]).

Example 8.1.7 ($\mathbf{L}_p$-space).   Call a function $s : [0; 1] \to \mathbb{R}$ a *step function* (Fig. 8.1), iff there are numbers $0 = a_0 < a_1 < \ldots < a_k = 1$ and $b_1, b_2, \ldots, b_k$ $(k \geq 1)$ such that

$$s(x) := \begin{cases} b_1 & \text{if} \ \ x = 0 \\ b_i & \text{if} \ \ a_{i-1} < x \leq a_i \ \ (i = 1, \ldots, k) \, . \end{cases}$$

For a step function s let $\int_0^1 s(x) \, \mathrm{d}x := \sum_{i=1}^k b_i \cdot (a_i - a_{i-1})$ be the integral as usual. Let RSF be the set of rational step functions, that is, the set of all step functions definable by rational numbers a_i, b_i, and let $\alpha_s :\subseteq \Sigma^* \to \text{RSF}$ be some canonical notation of the set RSF. For $1 \leq p < \infty$ define a metric d_p on RSF by

$$d_p(s, s') := \left(\int_0^1 |s(x) - s'(x)|^p \mathrm{d}x \right)^{1/p} \, .$$

It is well-known that d_p is a metric on RSF and that the completion of (RSF, d_p) is (isometric to) the space $\mathrm{L}_p[0; 1]$.
Let p be a computable real number. Then

$$\{(t, u, v, w) \in (\Sigma^*)^4 \mid \nu_{\mathbb{Q}}(t) < d_p(\alpha_s(u), \alpha_s(v)) < \nu_{\mathbb{Q}}(w)\}$$

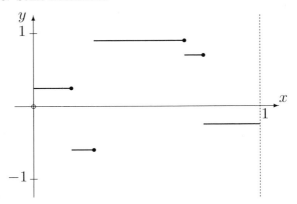

Fig. 8.1. A step function

is r.e. (see 8.7), and so $\mathbf{A}_p := (\mathrm{RSF}, d_p, \alpha_s)$ is a computable, noted metric space. Therefore, its constructive completion $(\mathrm{L}_p[0;1], d, A, \alpha)$ is a computable metric space. The associated Cauchy representation $\delta_{\mathbf{M}}$ induces a "natural" computability concept on $\mathrm{L}_p[0;1]$. Notice that we have constructed a new space together with a computability concept on it explicitly from a simple countable metric space.   □

Very often, continuous functions on metric spaces are defined by continuous extension. Let M, M' be metric spaces, let $A \subseteq M$ be dense and let M' be complete. If $f : A \to M'$ is a uniformly continuous function, then it has a unique continuous extension $\overline{f} : M \to M'$. Examples are addition and multiplication which can be extended from the rational numbers to the real numbers.

Example 8.1.8 (integration on $\mathrm{L}_1[0;1]$). Consider $p = 1$ in Example 8.1.7. Extend the integral $\int_0^1 s(x)\,\mathrm{d}x$ from RSF to $\mathrm{L}_1[0;1]$ by

$$\int_0^1 \lim_{i\to\infty} s_i \,\mathrm{d}x := \lim_{i\to\infty} \int_0^1 s_i(x)\,\mathrm{d}x$$

for all d_1-Cauchy sequences sequences $s_0, s_1, \ldots$ ($s_i \in \mathrm{RSF}$). The integration operator

$$F \mapsto \int_0^1 F\,\mathrm{d}x, \quad F \in \mathrm{L}_1[0;1] \quad \text{is} \quad (\delta_{\mathbf{M}}, \rho)\text{-computable},$$

where $\delta_{\mathbf{M}}$ is the canonical Cauchy representation from Example 8.1.7. For a proof assume $F = \lim_{i\to\infty} s_i$ and $\int_0^1 |s_i - s_k|\,\mathrm{d}x \le 2^{-i}$ for all $i < k$. Then

$$\left| \int_0^1 s_i(x)\,\mathrm{d}x - \int_0^1 s_k(x)\,\mathrm{d}x \right| = \left| \int_0^1 (s_i(x) - s_k(x))\,\mathrm{d}x \right|$$
$$\le \int_0^1 |s_i(x) - s_k(x)|\,\mathrm{d}x$$
$$\le 2^{-i}.$$

There is a Type-2 machine M which on input $\iota(u_0)\iota(u_1)\ldots \in \operatorname{dom}(\delta_{\mathbf{M}})$ produces a sequence $\iota(v_0)\iota(v_1)\ldots$ such that $\int_0^1 \alpha_s(u_i)(x)\,\mathrm{d}x = \nu_{\mathbb{Q}}(v_i)$. Then f_M is a $(\delta_{\mathbf{M}}, \rho_C)$-realization of integration on $\mathrm{L}_1[0;1]$. $\qquad\square$

The various definitions of representations of open, closed or compact subsets of $\mathbb{R}^n$ from Chap. 5 can be generalized to arbitrary computable metric spaces. As an example consider the definition of r.e. and of recursive open subsets of Σ^ω (Definition 2.4.9). Some of the equivalences proved in Chap. 5, however, may fail for other spaces [Pre99]. Also the representations for function spaces from Chap. 6 can be generalized to computable metric spaces in various ways. The details have not yet been investigated.

Exercises 8.1.

$\Diamond$ 1. Let $f : M \to M'$ be a surjective continuous function from a metric space M to a metric space M'. Show that $f[X]$ is dense in M', if X is dense in M.

$\Diamond$ 2. Show that (8.2) and (8.3) are equivalent.

$\Diamond$ 3. Let (Σ^ω, d) be the metric space from Lemma 2.2.5. Define $\alpha(w) := w0^\omega$. Show that $(\Sigma^\omega, d, \Sigma^*\{0^\omega\}, \alpha)$ is a computable metric space.

4. Prove Theorem 8.1.4.

5. Prove Lemma 8.1.6.

6. Let $\mathbf{M} = (M, d, A, \alpha)$ be a computable metric space. Show that $\mathbf{M}$ is (isometric to) a restriction of the constructive completion of the noted metric space $(A, d|_{A\times A}, \alpha)$.

7. Verify the statements from Example 8.1.7.

8. Let $\delta_{\mathbf{M}}$ be the Cauchy representation associated with an effective metric space $\mathbf{M}$.

   a) Show that the function

   $$(x_0, x_1, \ldots) \mapsto \lim_{i\to\infty} x_i$$

   for sequences $(x_0, x_1, \ldots)$ with $d(x_i, x_k) \leq 2^{-i}$ for $i < k$ is $([\nu_{\mathbb{N}} \to \delta_{\mathbf{M}}]_{\mathbb{N}}, \delta_{\mathbf{M}})$-computable (cf. Theorem 4.3.7).

   b) (computable Banach fixed point theorem) Let $\mathbf{M}$ be complete and computable. Let $X \subseteq M$ be a closed set containing a $\delta_{\mathbf{M}}$-computable element and let $f :\subseteq M \to M$ be a $(\delta_{\mathbf{M}}, \delta_{\mathbf{M}})$-computable function such that $X \subseteq \operatorname{dom}(f)$, $f[X] \subseteq X$ and f is contracting on X, that is, there is some $0 \leq c < 1$ such that $d(f(x), f(y)) \leq c \cdot d(x, y)$ for all $x, y \in X$. Show that f has a computable fixed point $x_f \in X$.

$\blacklozenge$ 9. (Computable Baire's theorem) Let $\mathbf{M}$ be a complete, computable metric space; consider ν and $\delta_{\mathbf{M}}$ for Theorem 8.1.4. Define a representation θ of the open subsets of M by $\theta(p) := \bigcup\{\nu(w) \mid w \lhd p\}$. Show that there is a $([\theta]^\omega, [\delta_{\mathbf{M}}]^\omega)$-computable multi-valued function mapping every sequence $U_0, U_1, \ldots$ of dense open sets to a sequence $x_0, x_1, \ldots$ of points which is dense and contained in every set U_i. This generalizes a result in [YMT99].

♦10. Let $\mathbf{M} = (M, d, A, \alpha)$ be a complete computable metric space. Let $\mathcal{K}^*$ be the set of all non-empty compact subsets of M, let d_H be the Hausdorff metric on $\mathcal{K}^*$, and let $\alpha_K : \iota(u_1)\iota(u_2)\ldots\iota(u_k) \mapsto \{\alpha(u_1), \alpha(u_2), \ldots, \alpha(u_k)\}$ be the standard notation of the set $E^0(A)$ of the non-empty finite subsets of A. Let $(f_1, \ldots, f_k)$ be a *hyperbolic iterated function system*(that is, a list of contracting functions on M [Bar93] (see Exercise 8 above)) such that $f_1, \ldots, f_k$ are $(\delta_\mathbf{M}, \delta_\mathbf{M})$-computable.

a) Show that $\mathbf{K} := (\mathcal{K}^*, d_H, E^0(A), \alpha_K)$ is a complete computable metric space. Let δ_K be the associated Cauchy representation.

b) Show that $f : \mathcal{K}^* \to \mathcal{K}^*$,

$$f(K) := f_1[K] \cup f_2[K] \cup \ldots \cup f_k[K]$$

is a contracting (δ_K, δ_K)-computable function.

c) Show that the fixed point of f (the *attractor*of the iterated function system) is a δ_K-computable compact set (apply Exercise 8), see [KK99].

11. Consider Examples 8.1.7 and 8.1.8. Show that

$$(F, a, b) \mapsto \int_a^b F \, dx, \quad 0 \leq a \leq b \leq 1, \quad F \in L_1[0; 1] ,$$

is $(\delta_\mathbf{M}, \rho, \rho, \rho)$-computable.

♦12. Consider Examples 8.1.7 and 8.1.8. Show that

$$(f, F) \mapsto f \cdot F, \quad f \in C[0; 1], \quad F \in L_1[0; 1],$$

is $(\delta_\to^{[0;1]}, \delta_\mathbf{M}, \delta_\mathbf{M})$-computable.

♦13. Effectivity on the space of all complete separable metric spaces [Wei93]:

a) Show that every complete effective metric space is isometric to the constructive completion of a noted pseudometric space $\mathbf{A} = (A, d, \alpha)$ such that $A \subseteq \Sigma^*$ and $\alpha(w) = w$.

b) Let NPS be the set of all noted pseudometric spaces from 13a above. Define a notation ν of subsets of NPS by

$$(A, d, \alpha) \in \nu\langle t, u, v, w\rangle : \Longleftrightarrow u, v \in A \text{ and } \nu_\mathbb{Q}(t) < d(u, v) < \nu_\mathbb{Q}(w) ,$$

cf. (8.7). Show that $\mathbf{S} := (\text{NPS}, \sigma, \nu)$, where $\sigma := \text{range}(\nu)$, is a computable topological space.

Let $\delta_M(p)$ be the constructive completion of $\delta_\mathbf{S}(p)$ ($\mathbf{S}$ from 13b above). Then δ_M is a canonical representation of (essentially all) complete effective metric spaces.

8.2 Degrees of Discontinuity

In recursion theory, the difficulty of non-recursive sets and functions can be compared by means of reducibilities [Rog67, Wei87, Odi89]; for example, many-one reducibility or Turing reducibility. In TTE we have two kinds of

effectivity: continuity and computability. In this section we will compare the
difficulty of problems by continuous reductions.

In recursion theory, $A \subseteq \mathbb{N}$ is called "many-one reducible" to $B \subseteq \mathbb{N}$,
$A \leq_m B$ for short, iff $A = f^{-1}[B]$ for some (total) recursive function $f : \mathbb{N} \to \mathbb{N}$. We transfer this definition to subsets of Σ^ω as follows.

Definition 8.2.1 (Wadge reducibility).   *For subsets $A, B \subseteq \Sigma^\omega$, define
Wadge reducibility and equivalence by*

$$A \leq_w B : \iff A = f^{-1}[B] \text{ for some continuous function } f : \Sigma^\omega \to \Sigma^\omega ,$$

$$A \equiv_w B : \iff A \leq_w B \text{ and } B \leq_w A .$$

The equivalence classes $A/_{\equiv_w}$ $(A \subseteq \Sigma^\omega)$ are called "Wadge degrees".

Obviously, $\leq_w$ is a pre-order and $\equiv_w$ is an equivalence relation on the power
set of Σ^ω, and the equivalence classes are partially ordered by $\leq_w/_{\equiv_w}$ which
we abbreviate by $\leq_w$.

Example 8.2.2.
1. $\emptyset \leq_w B$, iff $B \neq \Sigma^\omega$,
2. A is clopen, iff $A \leq_w 0\Sigma^\omega$,
3. A is open, iff $A \leq_w \Sigma^\omega \setminus \{0^\omega\}$ (Exercise 2.4.5),
4. $\{0^\omega\} \not\leq_w \Sigma^\omega \setminus \{0^\omega\}$,
5. A is a G_δ-set (Definition 2.2.1), iff $A \leq_w G_0$ where $G_0 := \{p \in \Sigma^\omega \mid p(i) = 0 \text{ infinitely often}\}$ (Exercise 2.4.5),
6. $\Sigma^\omega \setminus G_0 \not\leq_w G_0$ (Exercise 2.3.2). $\qquad\qquad\qquad\qquad\qquad$ $\square$

While the partial order on the many-one degrees of subsets of $\mathbb{N}$ is very
complicated and still subject of investigations, the order of the Wadge degrees is surprisingly simple, at least on the Borel sets (where the set of *Borel
subsets* of Σ^ω is the smallest class of subsets of Σ^ω which contains the open
sets and is closed under complement and countable union).

Theorem 8.2.3 (Wadge hierarchy).   There are an ordinal number α_w
and three families $(S_\beta)_{\beta < \alpha_w}$, $(P_\beta)_{\beta < \alpha_w}$ and $(E_\beta)_{\beta < \alpha_w}$ of subsets of Σ^ω such
that
1. $S_\beta <_w E_\beta$, $P_\beta <_w E_\beta$, $S_\beta \not\leq_w P_\beta$ and $P_\beta \not\leq_w S_\beta$ for all $\beta < \alpha_w$,
2. $E_\beta <_w S_\gamma$ and $E_\beta <_w P_\gamma$ for $\beta < \gamma < \alpha_w$,
3. For every Borel set $A \subseteq \Sigma^\omega$ there is some $\beta < \alpha_w$ such that $A \equiv_w S_\beta$,
   $A \equiv_w P_\beta$, or $A \equiv_w E_\beta$,

Fig. 8.2 shows the Wadge hierarchy on the Borel sets. If every set is replaced by its equivalence class, we obtain the hierarchy of the Wadge degrees.
For proofs and more details see [Wad72, Wad83, Eng86].

Let ω be the first infinite ordinal and ω_1 the first uncountable ordinal
[Sch77]. The Wadge degrees with index $\beta \leq \omega_1$ can be constructed easily.

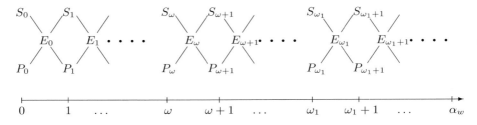

Fig. 8.2. The Wadge hierarchy

We use the following fact: for every ordinal $\gamma < \omega_1$, there is an increasing sequence $\alpha_\gamma(0), \alpha_\gamma(1), \ldots$ of ordinals such that $\sup_i \alpha_\gamma(i) = \gamma$.

Theorem 8.2.4 (ω_1-segment of the Wadge hierarchy).   Consider Theorem 8.2.3. For $\beta \leq \omega_1$ suitable sets S_β, P_β and E_β can be defined inductively as follows.

$$S_0 := \emptyset, \quad S_1 := \Sigma^\omega \setminus \{0^\omega\},$$
$$S_{\omega_1} := G_0 := \{p \in \Sigma^\omega \mid p(i) = 0 \text{ infinitely often}\},$$
$$P_\beta := \Sigma^\omega \setminus S_\beta, \quad E_\beta := S_\beta \oplus P_\beta,$$
$$\langle p, q \rangle \in S_{\beta+1} \; :\Longleftrightarrow \quad p \neq 0^\omega \text{ and } q \notin S_\beta,$$
$$S_\gamma := \bigcup_{i \in \mathbb{N}} 0^i 1 S_{h_\gamma(i)}$$

for all $1 \leq \beta < \omega_1$ and all limit numbers $\gamma < \omega_1$ (where $A \oplus B := 0A \cup 1B$).

For more details see Exercise 8.2.2 and [Wei91a]. We extend the Wadge reducibility to functions and sets of functions as follows.

Definition 8.2.5 (Wadge reducibility for functions). *Let W, X, Y, Z be finite Cartesian products of Σ^* and Σ^ω. Let $f :\subseteq W \to X$ and $g :\subseteq Y \to Z$ be partial functions, let S be a set of partial functions $h :\subseteq W \to X$ and let T be a set of partial functions $h :\subseteq Y \to Z$. Define Wadge reducibility and equivalence by*

$$f \leq_{\mathrm{w}} g \iff \begin{cases} \text{there are continuous functions } F, G \text{ such that} \\ f(w) = F(w, g \circ G(w)) \text{ for all } w \in \mathrm{dom}(f), \end{cases}$$
$$S \leq_{\mathrm{w}} T \iff \begin{cases} \text{there are continuous functions } F, G \text{ such that} \\ (\forall g \in T)(w \mapsto F(w, g \circ G(w)) \text{ extends some } f \in S), \end{cases}$$
$$f \equiv_{\mathrm{w}} g \iff f \leq_{\mathrm{w}} g \text{ and } g \leq_{\mathrm{w}} f,$$
$$S \equiv_{\mathrm{w}} T \iff S \leq_{\mathrm{w}} T \text{ and } T \leq_{\mathrm{w}} S.$$

If $f \leq_{\mathrm{w}} g$, then f can be determined from g and two additional continuous functions by composition such that for determining $f(w)$, $g(y)$ must be evaluated only for a single $y \in Y$. For subsets $A, B \subseteq \Sigma^\omega$,

$$\mathrm{cf}_A \leq_{\mathrm{w}} \mathrm{cf}_B \iff A \oplus (\Sigma^\omega \setminus A) \leq_{\mathrm{w}} B \oplus (\Sigma^\omega \setminus B)$$

(Exercise 8.2.4), and so Wadge reducibility on functions generalizes Wadge reducibility on sets. There are problems which do not have a single but instead a set of (possibly non-continuous) solutions, for example the problem of translating $\rho_{b,2}$ to $\rho_{b,10}$ (Theorem 4.1.13). If $S \leq_w T$, then there is a uniform method defined by two continuous functions F an G for transforming every "solution" $g \in T$ to a "solution" $f \in S$, where $f(w) = F(w, g \circ G(w))$.

By the next lemma the characteristic function of $\{0^\omega\}$ is the "easiest" discontinuous function.

Lemma 8.2.6. For any function $g :\subseteq \Sigma^\omega \to \Sigma^*$ or $g :\subseteq \Sigma^\omega \to \Sigma^\omega$,

$$\mathrm{cf}_{\{0^\omega\}} \leq_w g \iff g \text{ is discontinuous.}$$

Proof: See Exercise 8.2.8.   □

Several degrees are characterized in [Ste89, Myl92, Wei92b, Wei92c]. There is a close relation between the degrees of discontinuity of functions and the number of branchings in flowchart programs operating on the real numbers (real-RAMS, Sect. 9.7) which compute them [HW94, Her96].

Exercises 8.2.

Let $\mathrm{cf}_{\{0^\omega\}} : \Sigma^\omega \to \Sigma^*$ be the characteristic function of $\{0^\omega\}$ and let EC : $\Sigma^\omega \to \Sigma^\omega$ be the translation from the enumeration representation En to the characteristic function representation Cf of $2^\mathbb{N}$ (Definition 3.1.2), which are not continuous (Exercise 2.2.2, Example 3.1.4.9).
1. Verify the statements from Example 8.2.2.
◆ 2. Consider Theorems 8.2.3 and 8.2.4.
   a) Prove 1 and 2 of Theorem 8.2.3 for finite β, γ.
   b) Prove $A \leq_w S_n$ or $P_n \leq_w A$ for all $n < \omega$ and all $A \subseteq \Sigma^\omega$.
   c) Show that $A \subseteq \Sigma^\omega$ is a Boolean combination of open sets, iff $A \equiv_w S_n$ or $A \equiv_w P_n$ or $A \equiv_w E_n$ for some $n < \omega$.
◇ 3. Show that for functions f, g, h,

$$f \leq_w g \text{ and } g \leq_w h \implies f \leq_w h.$$

4. Show that for subsets $A, B \subseteq \Sigma^\omega$,

$$\mathrm{cf}_A \leq_w \mathrm{cf}_B \iff A \oplus (\Sigma^\omega \setminus A) \leq_w B \oplus (\Sigma^\omega \setminus B).$$

5. Let $f :\subseteq \Sigma^\omega \to \Sigma^*$ such that $f(0^\omega) := 0$ and $f(p) := \mathrm{div}$ for $p \neq 0^\omega$. Prove $f \equiv_w \mathrm{cf}_{\{0^\omega\}}$.
6. Define $f : \Sigma^\omega \times \Sigma^\omega \to \Sigma^*$ by $\mathrm{dom}(f) := \mathrm{dom}(\rho) \times \mathrm{dom}(\rho)$ and

$$f(p,q) := \begin{cases} 0 & \text{if } \rho(p) = \rho(q) \\ 1 & \text{otherwise} \end{cases}$$

for $p, q \in \mathrm{dom}(\rho)$. Prove $f \equiv_w \mathrm{cf}_{\{0^\omega\}}$.

♦ 7. Prove $\mathrm{cf}_{\{0^\omega\}} <_\mathrm{w} \mathrm{EC}$.

8. Prove Lemma 8.2.6.

9. Consider the problem of finding an index $i \in \{0,1\}$ such that $p_i = 0^\omega$ for any pair $(p_0, p_1) \in \Sigma^\omega \times \Sigma^\omega$ such that $p_0 = 0^\omega$ or $p_1 = 0^\omega$ (or both). The following functions f_0 and f_1 solve this problem:

$$f_i(p_0, p_1) := \begin{cases} i & \text{if} \quad p_0 = 0^\omega \quad \text{and} \quad p_1 = 0^\omega\,, \\ 0 & \text{if} \quad p_0 = 0^\omega \quad \text{and} \quad p_1 \neq 0^\omega\,, \\ 1 & \text{if} \quad p_0 \neq 0^\omega \quad \text{and} \quad p_1 = 0^\omega\,, \\ \mathrm{div} & \text{if} \quad p_0 \neq 0^\omega \quad \text{and} \quad p_1 \neq 0^\omega\,. \end{cases}$$

Prove $f_0 \equiv_\mathrm{w} \mathrm{cf}_{\{0^\omega\}}$, $f_1 \equiv_\mathrm{w} \mathrm{cf}_{\{0^\omega\}}$ and $\{f_0, f_1\} <_\mathrm{w} \{\mathrm{cf}_{\{0^\omega\}}\}$. (Therefore, $\mathrm{cf}_{\{0^\omega\}}$ can be reduced to f_0 and to f_1 but not to the set $\{f_0, f_1\}$.)

10. If $x_0 \cdot x_1 = 0$ for two real numbers x_0, x_1, then $x_i = 0$ for some $i \in \{0,1\}$. Let S be the set of all functions $g :\in \Sigma^\omega \times \Sigma^\omega \to \Sigma^*$ such that for all $p_0, p_1 \in \mathrm{dom}(\rho)$,

$$g(p_0, p_1) \begin{cases} \in \{0,1\} & \text{if} \quad \rho(p_0) = 0 \quad \text{and} \quad \rho(p_1) = 0 \\ = 0 & \text{if} \quad \rho(p_0) = 0 \quad \text{and} \quad \rho(p_1) \neq 0 \\ = 1 & \text{if} \quad \rho(p_0) \neq 0 \quad \text{and} \quad \rho(p_1) = 0\,. \end{cases}$$

Prove $S \equiv_\mathrm{w} \{f_0, f_1\}$, where f_0, f_1 are from Exercise 9. above.

11. Let T be the set of all functions $h :\subseteq \Sigma^\omega \times \Sigma^\omega \to \Sigma^*$ such that for all $p_0, p_1 \in \mathrm{dom}(\rho)$,

$$x_{h(p_0, p_1)} = \min(x_0, x_1) \quad \text{if} \quad x_0 = \rho(p_0) \quad \text{and} \quad x_1 = \rho(p_1).$$

The set T contains no continuous function (Exercise 4.3.15). Prove $S \equiv_\mathrm{w} \{f_0, f_1\}$, where f_0, f_1 are from Exercise 9 above.

♦ 12. For representations δ, δ' let $(\delta \to \delta')$ be the set of all translations $f :\subseteq \Sigma^\omega \to \Sigma^\omega$ from δ to δ' (Definition 2.3.2). Prove

$$\{\mathrm{EC}\} \equiv_\mathrm{w} (\rho_{\mathrm{Cn}} \to \rho) \equiv_\mathrm{w} (\rho_< \to \rho_>)$$

(Definitions 4.1.4 and 4.1.14.1).

♦ 13. Let S be the set of all $(\delta_\to^{[0;1]}, \rho)$-realizations of the function

$$f \mapsto f'(1/2)$$

and let S' be the set of all $(\delta_\to^{[0;1]}, \delta_\to^{[0;1]})$-realizations of

$$f \mapsto f'$$

(cf. Theorem 6.4.3). Prove $\{\mathrm{EC}\} \equiv_\mathrm{w} S \equiv_\mathrm{w} S'$.

9. Other Approaches to Computable Analysis

In the preceeding chapters we have developed Type-2 Theory of effectivity, TTE, which provides tools for studying various aspects of computability in analysis and, in particular, offers a "natural" definition of computable real functions. In contrast to the computable number functions where numerous definitions coincide (Church's Thesis) and alternatives are no longer considered seriously [Rog67, Odi89] several partly non-equivalent approaches to computable analysis have been proposed and are still being investigated. In this chapter we outline the concepts of some of these approaches and compare them with TTE. For avoiding additional technical framework we will express the original definitions in the terms used in this book whenever advisable.

9.1 Banach/Mazur Computability

On the basis of joint work with S. Banach in about 1950, S. Mazur [Maz63] introduced a weak kind of computability for functions operating on the set $\mathbb{R}_c$ of computable real numbers. As we know, every (ρ, ρ)-computable real function f maps computable real numbers to computable real numbers (Theorem 3.1.6) and, more uniformly, maps every $(\nu_{\mathbb{N}}, \rho)$-computable sequence $i \mapsto x_i$ in $\mathrm{dom}(f)$ to a $(\nu_{\mathbb{N}}, \rho)$-computable sequence $i \mapsto f(x_i)$ (Exercise 9.1.3). This second property is interesting, since by Theorem 4.2.6, no $(\nu_{\mathbb{N}}, \rho)$-computable sequence $i \mapsto x_i$ of real numbers lists *all* computable real numbers. Banach and Mazur take it as the axiom of computability.

Definition 9.1.1 (Banach/Mazur computable). *A real function $f :\subseteq \mathbb{R} \to \mathbb{R}$ with $\mathrm{dom}(f) \subseteq \mathbb{R}_c$ is Banach/Mazur computable (BM-computable) or sequentially computable, iff $i \mapsto f(x_i)$ is $(\nu_{\mathbb{N}}, \rho)$-computable for every $(\nu_{\mathbb{N}}, \rho)$-computable sequence $i \mapsto x_i$ in $\mathrm{dom}(f)$.*

The remarkable result this uniformity condition induces is sequential continuity restricted to computable sequences. Remember that a function on a metric space is continuous, iff it is sequentially continuous: if $i \mapsto x_i$ is *any* sequence in $\mathrm{dom}(f)$ converging to $x \in \mathrm{dom}(f)$, then the sequence $i \mapsto f(x_i)$ converges to $f(x)$.

Theorem 9.1.2. Let $f :\subseteq \mathbb{R} \to \mathbb{R}$ be a BM-computable function. If $i \mapsto x_i$ is a computable sequence in $\mathrm{dom}(f)$ converging to $x \in \mathrm{dom}(f)$, then the sequence $i \mapsto f(x_i)$ converges to $f(x)$.

Proof: See [Maz63].   □

A BM-computable real function f is continuous, if its domain has a "dense r.e. subset" (Exercise 9.1.2). On the other hand, there is an infinite subset $Y \subseteq \mathbb{R}_c$ such that *every* function $f :\subseteq \mathbb{R} \to \mathbb{R}$ with $\mathrm{dom}(f) = Y$ and $\mathrm{range}(f) \subseteq \mathbb{R}_c$ is BM-computable, in particular, a discontinuous one, which cannot be (ρ, ρ)-computable (Exercise 9.1.4). Therefore, the uniformity condition in Definition 9.1.1 is not strong enough to exclude such strange examples.

Exercises 9.1.

1. Show that there is a real function $f :\subseteq \mathbb{R} \to \mathbb{R}$ with $\mathrm{dom}(f) = \mathbb{R}_c$ mapping every computable number to a computable number which is not BM-computable.
♦ 2. Let $f :\subseteq \mathbb{R} \to \mathbb{R}$ with $\mathrm{dom}(f) \subseteq \mathbb{R}_c$ be a BM-computable function such that $\mathrm{range}(s)$ is dense in $\mathrm{dom}(f)$ for some $(\nu_\mathbb{N}, \rho)$-computable sequence $s : \mathbb{N} \to \mathbb{R}$. Show that f is continuous.
3. Show that every (ρ, ρ)-computable function f with $\mathrm{dom}(f) \subseteq \mathbb{R}_c$ is BM-computable.
4. Let $A \subseteq \mathbb{N}$ be immune (that is, A is infinite and has no infinite r.e. subset [Rog67, Odi89]). Define $Y := \{2^{-i} \mid i \in A\} \cup \{0\}$.
   a) Show that *every* function $f :\subseteq \mathbb{R} \to \mathbb{R}$ with $\mathrm{dom}(f) = Y$ and $\mathrm{range}(f) \subseteq \mathbb{R}_c$ is BM-computable.
   b) Show that there is a BM-computable real function which is not continuous and so not (ρ, ρ)-computable.

9.2 Grzegorczyk's Characterizations

In his paper [Grz55] A. Grzegorczyk proposes a definition of the computable real functions which can be written in our terms as follows:

Definition 9.2.1 (Grzegorczyk computable).   *A real function* $f :\subseteq \mathbb{R} \to \mathbb{R}$ *is G-computable, iff it is* (ρ_G, ρ_G)-*computable, where*

$$\rho_G(p) = x : \Longleftrightarrow \begin{cases} \textit{there are words } w_0, w_1 \ldots \in \mathrm{dom}(\nu_\mathbb{Z}) \\ \textit{such that } p = \iota(w_0)\iota(w_1)\iota(w_2)\ldots \\ \textit{and } \left| x - \frac{\nu_\mathbb{Z}(w_i)}{i+1} \right| < \frac{1}{i+1} \textit{ for all } i . \end{cases}$$

Since the representations ρ and ρ_G are equivalent (Exercises 4.1.7, 9.2.1), a real function is computable, iff it is G-computable. In [Grz57] the G-computable functions are characterized in various ways, two of which are given below. For the sake of simplicity, we will consider real functions defined on the interval $[0; 1]$.

Theorem 9.2.2. A real function $f :\subseteq \mathbb{R} \to \mathbb{R}$ with domain $[0; 1]$ is G-computable, iff 1 and 2 hold.
1. The restriction of f to $[0; 1] \cap \mathbb{Q}$ is BM-computable.
2. f has a computable modulus of continuity on $[0; 1]$.

Proof: See Exercise 9.2.2 for "if" and Theorem 6.2.7 for "only if".   □

The condition "f is continuous" replacing Condition 2 is not sufficient for computability of f: Pour-El and Richards [PER89] have constructed a continuous function satisfying 1 which is not computable.

The other characterization from [Grz57] we mention here considers functions on rational intervals (Definition 4.1.2). It is similar to the characterization of the computable functions $f :\subseteq \Sigma^\omega \to \Sigma^\omega$ by monotone word functions from Lemma 2.1.11.

Theorem 9.2.3. A real function $f : \mathbb{R} \to \mathbb{R}$ is G-computable, iff there is an (I^1, I^1)-computable function $F : \mathrm{Cb}^1 \to \mathrm{Cb}^1$ such that
1. $f[K] \subseteq F(K)$,
2. $F(K) \subseteq F(L)$, if $K \subseteq L$ (F is monotone) and
3. $\{f(x)\} = \bigcap_{x \in K} F(K)$ for all $x \in \mathbb{R}$.

Exercises 9.2.

1. Prove that ρ_G from Definition 9.2.1 is equivalent to ρ.
2. Prove the "if" part of Theorem 9.2.2.
3. Consider Theorem 9.2.3.
   a) Prove the "if" part.
   ◆ b) Prove the "only if" part for a bounded function f.
      Hint: Let J_0 be a closed interval with rational endpoints such that range$(f) \subseteq J_0$. Define the "open extension" of a closed interval $[a; b]$ by $\mathrm{ext}[a; b] := (a - (b - a); b + (b - a))$. By Lemma 2.1.11 there is a monotone function $h : \Sigma^* \to \Sigma^*$ such that $h_\omega :\subseteq \Sigma^\omega \to \Sigma^\omega$ is a (ρ_{sd}, ρ_{sd})-realization of f. For $J \in \mathrm{Cb}^1$ let $F(J)$ be the open extension of the intersection of J_0 and

   $$\bigcap \left\{ \rho_{sd}[x \bullet y \Sigma^\omega] \mid (\exists\, u, v \in \Sigma^*)(J \subseteq \rho_{sd}[u \bullet v \Sigma^\omega] \text{ and } h(u \bullet v) = x \bullet y) \right\}.$$

   c) Prove the general "only if" part by modifying the above proof.

9.3 The Pour-El/Richards Approach

In their book "Computability in Analysis and Physics" [PER89] (also [PE99])
M. Pour-El and J. Richards introduce computable real functions according to Grzegorczyk's [Grz57] characterization given in Theorem 9.2.2 (Ba-
nach/Mazur computable + computable uniform modulus of continuity). Part
I deals in a systematic way with the computability theory of real numbers,
real sequences, continuous functions, uniform convergence, integration, max-
ima and minima, the intermediate value theorem and several other topics.

Part II of the book generalizes the concept of a computable sequence
of real numbers to Banach spaces. The set of computable sequences $\mathcal{CS}$ of
a Banach space X is introduced axiomatically and the behaviour of linear,
possibly non-continuous operators on computable elements and computable
sequences is investigated.

Definition 9.3.1 (computability structure [PER89]). *A computabil-
ity structure on a Banach space X is a pair $(X, \mathcal{CS})$ where $\mathcal{CS}$ is a set of
sequences on X satisfying the following axioms:*

1. *(Linear Forms) Let $n \mapsto x_n$ and $n \mapsto y_n$ be sequences in $\mathcal{CS}$, let $n \mapsto \alpha_n$
   and $n \mapsto \beta_n$ be computable sequences of real (or complex) numbers and
   let $d : \mathbb{N} \to \mathbb{N}$ be a computable function. Then the sequence*

$$n \mapsto s_n := \sum_{k=0}^{d(n)} \left(\alpha_{\langle n,k \rangle} x_k + \beta_{\langle n,k \rangle} y_k \right)$$

   is in $\mathcal{CS}$.

2. *(Limits) Let $n \mapsto x_n$ be a sequence in $\mathcal{CS}$, let $n \mapsto y_n$ be a sequence on
   X and let $e : \mathbb{N}^2 \to \mathbb{N}$ be a computable function such that*

$$\|x_{\langle n,k \rangle} - y_n\| \leq 2^{-N} \quad \text{for all} \ \ k \geq e(n, N) .$$

   Then $n \mapsto y_n$ is a sequence in $\mathcal{CS}$.

3. *(Norms) If $n \mapsto x_n$ is a sequence in $\mathcal{CS}$, then $n \mapsto \|x_n\|$ is a computable
   sequence of real numbers.*

*An element $x \in X$ is called PER-computable, iff the constant sequence $n \mapsto x$
is in $\mathcal{CS}$.*

*A sequence $n \mapsto e_n$ generates ([PER89]: is an effective generating set for)
the computability structure $(X, \mathcal{CS})$, iff it is in $\mathcal{CS}$ and the linear span (that
is, the set of finite linear combinations) of $\{e_n \mid n \in \mathbb{N}\}$) is dense in X.*

Examples of computability structures are the computable sequences of
real numbers generated by $n \mapsto 1$ (Exercise 9.3.1) and the $(\nu_\mathbb{N}, \delta_\to^{[0;1]})$-
computable sequences on C[0; 1] generated by $n \mapsto x^n$ (Exercise 9.3.2). The
following "First Main Theorem" has many interesting applications.

Theorem 9.3.2 (First Main Theorem [PER89]). Let $(X, \mathcal{CS})$ and $(Y, \mathcal{CS}')$ be computability structures and let $n \mapsto e_n$ generate $(X, \mathcal{CS})$. Let $T :\subseteq X \to Y$ be a linear operator such that graph(T) is closed, $\{e_n \mid n \in \mathbb{N}\} \subseteq \text{dom}(T)$ and $n \mapsto T(e_n)$ is in $\mathcal{CS}'$. Then T is continuous, iff $T(x)$ is computable for every computable element $x \in \text{dom}(T)$.

The differentiation operator on $D :\subseteq C[0; 1] \to C[0; 1]$ is an example of a linear discontinuous operator with closed graph. By the First Main Theorem it maps some computable continuously differentiable function to a noncomputable one (Example 6.4.9). As another example, the solution operator of the wave equation maps some computable initial condition at time 0 to a non-computable solution at time 1. This surprising result has caused numerous discussions. For a detailed discussion in terms of TTE see [WZ98b]. For a more general discussion of computable invariance see [Bra99a].

A computable metric space (Definition 8.1.2) can be related to every computability structure with a generating sequence such that the computable sequences in both contexts coincide:

Theorem 9.3.3. Let $n \mapsto e_n$ generate the computability structure $(X, \mathcal{CS})$. Let A be the set of all finite rational linear combinations of elements of $\{e_n \mid n \in \mathbb{N}\}$, let α be a standard notation of A by means of terms and let d be the metric on X induced by the norm. Then
1. $\mathbf{M} := (X, d, A, \alpha)$ is a computable metric space,
2. $\mathcal{CS}$ is the set of all $(\nu_{\mathbb{N}}, \delta_{\mathbf{M}})$-computable sequences $n \mapsto x_n$ on X where $\delta_{\mathbf{M}}$ is the Cauchy representation associated with M.

Proof: See Exercise 9.3.4.   □

For computability structures with generating sequences, this theorem embeds the Pour-El/Richards approach in TTE (see also [SHT99]).

Part III of [PER89] deals with *effectively determined operators* on *effectively generated* Hilbert spaces and studies computability of the spectrum and the eigenvalues.

Exercises 9.3.

1. Show that the computable sequences of real numbers are a computability structure generated by $n \mapsto 1$.
2. Show that the $(\nu_{\mathbb{N}}, \delta_{\to}^{[0;1]})$-computable sequences on $C[0; 1]$ are a computability structure generated by $n \mapsto x^n$.
3. For representations $\delta :\subseteq \Sigma^\omega \to M$ and $\delta' :\subseteq \Sigma^\omega \to M'$ let $f :\subseteq M \to M'$ be (δ, δ')-computable. Show that f maps every $(\nu_{\mathbb{N}}, \delta)$-computable sequence to a $(\nu_{\mathbb{N}}, \delta')$-computable sequence.
◆ 4. ·Prove Theorem 9.3.3.

9.4 Ko's Approach

In his book "Complexity Theory of Real Functions" [Ko91] (see also the survey [Ko98]), K. Ko applies the famous **NP**-completeness theory from discrete complexity to prove lower bounds for basic numerical operations, such as maximization and integration. His definitions of computability and complexity of real functions are essentially equivalent to the definitions in TTE. Ko represents real numbers by *Cauchy functions*.

Definition 9.4.1 (Cauchy functions of a real number).
1. For $n \in \mathbb{N}$ let

$$\mathrm{D}_n := \{z \cdot 2^{-n} \mid z \in \mathbb{Z}\} \quad and$$

$$\mathrm{WD}_n := \{su{\scriptstyle\bullet}v \mid s \in \{-, \lambda\}, u \in 1\{0, 1\}^* \cup \{\lambda\}, v \in \{0, 1\}^n\} \, .$$

Let $\nu_{\mathrm{b},2} :\subseteq \Sigma^* \to \mathrm{D} := \bigcup\{\mathrm{D}_n \mid n \in \mathbb{N}\}$ be the base-2 notation from Definition 4.1.12 with domain $\mathrm{WD} := \bigcup\{\mathrm{WD}_n \mid n \in \mathbb{N}\}$.
2. A Cauchy function for $x \in \mathbb{R}$ is a function $\phi : \mathbb{N} \to \mathrm{WD}$ such that

$$\phi(n) \in \mathrm{WD}_n \quad and \quad |\nu_{\mathrm{b},2} \circ \phi(n) - x| \le 2^{-n} \qquad (9.1)$$

for all $n \in \mathbb{N}$. Let CF_x be the set of all Cauchy functions of x and let $\delta_{\mathrm{K}}(\phi) = x$, iff $\phi \in \mathrm{CF}_x$.

In terms of TTE, the function δ_{K} is a "representation" of the real numbers, where sequences of *words* are used as names. This representation δ_{K} and our signed digit representation ρ_{sd} (Definition 7.2.4) are very similar. If $\delta_{\mathrm{K}}(\phi) = x$, then by (9.1), $\nu_{\mathrm{b},2} \circ \phi(n) \in \mathrm{D}_n$ is the center of a closed interval J of length $2 \cdot 2^{-n}$ such that $x \in J$. Correspondingly, if $\rho_{\mathrm{sd}}(p) = x$, then also $\nu_{\mathrm{sd}}(p[n])$ (Definition 7.2.4) is the center of a closed interval J of length $2 \cdot 2^{-n}$ such that $x \in J$ (cf. the discussion after Definition 7.2.4). Therefore, the prefix $p[n]$ of p gives the same kind and amount of information as the sequence element $\phi(n)$.

Instead of Type-2 machines transforming infinite sequences $p \in \Sigma^\omega$ of symbols, Ko uses *function-oracle Turing machines* [HU79] (oracle machines, for short) mapping Cauchy functions to Cauchy functions. An oracle machine M (of the type we are considering here) is an ordinary Turing machine with one input and one output tape equipped with an additional distinguished work tape (the query tape) in which a new statement "call oracle" can occur arbitrarily often. Supplied with an oracle $\phi : \mathbb{N} \to \Sigma^*$, the machine M replaces the current inscription w of the query tape by $\phi(|w|)$ in one step, whenever the statement "call oracle" has been reached. The result of the computation of M on input 0^n supplied with the oracle ϕ is denoted by $M^\phi(n)$.

Definition 9.4.2 (computable real function, complexity [Ko91]).
Consider an oracle machine M, a real function $f :\subseteq \mathbb{R} \to \mathbb{R}$ with $\mathrm{dom}(f) = [a; b]$ and a function $t : \mathbb{N} \to \mathbb{N}$.

1. *The machine M computes f, iff*

$$\phi \in \mathrm{CF}_x \Longrightarrow M^\phi \in \mathrm{CF}_{f(x)} \quad \text{(for all } x \in [a;b] \text{ and } \phi : \mathbb{N} \to \Sigma^*) .$$

2. *The machine M computes f in time t, iff it computes f and additionally, on input 0^n with oracle ϕ, it halts after at most $t(n)$ steps.*

Notice that the formula in 1 can be written as "$f \circ \delta_K(\phi) = \delta_K(M^\phi)$, if $\phi \in \mathrm{dom}(f \circ \delta_K)$" (cf. Definition 3.1.3.3). Informally, a Cauchy function ϕ for x can be easily translated to a ρ_{sd}-name of x and vice versa. Therefore, oracle machines can be simulated by Type-2 machines and vice versa without much loss of time. More precisely we obtain:

Theorem 9.4.3 (Ko's definitions versus TTE).   A real function $f :\subseteq \mathbb{R} \to \mathbb{R}$ with $\mathrm{dom}(f) = [a;b]$
1. is computable according to Ko's definition, iff it is (ρ_{sd}, ρ_{sd})-computable,
2. is computable in polynomial time according to Ko's definition, iff it is computable in polynomial time according to Definition 7.2.6.

Proof : See Exercise 9.4.1.   □

From the large number of very interesting theorems in [Ko91, Ko98] we select two examples. In the following let **P** be the class of languages accepted by deterministic Turing machines in polynomial time, **NP** the class of languages accepted by non-deterministic Turing machines in polynomial time, **FP** the class of functions computed by deterministic Turing machines in polynomial time, and #**P** the class of functions that enumerate the number of accepting computations of polynomial-time nondeterministic Turing machines.

By Theorem 6.2.5, the maximum value of a computable function on a recursive compact set is a computable number, and various max operators map computable functions to computable ones (Exercises 6.2.6 and 6.2.7). By the next theorem it is unknown whether two important max operators map polynomially computable functions to polynomially computable functions.

Theorem 9.4.4 (maximum of pol-computable functions [Ko91]). The following properties are equivalent:
1. **P** = **NP**,
2. $g(x) := \max\{f(x,y) \mid 0 \le y \le 1\}$ is computable in polynomial time for every function $f \in C([0;1]^2)$ computable in polynomial time,
3. $h(x) := \max\{f(y) \mid 0 \le y \le x\}$ is computable in polynomial time for every function $f \in C[0;1]$ computable in polynomial time.

In addition, integration of functions computable in polynomial time may be difficult:

Theorem 9.4.5 (integration [Ko91]). The following properties are equivalent:

1. $\mathbf{FP} = \#\mathbf{P}$,
2. the function $h(x) := \int_0^x f(y)\, dy$ is computable in polynomial time for every function $f \in C[0;1]$ computable in polynomial time.

Exercises 9.4.

1. Prove Theorem 9.4.3.

9.5 Domain Theory

Domain theory [Wei87, AJ94, SHLG94] has been developed as a tool for the mathematical semantics of higher order programming languages and to understand computability of higher type functionals [Sco70, Erš72, Wei87, AJ94, SHLG94]. In his early paper Scott [Sco70] has already suggested to embed the real numbers in a domain consisting of real numbers and intervals of real numbers. Later this idea has been realized in various ways (for example [WD80, WS81, SHT95, DG96, Eda95a, Eda96, Eda97, Esc97, ES99b, SHT99, Bla99]).

In the following, we define a domain containing the real numbers such that the computable real functions are the restrictions of the computable functions on this domain [DG99]. We will tacitly use some elementary concepts from domain theory (for example [Wei87, AJ94, SHLG94]). In particular, for a partial order $(M, \leq)$ a subset $X \subseteq M$, is an *ideal*, iff it is *directed* (that is, $(\forall a, b \in X)(\exists c \in X)(a \leq c$ and $b \leq c))$ and *closed downwards* (that is, $a \in X$, if $a \leq b$ and $b \in X$).

Definition 9.5.1 (a domain extending the real numbers). *Define a noted partial order* $\overline{B}_0 := (B_0, \leq, \beta_0)$ *by*

1. $B_0 := \{(z \cdot 2^{-n}; (z+2) \cdot 2^{-n}) \subseteq \mathbb{R} \mid z \in \mathbb{Z}, n \in \mathbb{N}\} \cup \{\mathbb{R}\}$,
2. *the partial order on* B_0, $a \leq b : \Longleftrightarrow b \subseteq a$
3. *and a standard notation* $\beta_0 :\subseteq \Sigma^* \to B_0$ *of* B_0.

Let $\overline{D} := (D, \sqsubseteq, \bot, B, \beta)$ *be the ideal completion of* $\overline{B}_0$, *that is,* $(D, \sqsubseteq)$ *is the set of all non-empty ideals of the partial order* $(B_0, \leq)$, *ordered by set inclusion,* $\bot$ *is the ideal* $\{\mathbb{R}\}$ *and* β *is a notation of* $B \subseteq D$ *such that* $\beta(w) := \{b \in B_0 \mid b \leq \beta_0(w)\}$.

Embed B_0 *in* D *by* $c \mapsto \hat{c} := \{b \in B_0 \mid b \leq c\}$.
Embed the real numbers in D *by* $e(x) := \{b \in B_0 \mid x \in b\} \in D$.

A function $f : D \to D$ is *continuous*, iff it is monotone and commutes with sup on directed sets, that is, $f(\sup X) = \sup f[X]$ for every directed set $X \subseteq D$. Every monotone function $f : B \to D$ has a unique continuous extension $\bar{f} : D \to D$. A continuous function $f : D \to D$ is *computable*, iff the set $\{(u, v) \mid \beta(v) \sqsubseteq f \circ \beta(u)\}$ is r.e.

Call a (γ, γ_0)-realization g of a function $f :\subseteq M \to M_0$ (Definition 3.1.3) *strong*, iff $g(y) = \text{div}$ for all $y \in \text{dom}(\gamma)$ such that $\gamma(y) \notin \text{dom}(f)$ (Exercise 3.1.5).

By the following theorem, TTE continuity and computability on the real numbers can be characterized by means of domain continuity and computability, respectively.

Theorem 9.5.2 (TTE effective versus domain effective). Let $\overline{D}$ be the effective domain from Definition 9.5.1. Then for every partial function $f :\subseteq \mathbb{R} \to \mathbb{R}$,

1. f is strongly (ρ, ρ)-continuous, iff $f = e^{-1} \circ \bar{f} \circ e$ for some continuous function $\bar{f} : D \to D$,
2. f is strongly (ρ, ρ)-computable, iff $f = e^{-1} \circ \bar{f} \circ e$ for some computable function $\bar{f} : D \to D$,

---

Proof: We merely outline a proof of Part 2. The proof of Part 1 is similar. If $\bar{f}$ is computable, then from the r.e. set $\{(u, v) \mid \beta(v) \sqsubseteq \bar{f} \circ \beta(u)\}$ a program for a strong (ρ, ρ)-realization of $e^{-1} \circ \bar{f} \circ e$ can be programmed straightforwardly.

On the other hand, let f be strongly $(\rho_{\text{sd}}, \rho_{\text{sd}})$-computable. By Lemma 2.1.11 there is a computable monotone function $h : \Sigma^* \to \Sigma^*$ such that $h_\omega :\subseteq \Sigma^\omega \to \Sigma^\omega$ is a strong $(\rho_{\text{sd}}, \rho_{\text{sd}})$-realization of f. For $b \in B_0$ let $\bar{f}(\hat{b}) := \hat{a}$ where a is the shortest interval in B_0 such that

$$\bigcap \{\rho_{\text{sd}}[x \bullet y \Sigma^\omega] \mid (\exists\, u, v \in \Sigma^*)(b \subseteq \rho_{\text{sd}}[u \bullet v \Sigma^\omega] \text{ and } h(u \bullet v) = x \bullet y)\} \subseteq a$$

where $\bigcap \emptyset = \mathbb{R}$ (cf. Exercise 9.2.3). From domain theory it is known that $\bar{f}(d) = \sup\{\bar{f}(\hat{b}) \mid b \in B_0, \hat{b} \sqsubseteq d\}$ for all $d \in D$. To verify $f = e^{-1} \circ \bar{f} \circ e$ use the fact that $\rho_{\text{sd}}^{-1}[\{x\}]$ is compact for every $x \in \mathbb{R}$ (Exercise 7.2.9). □

Notice that in the above construction $\bar{f}(a) \in B$, if $a \in B$, cf. Theorem 9.2.3. Here we have considered merely algebraic domains. Sometimes more general domains are useful [Eda95a, Eda95b, Eda96, ES99a]. Since all of these domains have T_0-topologies with countable base, computability on them can be defined also via representations (see [WS83] for the algebraic case) and, therefore, can be expressed in TTE ([SHT99]).

Exercises 9.5.

1. Complete the proof of Theorem 9.5.2.

9.6 Markov's Approach

The *Russian School of Constructive Mathematics* was founded by A.A. Markov, Jr. in the late 1940s. Besides a restriction to a special constructive logic as a main feature, only computable real numbers are considered in this approach. They are encoded by *Markov algorithms* (*normal algorithms*) computing fast converging Cauchy sequences of rational numbers, and computable real functions are defined by *algorithms* transforming such programs [Kuš84, Kuš99].

We embed the recursion theoretical component of Markov's approach to TTE. For this purpose we use the canonical notation

$$w \mapsto \xi_w^{*\omega}(\lambda)$$

of the computable elements of Σ^ω.

Definition 9.6.1 (canonical notation of computable points).
Let $\delta :\subseteq \Sigma^\omega \to M$ be a representation. Define a notation $\nu_\delta : \Sigma^* \to M_c$ of the set of the δ-computable elements of M by

$$\nu_\delta(w) := \delta \circ \xi_w^{*\omega}(\lambda) \ .$$

Therefore, a word w is a ν_δ-name of a computable element $x \in M$, iff on input λ the Turing machine with code w computes an infinite sequence $p \in \Sigma^\omega$ which is a δ-name of x. If for two representations $\gamma \leq \delta$, then $\nu_\gamma \leq \nu_\delta$ (Exercise 9.6.1). In terms of TTE, the Markov computable real functions can be defined as follows.

Definition 9.6.2 (Markov computable real function). *A partial function $f :\subseteq \mathbb{R}_c^n \to \mathbb{R}_c$ is Markov computable, iff it is $(\nu_\rho, \ldots, \nu_\rho)$-computable.*

Therefore, a real function on the computable real numbers is Markov computable, iff it can be realized by a computable function transforming Turing machine codes appropriately. Computability induced by δ and computability induced by ν_δ are closely related. For the sake of simplicity we consider only functions with a single argument.

Theorem 9.6.3 (δ-computable implies ν_δ-computable).
Let $\gamma :\subseteq \Sigma^\omega \to M$ and $\delta :\subseteq \Sigma^\omega \to N$ be representations.
If $f :\subseteq M \to N$ is (γ, δ)-computable, then the restriction $f_0 :\subseteq M_c \to N_c$ of f to the computable elements of M and N, respectively, is (ν_γ, ν_δ)-computable.

Proof: Let $g :\subseteq \Sigma^\omega \to \Sigma^\omega$ be a computable (γ, δ)-realization of f. Let $u^{*\omega} :\subseteq \Sigma^* \times \Sigma^* \to \Sigma^\omega$ be the computable universal function of $\xi^{*\omega}$ (Theorem 2.3.5). Since $g \circ u^{*\omega}$ is computable, by the smn-theorem for $\xi^{*\omega}$

(Theorem 2.3.5) there is a computable function $r : \Sigma^* \to \Sigma^*$ such that $g \circ u^{*\omega}(w, v) = \xi^{*\omega}_{r(w)}(v)$ for $w \in \mathrm{dom}(f_0 \circ \nu_\gamma)$. We obtain

$$f_0 \circ \nu_\gamma(w) = f \circ \gamma \circ \xi^{*\omega}_w(\lambda) = \delta \circ g \circ \xi^{*\omega}_w(\lambda) = \delta \circ \xi^{*\omega}_{r(w)}(\lambda) = \nu_\delta \circ r(w) ,$$

and so f_0 is (ν_γ, ν_δ)-computable. □

Corollary 9.6.4 (TTE-computable implies Markov computable).
The restriction of every TTE-computable real function to the computable real numbers is Markov computable.

But not every Markov computable real function is the restriction of a TTE-computable real function. The following counter-example by P. Hertling [Her94] is a variant of an example by Myhill ([KLS59]).

Example 9.6.5. We construct a (ν_ρ, ν_ρ)-computable function f which is not continuous. Since every (ρ, ρ)-computable function is continuous, f cannot be the restriction of such a function. We may use the Cauchy representation ρ_C instead of ρ, since $\nu_{\rho_C} \equiv \nu_\rho$.
    For $w \in \mathrm{dom}(\nu_{\rho_C})$ let $\xi^{*\omega}_w(\lambda) = \iota(v_{w0})\iota(v_{w1})\iota(v_{w2})\dots$. Define a function $f :\subseteq \mathbb{R}_c \to \mathbb{R}_c$ as follows.

$$f(x) := \begin{cases} 0 & \text{if } x = 0 \\ 1 & \text{if } x \in M \\ \mathrm{div} & \text{otherwise,} \end{cases}$$

where M is the set of all $x \in \mathbb{R}_c$ such that

$$(\forall w \in \nu_{\rho_C}^{-1}[\{x\}]) \ (\exists j \leq |w|) \ |\nu_\mathbb{Q}(v_{wj})| > 2^{-j},$$

that is, for every name w of x among the first $|w|$ elements of the related Cauchy sequence there is a witness for $x \neq 0$. Obviously, $x \neq 0$ for all $x \in M$.
Proposition 1: $(\forall n_0 \in \mathbb{N}) \ (\exists x_0 \in M) \ |x_0| < 2^{-n_0}$
Proof : Let $n_0 \in \mathbb{N}$. Since $\mathbb{R}_c$ is dense in $\mathbb{R}$, also its finite variant

$$T := \mathbb{R}_c \setminus \big\{ x \in \mathbb{R} \mid (\exists w) \ (\nu_{\rho_C}(w) = x \text{ and } |w| \leq n_0 + 1) \big\}$$

is dense in $\mathbb{R}$.
Therefore, there is some number $x_0 \in T$ such that $2^{-n_0-1} < x_0 < 2^{-n_0}$.
Assume $\nu_{\rho_C}(u) = x_0$. Then $|u| \geq n_0 + 2$ by the definition of T. Since $|\nu_\mathbb{Q}(v_{u(n_0+2)}) - x_0| \leq 2^{-n_0-2}$, $|\nu_\mathbb{Q}(v_{u(n_0+2)})| > 2^{-n_0-2}$. Therefore, $x_0 \in M$ and $x_0 < 2^{-n_0}$ and so the proposition is proved. Since $f(0) = 0$ and $f(x) = 1$ for numbers $x \in \mathbb{R}_c$ arbitrarily close to 0, the function f cannot be continuous.
Proposition 2: f is $(\nu_{\rho_C}, \nu_{\rho_C})$-computable.

Proof : Let $w_0, w_1 \in \Sigma^*$ such that $\nu_{\rho_C}(w_0) = 0$ and $\nu_{\rho_C}(w_1) = 1$. Define a computable function $g :\subseteq \Sigma^* \to \Sigma^*$ by

$$g(w) := \begin{cases} w_0 & \text{if } (\forall j \leq |w|) \ |\nu_{\mathbb{Q}}(v_{wj})| \leq 2^{-j} \\ w_1 & \text{if } (\exists j \leq |w|) \ |\nu_{\mathbb{Q}}(v_{wj})| > 2^{-j} \\ \text{div} & \text{otherwise.} \end{cases}$$

Then g is a $(\nu_{\rho_C}, \nu_{\rho_C})$-realization of f. $\qquad\qquad\square$

In the above counter-example, $\text{dom}(f)$ is a rather complicated set. By a famous result of Ceitin [Ceĭ59], the converse of Theorem 9.6.4 holds, if $\text{dom}(f)$ is sufficiently simple.

Theorem 9.6.6 (Ceitin's theorem). Let $f :\subseteq \mathbb{R}_c \to \mathbb{R}_c$ be (ν_ρ, ν_ρ)-computable. Let $\text{range}(h)$ be dense in $\text{dom}(f)$ for some $(\nu_{\mathbb{N}}, \rho)$-computable function $h : \mathbb{N} \to \mathbb{R}$. Then the function $\bar{f} :\subseteq \mathbb{R} \to \mathbb{R}$, defined by $\text{graph}(\bar{f}) := \text{graph}(f)$, is (ρ, ρ)-computable.

For a proof, generalizations and further references see [Her97, Spr99]. Markov's approach to computable analysis is also used, for example in [Abe80, SHT95]. In most applications, Markov computability is proved (directly or indirectly) by reduction to TTE-computability (Theorem 9.6.3), although by Example 9.6.5 this is not possible in general.

Exercises 9.6.

1. Prove that for two representations, $\gamma \leq \delta$ implies $\nu_\gamma \leq \nu_\delta$.
2. Let (M, d, M, α) be a discrete (that is, $d(x, y) = 1$, if $x \neq y$) effective metric space (Definition 8.1.2). Let $\delta_{\mathbf{M}}$ be its Cauchy representation and let $\nu_{\delta_{\mathbf{M}}}$ be the notation of the $\delta_{\mathbf{M}}$-computable points according to Definition 9.6.1. Show

$$\alpha \equiv \delta_{\mathbf{M}} \equiv \nu_{\delta_{\mathbf{M}}} .$$

9.7 The real-RAM and Related Models

Theories like algebraic complexity (for example of real matrix multiplication) [Str84] or computational geometry [PS85] use models of computation where real numbers are considered as entities which can be added, multiplied, divided etc. or compared in a single step. As a mathematical model of computation we consider the "real-RAM", which generalizes the Random Access Machine [SS63] from the natural numbers to real numbers.

Informally, we define a real-RAM as a flowchart acting on a sequence of registers $N_0, N_1, \ldots$ for natural numbers and a sequence of registers $R_0, R_1, \ldots$ for real numbers. The following statements may occur for $i, j, k \in \mathbb{N}$.

Assignments: $N_i := N_i + 1$, $N_i := N_i - 1$, $R_i := N_j$ and $R_i := R_j$ op R_k
where op $\in \{+, -, \cdot, /\}$,
Branchings: $N_i = 0$ and $R_i < R_j$.
Instead of direct addresses $i, j, \ldots$, indirect addresses $N_i, N_j, \ldots$ are also allowed. For example, the statement $N_{N_5} := N_{N_5} + 1$ increments the register N_k by 1, where k is the actual content of the register N_5.

With canonical input and output encodings (all registers except the specified input registers contain 0 at the beginning) and flowchart semantics, a real-RAM can be used to compute functions $f :\subseteq \mathbb{R}^m \to \mathbb{R}^n$, $f :\subseteq \mathbb{R} \to \mathbb{N}$ or $f :\subseteq \mathbb{R}^* \to \mathbb{R}^*$, where $\mathbb{R}^* := \bigcup_{n \in \mathbb{N}} \mathbb{R}^n$. The complexity of a computation is its length.

Example 9.7.1 (real-RAM computability).

1. The Gauß staircase $x \mapsto \lfloor x \rfloor$ (Fig. 1.3) is real-RAM computable but not (ρ, ρ)-computable. The following real-RAM can be used:

   > input R_0
   > L_1 : if $R_0 < R_1$ then goto L_3, else goto L_2;
   > L_2 : $R_1 := R_1 + 1$, goto L_1;
   > L_3 : if $R_0 < R_1$ then goto L_4, else goto L_5;
   > L_4 : $R_1 := R_1 - 1$, goto L_3;
   > L_5 : HALT;
   > output R_1.

2. The characteristic function $\mathrm{cf}_X : \mathbb{R}^2 \to \mathbb{N}$ of the open square $X := \{z \in \mathbb{R}^2 \mid ||z|| < 1\}$ is real-RAM computable, and so X is decidable w.r.t. this model of computation. Remember that $\emptyset$ and $\mathbb{R}$ are the only ρ^2-decidable subsets of $\mathbb{R}^2$ (Corollary 4.3.16, Theorem 4.1.16).
3. Neither the square root $\sqrt{\ }$ nor the exponential function $\exp : \mathbb{R} \to \mathbb{R}$ are real-RAM computable (Exercise 9.7.5). Remember that both functions are (ρ, ρ)-computable.
4. There is some real-RAM computable function $f :\subseteq \mathbb{R}^2 \to \mathbb{R}$ such that

$$|f(x, \varepsilon) - \exp(x)| < \varepsilon \text{ for all } x, \varepsilon > 0 .$$

   (Exercise 9.7.3).
5. The epigraph $\{(x, y) \mid y \le \sqrt{x}\} \subseteq \mathbb{R}^2$ of the square root function is a recursive closed set (Exercise 5.1.31) and its characteristic function is real-RAM computable.
6. The epigraph $\{(x, y) \mid y \le \exp(x)\} \subseteq \mathbb{R}^2$ of the exponential function is a recursive closed set (Exercise 5.1.31), but its characteristic function is not real-RAM computable [Bra99b].

These examples already show that real-RAM computability and TTE-computability are incomparable concepts. Since every (non-trivial) branching $R_i < R_j$ in a real-RAM causes a point of discontinuity, most real-RAM computable real functions are discontinuous, and so not TTE-computable. On the other hand, important computable functions like the exponential function are not real-RAM computable.

Many TTE-computable real functions can be approximated by real-RAMs (Example 9.7.1.4). But approximate computation is an insufficient substitute for exact computation: in programming languages as well as models of computation, usually programs for functions f, g can be combined easily to a program of the composition $g \circ f$ by using the output of the first program as the input for the second one. This composition method fails for approximate algorithms in the real-RAM model.

Turing machines which can store a real number in every tape cell can be used as a model of computation [BSS89, Nov95]. They are equivalent to real-RAMs. Occasionally, further functions are considered as primitive, for example computable real constants, $\sqrt{\ }$ or the exponential function exp. But adding finitely many computable functions to the basic assignments of a real-RAM is not sufficient, since there still remain (ρ, ρ)-computable functions which are not real-RAM computable (Exercise 9.7.4).

Computability and complexity in the real-RAM model of computation, often called the "BSS-model", have been investigated in detail by L. Blum et al. [BCSS98]. *Information Based Complexity* [TWW88], "IBC" for short, uses the real-RAM model [Nov95].

Real-RAMs cannot be realized by physical machines, that is, they are unrealistic, for the following reason: in a finite amount of time every physical information channel can transfer only finitely many bits of information and every physical memory is finite. Since there are uncountably many real numbers, it is impossible to identify an arbitrary real number by a finite amount of information. Therefore, it is impossible to transfer an arbitrary real number to or from a computer in a finite amount of time or to store a real number in a computer.

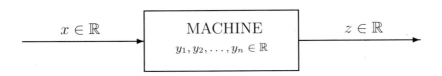

Fig. 9.1. An unrealistic machine model

In scientific computation, often (or even mostly?) PASCAL, FORTRAN, real-RAM, ... programs for real functions are designed such that they are (mathematically) correct with respect to the real number semantics, and then they are realized on computers where the data type REAL means floating-point semantics. Although the real-RAM model is unrealistic, in many applications real-RAM computations can be "approximated" reasonably by corresponding computations on *floating-point numbers*. Fig. 9.2 shows all pairs (x, y) of floating-point numbers with basis 2 for which mantissa and exponent have 2 digits.

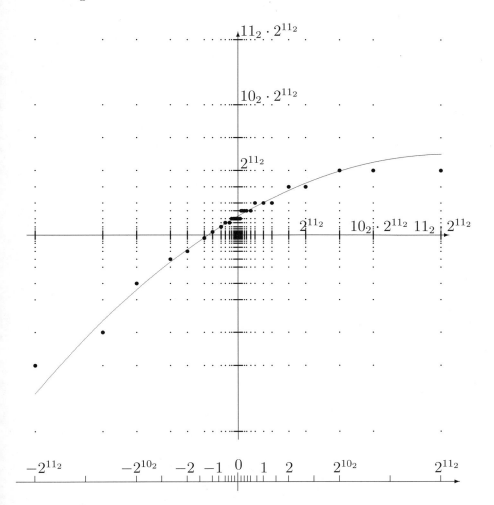

Fig. 9.2. Top: all pairs (x, y) of floating-point numbers (basis 2, mantissa and exponent 2 digits); a smooth function and the best approximation of its graph by pairs of floating-point numbers. Bottom: zoom in on the x-axis, amplification factor 3

A typical set of computer floating-point numbers is

$$\text{FN} := \left\{ m \cdot 2^{e-64} \mid m, e \in \mathbb{Z}, \ -2^{64} < m < 2^{64}, \ -(2^{14} - 1) < e < 2^{14} \right\}.$$

Well-behaved real functions can be approximated reasonably by functions on a given (finite) set FP of floating-point numbers. Fig. 9.2 shows the graph of a "smooth" real function and its approximation by pairs of floating-point numbers. However, a function which varies several times between 2^{11_2} and -2^{11_2} in the interval $(2^{11_2}; 11_2 \cdot 2^{10_2})$ (in decimal notation: between 8 and -8 in the interval $(8; 12)$) cannot be approximated reasonably by a function on these floating-point numbers. Often the quality of approximation can be improved by using multiple precision floating-point numbers. In the case of discontinuous functions even multiple precision does not help.

Example 9.7.2. The following real function is real-RAM computable but cannot be approximated by floating-point computations with longer and longer mantissas:

$$f(x) := \begin{cases} 1 & \text{if } x \in \mathbb{Q} \\ 0 & \text{if } x \notin \mathbb{Q} \text{ and } x^2 \in \mathbb{Q} \\ \text{div} & \text{otherwise.} \end{cases}$$

□

Therefore, only sometimes the floating-point realization of a real-RAM is a good approximation, and so only sometimes a theorem proved in the real-RAM theory has a meaning for computations on physical machines. As a negative example consider the following theorem [BCSS98]: "The characteristic function of the Mandelbrot set $M \subseteq \mathbb{R}^2$ is not real-RAM computable." The black part in Figure 9.3 is the Mandelbrot set.

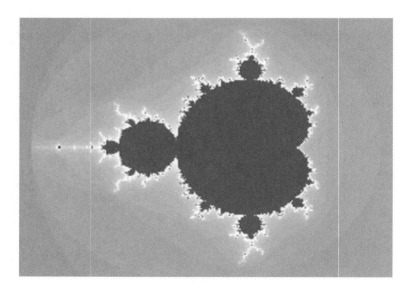

Fig. 9.3. The Mandelbrot set

Since real-RAMs cannot be realized anyway, seemingly this result is irrelevant for computations on machines.[1] There are attempts to explain why in many cases the realization of real-RAMs by floating point computations is so successful [Cuc99, Vig96, Woź99].

In [TW91] for "Information Based Complexity" (IBC) the "real-number model" is justified as follows:

> The rationale for the real-number model is that fixed-precision floating-point numbers are almost universally used for numerical calculations whether they occur in science, engineering, or economics. The cost of floating-point operations is independent of the size of operands. [...] The real-number model is used in IBC to decouple complexity from round-off issues. The numerical stability of optimal algorithms requires further study; [...]

For detailed discussions see also [Wei98, Woź99].

Exercises 9.7.

◇ 1. Prove the statement from Example 9.7.1.2.

2. Prove Example 9.7.2.

3. Show that there is some real-RAM computable function $f :\subseteq \mathbb{R}^2 \to \mathbb{R}$ such that
$$|f(x,\varepsilon) - \sqrt{x}| < \varepsilon \quad \text{for all} \quad x,\varepsilon > 0 \ .$$

♦ 4. Consider real-RAMs with assignments from a finite set $F = \{f_1, \ldots, f_m\}$ of computable real functions and the two types of branchings given above. Show that there is a (ρ, ρ)-computable real function $f : \mathbb{R} \to \mathbb{R}$ which is not computable by such a real-RAM.
Hint: The result of such a real-RAM on input 0 (if it exists) can be obtained by composition of the primitive functions. Show that there is a $(\nu_\mathbb{N}, \rho)$-computable sequence $i \mapsto x_i$ listing all such real numbers (Exercise 2.1.14) and apply Theorem 4.2.6. The argument can be extended to a computable sequence $f_0, f_1, \ldots$ of computable functions (Exercise 2.1.14).

♦ 5. Show that the exponential function and the square root $\sqrt{\ }$ are not real-RAM computable.
Hint: Let M be a real-RAM computing a function $f : \mathbb{R} \to \mathbb{R}$. Unwind the flowchart to a (generally infinite) tree. For every finite path p from the root to a "HALT" statement there are a set $X_p \subseteq \mathbb{R}$ and a rational function $f_p :\subseteq \mathbb{R} \to \mathbb{R}$, such that $f(x) = f_p(x)$ for $x \in X_p$ and $\{X_p \mid p$ is a finite halting path, $X_p \neq \emptyset\}$ is a partition of $\mathbb{R}$. Choose some p such that $X_p \cap [0; 1]$ is infinite. Remember that two analytic functions are equal, if they coincide on an infinite bounded set.

---

[1] The characteristic function of M is not (ρ, ρ, ρ)-computable (Corollary 4.3.16), the Mandelbrot set is co-r.e. (Definition 5.1.1, Exercise 5.1.32), it is an open problem whether it is a recursive closed set.

9.8 Comparison

Let us consider computable analysis as the mathematical theory of those
real functions (and other functions from analysis) which can be computed by
physical machines like digital computers. Since we do not know the precise
meaning of "computable by physical machines", every mathematical inves-
tigation in computable analysis must be based on a model of computation.
Such a model of computation is not "true" or "false" but can merely be
more or less realistic, powerful, expressive, illuticating or useful in practice
according to the specific situation. In this section, we compare the models
of computation presented in this chapter and TTE (which extends Grzegorczyk's definition and includes Ko's complexity theory) under various points
of view.

While TTE as well as the domain and the real-RAM approaches include
functions on *all* real numbers, the approaches by Banach/Mazur and Markov
consider only functions on the computable real numbers. Non-computable
real numbers do not exist in their settings. From a formal point of view Pour-
El and Richards also handle only computable real numbers and computable
points in their axiomatic approach (Definition 9.3.1).[2]

In most of the models, a computable function can be realized canonically
by a machine or program. The two exceptions are BM-computability and the
axiomatic approach by Pour-El and Richards (Definition 9.3.1). They differ
in their input (and output) conventions. In Markov's approach, the input
for a machine is a *program* computing a real number. In TTE, the input is
an infinite sequence of symbols considered as a name of a real number. In
the real-RAM model, the input is a real number. In the domain approach,
computable functions can be represented by recursively enumerable sets or
by enumeration operators [Rog67, Odi89] getting sets of natural numbers as
inputs.

Almost all the approaches mentioned in this chapter reduce computability
of real functions to discrete computations on Turing machines. Only the real-
RAM approach introduces computable real functions explicitly using merely
the register machine (or Turing machine) *control structure* as an ingredient
from ordinary computability theory. For further and more detailed compar-
isons of models of computation for analysis see [Zho98, SHT99, ES99b].

So far we have not mentioned "Constructive Analysis" by E. Bishop,
[BB85] who has developed large parts of classical and modern analysis within
a strictly constructive framework. It can be considered as a foundation for
computable analysis [Bri99]. His development is based on an informal, un-
specified notion of algorithm. This vagueness enables him to leave open the

---

[2] Since in their setting a computable function (on the computable numbers) must
have a uniform modulus of continuity, it has a unique continuous extension from
the dense set of computable numbers to the real numbers, and so it can be
assumed that they actually study computable functions on the real numbers.

possibility of interpreting his work within a variety of formal systems. In particular, constructive analysis by Bishop and his followers is a rich source of material for algorithms in TTE.

As a theory of those real functions (and other functions from analysis) which can be computed by physical machines, TTE (Type-2 Theory of Effectivity) seems to be the best choice.

- The Banach/Mazur definition considers only computable real numbers. It allows more than countably many computable real functions (Exercise 9.1.4).
- Although the Pour-El/Richards approach is consistent with TTE to a large extent, seemingly it is not as expressive and flexible. For example, it is not obvious, how $(\rho_<, \rho_>)$-computability, $(\delta_\to, \delta_\to)$-computability, computability of multi-valued functions or computational complexity can be formalized in that framework.
- The domain approach developed so far is consistent with TTE. Roughly speaking, a domain (for the real numbers) contains approximate objects as well as precise objects which are treated in separate sets in TTE. A computable domain function must map also all approximate objects reasonably. In many cases, constructing a domain which correspond to given representation still is a difficult task. Concepts for handling multi-valued functions and for computational complexity have not yet been developed for the domain approach. The elegant handling of higher type functions in domain theory can be simulated in TTE by means of function space representations $[\delta \to \delta']$ (Definition 3.3.13). To date, there seems to be no convincing reason to learn domain theory as a prerequisite for computable analysis.
- Markov's approach considers only computable real numbers, and seemingly it does not allow a definition of computational complexity.
- Most real-RAM computable real functions are not TTE-computable and most TTE-computable real functions are not real-RAM computable. Although some real-RAMs can be realized approximately by floating-point computations, real-RAMs cannot be realized (exactly) by physical machines, they are unrealistic. The theory does not separate realistic and unrealistic results.

On the other hand, Type-2 Theory of Effectivity, TTE, has a number of advantages:

- The model of computation (Type-2 machine) is realistic (see Fig. 2.2).
- The basic computability concept via naming systems and realizations (Definition 3.1.2) is very concrete.
- It allows definition of computable functions on the set of *all* real numbers.
- It combines the concepts of computation and approximation (topology) in a single theory.

- Various computability concepts on the real numbers ($\rho_<$-, $\rho_>$-, ...) and other spaces can be studied.
- It allows study of computability on many other sets, for example spaces of subsets and spaces of continuous functions.
- It allows a natural complexity theory.
- Computable multi-valued functions can be studied.

For a theoretical foundation of computability and computational complexity of real functions we have used Type-2 machines, that is, Turing machines with infinite input and output tapes as our basic model of computation. For more convenient programming they can be replaced by programs from some higher level programming language such as PASCAL (still with infinite input and output tapes).

The use of discrete computations for a foundation of scientific computation has been criticized repeatedly. In their "manifesto" [BCSS96, BCSS98] L. Blum et al. substantiate the real-RAM model as follows:

> A major obstacle in reconciling scientific computation and computer science is the present view of the machine, that is, the digital computer. As long as the computer is seen simply as a finite or discrete object, it will be difficult to systematize numerical analysis. We believe that the Turing machine as a foundation for real number algorithms can only obscure concepts.

The theory presented in this book refutes their statement. The authors continue:

> Toward resolving the problem we have posed, we are led to expanding the theoretical model of the machine to allow real number inputs. There has been a great hesitation to do this because of the digital nature of the computer. [...]

The arguments point to the real-RAM model of computation. However, there is an excellent alternative to their solution. V. Brattka and P. Hertling [BH95, BH98] propose the *feasible real*-RAM, a model of computation which uses "real numbers as entities of their own right" [Sma92] as in the real-RAM model and which allows computation of all and only all TTE-computable real functions. Furthermore, translation to Type-2 machines preserves polynomial time. V. Brattka [Bra99c] has generalized this concept of a higher level programming language for real functions to multi-valued functions on more general topological structures like spaces of sets or functions. J. Tucker and J. Zucker [TZ99] call such models as well as the real-RAM *abstract models of computation* in contrast to the *concrete models of computation* based on Turing machines.

References

[Abe80] Oliver Aberth. *Computable Analysis*. McGraw-Hill, New York, 1980.

[Ahl66] Lars Valerian Ahlfors. *Complex Analysis*. McGraw-Hill, New York, 1966.

[AHU83] Alfred V. Aho, John E. Hopcroft, and Jeffrey D. Ullman. *Data Structures and Algorithms*. Addison-Wesley, Reading, 1983.

[AJ94] S. Abramsky and A. Jung. Domain theory. In S. Abramsky, D.M. Gabbay, and T.S.E. Maibaum, editors, *Handbook of Logic in Computer Science, Volume 3*, pages 1–168, Oxford, 1994. Clarendon Press.

[Alt85] Helmut Alt. Multiplication is the easiest nontrivial arithmetic function. *Theoretical Computer Science*, 36:333–339, 1985.

[Avi61] Algirdas Avizienis. Signed-digit number representations for fast parallel arithmetic. *IRE Transactions on Electronic Computers*, EC-10:389–400, September 1961.

[Bar93] M.F. Barnsley. *Fractals Everywhere*. Academic Press, Boston, 1993.

[BB85] Errett Bishop and Douglas S. Bridges. *Constructive Analysis*, volume 279 of *Grundlehren der mathematischen Wissenschaften*. Springer, Berlin, 1985.

[BCSS96] Lenore Blum, Felipe Cucker, Mike Shub, and Steve Smale. Complexity and real computation: A manifesto. *International Journal of Bifurcation and Chaos*, 6:3–26, 1996.

[BCSS98] Lenore Blum, Felipe Cucker, Michael Schub, and Steve Smale. *Complexity and Real Computation*. Springer, New York, 1998.

[Bee93] Gerald Beer. *Topologies on Closed and Closed Convex Sets*, volume 268 of *Mathematics and Its Applications*. Kluwer Academic, Dordrecht, 1993.

[BH95] Vasco Brattka and Peter Hertling. Feasible real random access machines. Informatik Berichte 193, FernUniversität Hagen, Hagen, December 1995.

[BH98] Vasco Brattka and Peter Hertling. Feasible real random access machines. *Journal of Complexity*, 14(4):490–526, 1998.

[BH00] Vasco Brattka and Peter Hertling. Topological properties of real number representations. *Theoretical Computer Science*, 2000. accepted for publication.

[Bla97] Jens Blanck. Domain representability of metric spaces. *Annals of Pure and Applied Logic*, 83:225–247, 1997.

[Bla99] Jens Blanck. Effective domain representations of $H(X)$, the space of compact subsets. *Theoretical Computer Science*, 219:19–48, 1999.

[Bra97] Vasco Brattka. Order-free recursion on the real numbers. *Mathematical Logic Quarterly*, 43:216–234, 1997.

[Bra99a] Vasco Brattka. Computable invariance. *Theoretical Computer Science*, 210:3–20, 1999.

[Bra99b] Vasco Brattka. The emperor's new recursiveness. preprint, 1999.

[Bra99c] Vasco Brattka. Recursive and computable operations over topological structures. Informatik Berichte 255, FernUniversität Hagen, Fachbereich Informatik, Hagen, July 1999. Dissertation.

[Bre76] R.P. Brent. Fast multiple-precision evaluation of elementary functions. *Journal of the ACM*, 23(2):242–251, 1976.

[Bri99] Douglas S. Bridges. Constructive mathematics: a foundation for computable analysis. *Theoretical Computer Science*, 219:95–109, 1999.

[BSS89] Lenore Blum, Mike Shub, and Steve Smale. On a theory of computation and complexity over the real numbers: NP-completeness, recursive functions and universal machines. *Bulletin of the American Mathematical Society*, 21(1):1–46, July 1989.

[BW99] Vasco Brattka and Klaus Weihrauch. Computability on subsets of Euclidean space I: Closed and compact subsets. *Theoretical Computer Science*, 219:65–93, 1999.

[Ceĭ59] G.S. Ceĭtin. Algorithmic operators in constructive complete separable metric spaces. *Doklady Akad. Nauk*, 128:49–52, 1959. (in Russian).

[CPE75] J. Caldwell and Marian Boykan Pour-El. On a simple definition of computable functions of a real variable - with applications to functions of a complex variable. *Zeitschrift für Mathematische Logik und Grundlagen der Mathematik*, 21:1–19, 1975.

[Cuc99] Felipe Cucker. Real computations with fake numbers. In Jiří Wiedermann, Peter van Emde Boas, and Mogens Nielsen, editors, *Automata, Languages and Programming*, volume 1644 of *Lecture Notes in Computer Science*, pages 55–73, Berlin, 1999. Springer. 26th International Colloquium, ICALP'99, Prague, Czech Republic, July, 1999.

[Dei84] Thomas Deil. Darstellungen und Berechenbarkeit reeller Zahlen. Informatik Berichte 51, FernUniversität Hagen, Hagen, December 1984. Dissertation.

[DG96] Pietro Di Gianantonio. Real number computation and domain theory. *Information and Computation*, 127:11–25, 1996.

[DG99] Pietro Di Gianantonio. An abstract data type for real numbers. *Theoretical Computer Science*, 221:295–326, 1999.

[Die60] J. Dieudonné. *Foundations of Modern Analysis*. Academic Press, New York, 1960.

[Eda95a] Abbas Edalat. Domain theory and integration. *Theoretical Computer Science*, 151:163–193, 1995.

[Eda95b] Abbas Edalat. Dynamical systems, measures, and fractals via domain theory. *Information and Computation*, 120(1):32–48, 1995.

[Eda96] Abbas Edalat. Power domains and iterated function systems. *Information and Computation*, 124(2):182–197, 1996.

[Eda97] Abbas Edalat. Domains for computation in mathematics, physics and exact real arithmetic. *Bulletin of Symbolic Logic*, 3(4):401–452, 1997.

[Eng86] A.J.M. Engelen. *Homogeneous zero-dimensional absolute Borel sets*, volume 27 of *CWI Tracts*. Centrum voor Wiskunde en Informatica, Amsterdam, 1986.

[Eng89] Ryszard Engelking. *General Topology*, volume 6 of *Sigma series in pure mathematics*. Heldermann, Berlin, 1989.

[Erš72] Ju. L. Eršov. Computable functionals of finite type. *Algebra and Logic*, 11(4):203–242, 1972.

[Erš73] Ju. L. Eršov. Theorie der Numerierungen I. *Zeitschrift für Mathematische Logik und Grundlagen der Mathematik*, 19(4):289–388, 1973.

[Erš75] Ju. L. Eršov. Theorie der Numerierungen II. *Zeitschrift für Mathematische Logik und Grundlagen der Mathematik*, 21(6):473–584, 1975.

[Erš77] Ju. L. Eršov. Theorie der Numerierungen III. *Zeitschrift für Mathematische Logik und Grundlagen der Mathematik*, 23(4):289–371, 1977.

[ES99a] Abbas Edalat and Philipp Sünderhauf. Computable Banach spaces via domain theory. *Theoretical Computer Science*, 219:169–184, 1999.

[ES99b] Abbas Edalat and Philipp Sünderhauf. A domain-theoretic approach to computability on the real line. *Theoretical Computer Science*, 210:73–98, 1999.

[Esc97] Martín Hötzel Escardó. *PCF extended with real numbers: a domain-theoretic approach to higher-order exact real number computation*. PhD thesis, Imperial College, University of London, London, Great Britain, 1997.

[FS74] M.J. Fischer and L. Stockmeyer. Fast on-line integer multiplication. *Journal of Computer and Systems Sciences*, 9:317–331, 1974.

[Goo59] R.L. Goodstein. Recursive analysis. In A. Heyting, editor, *Constructivity in Mathematics*, Studies in Logic and The Foundations of Mathematics, pages 37–42, Amsterdam, 1959. North-Holland. Proc. Colloq., Amsterdam, Aug. 26–31, 1957.

[Grz55] Andrzej Grzegorczyk. Computable functionals. *Fundamenta Mathematicae*, 42:168–202, 1955.

[Grz57] Andrzej Grzegorczyk. On the definitions of computable real continuous functions. *Fundamenta Mathematicae*, 44:61–71, 1957.

[GS83] Dan Gordon and Eliahu Shamir. Computation of recursive functionals using minimal initial segments. *Theoretical Computer Science*, 23:305–315, 1983.

[Hau73] Jürgen Hauck. Berechenbare reelle Funktionen. *Zeitschrift für Mathematische Logik und Grundlagen der Mathematik*, 19:121–140, 1973.

[Hau78] Jürgen Hauck. Konstruktive Darstellungen reeller Zahlen und Folgen. *Zeitschrift für Mathematische Logik und Grundlagen der Mathematik*, 24:365–374, 1978.

[Hau80] Jürgen Hauck. Konstruktive Darstellungen in topologischen Räumen mit rekursiver Basis. *Zeitschrift für Mathematische Logik und Grundlagen der Mathematik*, 26:565–576, 1980.

[Hau81] Jürgen Hauck. Berechenbarkeit in topologischen Räumen mit rekursiver Basis. *Zeitschrift für Mathematische Logik und Grundlagen der Mathematik*, 27:473–480, 1981.

[Hau82] Jürgen Hauck. Stetigkeitseigenschaften berechenbarer Funktionale. *Zeitschrift für Mathematische Logik und Grundlagen der Mathematik*, 28:377–383, 1982.

[Hau87] Jürgen Hauck. Eine berechenbare Funktion mit rationalen Werten, die nicht rekursiv ist. *Zeitschrift für Mathematische Logik und Grundlagen der Mathematik*, 33:255–256, 1987.

[Her94] Peter Hertling. Der Satz von Ceitin. preprint, 1994.

[Her96] Peter Hertling. Unstetigkeitsgrade von Funktionen in der effektiven Analysis. Informatik Berichte 208, FernUniversität Hagen, Hagen, November 1996. Dissertation.

[Her97] Peter Hertling. Effectivity and effective continuity of functions between computable metric spaces. In Douglas S. Bridges, Cristian S. Calude, Jeremy Gibbons, Steve Reeves, and Ian H. Witten, editors, *Combinatorics, Complexity, and Logic*, Discrete Mathematics and Theoretical Computer Science, pages 264–275, Singapore, 1997. Springer. Proceedings of DMTCS'96.

[Her99a] Peter Hertling. An effective Riemann Mapping Theorem. *Theoretical Computer Science*, 219:225–265, 1999.

[Her99b] Peter Hertling. A real number structure that is effectively categorical. *Mathematical Logic Quarterly*, 45(2):147–182, 1999.

[HU79] John E. Hopcroft and Jeffrey D. Ullman. *Introduction to Automata Theory, Languages and Computation*. Addison-Wesley, Reading, 1979.

[HW94] Peter Hertling and Klaus Weihrauch. Levels of degeneracy and exact
lower complexity bounds for geometric algorithms. In *Proceedings of the Sixth
Canadian Conference on Computational Geometry*, pages 237–242. University
of Saskatchewan, 1994. Saskatoon, Saskatchewan, August 2–6, 1994.

[KK99] Hiroyasu Kamo and Kiko Kawamura. Computability of self-similar sets.
Mathematical Logic Quarterly, 45:23–30, 1999.

[Kla61] D. Klaua. *Konstruktive Analysis*. Deutscher Verlag der Wissenschaften,
Berlin, 1961.

[KLS59] G. Kreisel, D. Lacombe, and J.R. Shoenfield. Partial recursive functionals
and effective operations. In A. Heyting, editor, *Constructivity in Mathematics*,
Studies in Logic and The Foundations of Mathematics, pages 290–297, Amsterdam, 1959. North-Holland. Proc. Colloq., Amsterdam, Aug. 26–31, 1957.

[Ko91] Ker-I Ko. *Complexity Theory of Real Functions*. Progress in Theoretical
Computer Science. Birkhäuser, Boston, 1991.

[Ko98] Ker-I Ko. Polynomial-time computability in analysis. In Yu. L. Ershov,
S.S. Goncharov, A. Nerode, and J.B. Remmel, editors, *Handbook of Recursive
Mathematics*, volume 139 of *Studies in Logic and The Foundations of Mathematics*, pages 1271–1317, Amsterdam, 1998. Elsevier. Volume 2, Recursive
Algebra, Analysis and Combinatorics.

[Kur66] K. Kuratowski. *Topology, Volume 1*. Academic Press, New York, 1966.

[Kuš84] Boris Abramovich Kušner. *Lectures on Constructive Mathematical Analysis*, volume 60. American Mathematical Society, Providence, 1984.

[Kuš99] Boris Abramovich Kušner. Markov's constructive analysis; a participant's
view. *Theoretical Computer Science*, 219:267–285, 1999.

[KW85] Christoph Kreitz and Klaus Weihrauch. Theory of representations. *Theoretical Computer Science*, 38:35–53, 1985.

[Lac55] Daniel Lacombe. Extension de la notion de fonction récursive aux fonctions
d'une ou plusieurs variables réelles III. *Comptes Rendus Académie des Sciences
Paris*, 241:151–153, July 1955. Théorie des fonctions.

[Mal71] Anatolií Ivanovič Mal'cev. Constructive algebras. I. In *The Metamathematics of Algebraic Systems*, pages 148–214, Amsterdam, 1971. North-Holland.
Collected papers: 1936–1967.

[Maz63] S. Mazur. *Computable Analysis*, volume 33. Razprawy Matematyczne,
Warsaw, 1963.

[Mos57] A. Mostowski. On computable sequences. *Fundamenta Mathematicae*,
44:37–51, 1957.

[MTY97] Takakazu Mori, Yoshiki Tsujii, and Mariko Yasugi. Computability structures on metric spaces. In Douglas S. Bridges, Cristian S. Calude, Jeremy
Gibbons, Steve Reeves, and Ian H. Witten, editors, *Combinatorics, Complexity, and Logic*, Discrete Mathematics and Theoretical Computer Science, pages
351–362, Singapore, 1997. Springer. Proceedings of DMTCS'96.

[Mül86a] Norbert Th. Müller. Computational complexity of real functions and real
numbers. Informatik Berichte 59, FernUniversität Hagen, Hagen, June 1986.

[Mül86b] Norbert Th. Müller. Subpolynomial complexity classes of real functions
and real numbers. In Laurent Kott, editor, *Proceedings of the 13th International
Colloquium on Automata, Languages, and Programming*, volume 226 of *Lecture
Notes in Computer Science*, pages 284–293, Berlin, 1986. Springer.

[Mül87] Norbert Th. Müller. Uniform computational complexity of Taylor series.
In Thomas Ottmann, editor, *Proceedings of the 14th International Colloquium
on Automata, Languages, and Programming*, volume 267 of *Lecture Notes in
Computer Science*, pages 435–444, Berlin, 1987. Springer.

[Mül95] Norbert Th. Müller. Constructive aspects of analytic functions. In Ker-I Ko and Klaus Weihrauch, editors, *Computability and Complexity in Analysis*, volume 190 of *Informatik Berichte*, pages 105–114. FernUniversität Hagen, September 1995. CCA Workshop, Hagen, August 19–20, 1995.

[Myh71] J. Myhill. A recursive function defined on a compact interval and having a continuous derivative that is not recursive. *Michigan Math. J.*, 18:97–98, 1971.

[Myl92] Uwe Mylatz. Vergleich unstetiger Funktionen in der Analysis. Diplomarbeit, Fachbereich Informatik, FernUniversität Hagen, 1992.

[Nov95] Erich Novak. The real number model in numerical analysis. *Journal of Complexity*, 11(1):57–73, 1995.

[Odi89] Piergiorgio Odifreddi. *Classical Recursion Theory*, volume 129 of *Studies in Logic and the Foundations of Mathematics*. North-Holland, Amsterdam, 1989.

[PE99] Marian Boykan Pour-El. The structure of computability in analysis and physical theory: An extension of Church's thesis. In Edward R. Griffor, editor, *Handbook of Computability Theory*, volume 140 of *Studies in Logic and The Foundations of Mathematics*, pages 449–471, Amsterdam, 1999. Elsevier.

[Pen89] Roger Penrose. *The Emperor's New Mind. Concerning Computers, Minds and The Laws of Physics*. Oxford University Press, New York, 1989.

[PER89] Marian B. Pour-El and J. Ian Richards. *Computability in Analysis and Physics*. Perspectives in Mathematical Logic. Springer, Berlin, 1989.

[Pre99] Gero Presser. Effektive Teilmengen metrischer Räume. Diplomarbeit, Fachbereich Informatik, Universität Dortmund, 1999.

[PS85] Franco P. Preparata and Michael Ian Shamos. *Computational Geometry*. Texts and Monographs in Computer Science. Springer, New York, 1985.

[Rog67] Hartley Rogers. *Theory of Recursive Functions and Effective Computability*. McGraw-Hill, New York, 1967.

[Rud64] Walter Rudin. *Principles of Mathematical Analysis*. McGraw-Hill, New York, 1964.

[Sch77] Kurt Schütte. *Proof Theory*, volume 225 of *Grundlehren der mathematischen Wissenschaften*. Springer, Berlin, 1977. Translated by J.N. Crossley.

[Sch90] Arnold Schönhage. Numerik analytischer Funktionen und Komplexität. *Jahresbericht der Deutschen Mathematiker-Vereinigung*, 92:1–20, 1990.

[Sch97] Matthias Schröder. Fast online multiplication of real numbers. In Rüdiger Reischuk and Michel Morvan, editors, *STACS 97*, volume 1200 of *Lecture Notes in Computer Science*, pages 81–92, Berlin, 1997. Springer. 14th Annual Symposium on Theoretical Aspects of Computer Science, Lübeck, Germany, February 27 – March 1, 1997.

[Sch98] Matthias Schröder. Effective metrization of regular spaces. In Ker-I Ko, Anil Nerode, Marian B. Pour-El, Klaus Weihrauch, and Jiří Wiedermann, editors, *Computability and Complexity in Analysis*, volume 235 of *Informatik Berichte*, pages 63–80. FernUniversität Hagen, August 1998. CCA Workshop, Brno, Czech Republik, August, 1998.

[Sch99] Matthias Schröder. Online computations of differentiable functions. *Theoretical Computer Science*, 219:331–345, 1999.

[Sch00] Matthias Schröder. Extended admissibility. Submitted, 2000.

[Sco70] Dana Scott. Outline of a mathematical theory of computation. Technical Monograph PRG-2, Oxford University, Oxford, November 1970.

[SGV94] A. Schönhage, A.F.W. Grotefeld, and E. Vetter. *Fast Algorithms*. B.I. Wissenschaftsverlag, Mannheim, 1994.

[SHLG94] V. Stoltenberg-Hansen, I. Lindström, and E.R. Griffor. *Mathematical Theory of Domains*, volume 22 of *Cambrige Tracts in Theoretical Computer Science*. Cambridge University Press, Cambridge, 1994.

274   References

[SHT95] Viggo Stoltenberg-Hansen and John V. Tucker. Effective algebras. In S. Abramsky, D.M. Gabbay, and T.S.E. Maibaum, editors, *Handbook of Logic in Computer Science, Volume 4*, pages 357–527, Oxford, 1995. Clarendon Press.

[SHT99] Viggo Stoltenberg-Hansen and John V. Tucker. Concrete models of computation for topological algebras. *Theoretical Computer Science*, 219:347–378, 1999.

[Sma92] Steve Smale. Theory of computation. In A. et al. Casacuberta, editor, *Mathematical Research Today and Tomorrow*, pages 60–69, Berlin, 1992. Springer.

[Spe49] Ernst Specker. Nicht konstruktiv beweisbare Sätze der Analysis. *The Journal of Symbolic Logic*, 14(3):145–158, 1949.

[Spe59] Ernst Specker. Der Satz vom Maximum in der rekursiven Analysis. In A. Heyting, editor, *Constructivity in mathematics*, Studies in Logic and The Foundations of Mathematics, pages 254–265, Amsterdam, 1959. North-Holland. Proc. Colloq., Amsterdam, Aug. 26–31, 1957.

[Spr99] Dieter Spreen. Representations versus numberings: On the relationship of two computability notions. Informatik Berichte 99-02, Universität Siegen, Siegen, 1999.

[SS63] J.C. Shepherdson and H.E. Sturgis. Computability of recursive funtions. *Journal of the ACM*, 10:217–255, 1963.

[Ste89] Thorsten von Stein. Vergleich nicht konstruktiv lösbarer Probleme in der Analysis. Diplomarbeit, Fachbereich Informatik, FernUniversität Hagen, 1989.

[Str84] V. Strassen. Algebraische Berechnungskomplexität. In W. Jäger, J. Moser, and R. Remmert, editors, *Perspectives in Mathematics*, pages 509–550, Basel, 1984. Birkhäuser.

[Tur36] Alan M. Turing. On computable numbers, with an application to the "Entscheidungsproblem". *Proceedings of the London Mathematical Society*, 42(2):230–265, 1936.

[Tur37] Alan M. Turing. On computable numbers, with an application to the "Entscheidungsproblem". A correction. *Proceedings of the London Mathematical Society*, 43(2):544–546, 1937.

[TW91] Joseph F. Traub and H. Woźniakowski. Information-based complexity: New questions for mathematicians. *The Mathematical Intelligencer*, 13(2):34–43, 1991.

[TWW88] Joseph F. Traub, G.W. Wasilkowski, and H. Woźniakowski. *Information-Based Complexity*. Computer Science and Scientific Computing. Academic Press, New York, 1988.

[TZ99] J.V. Tucker and J.I. Zucker. Computation by 'While' programs on topological partial algebras. *Theoretical Computer Science*, 219:379–420, 1999.

[Vig96] Jean Vignes. A stochastic approach to the analysis of round-off error propagation - a survey of the CESTAC method. In Jean-Paul Allouche, Jean-Claude Bajard, Laurent-Stephane Didier, and Pierre Liardet, editors, *Second Real Numbers and Computer Conference*, pages 233–251. Centre International de Rencontres Mathématiques, 1996. Marseille, April 9–11, 1996.

[Wad72] W.W. Wadge. Degrees of complexity of subsets of the Baire space. *Notices of the Amer. Math. Soc.*, A:714–715, 1972.

[Wad83] W.W. Wadge. *Reducibility and determinateness on the Baire space*. Thesis, University of California, Berkeley, 1983.

[WD80] Klaus Weihrauch and Thomas Deil. Berechenbarkeit auf cpo-s. Schriften zur Informatik und angewandten Mathematik 63, Technische Hochschule Aachen, Aachen, 1980. Eine Vorlesung von Klaus Weihrauch ausgearbeitet von Thomas Deil.

[Wei85]   Klaus Weihrauch. Type 2 recursion theory. *Theoretical Computer Science*, 38:17–33, 1985.

[Wei87]   Klaus Weihrauch. *Computability*, volume 9 of *EATCS Monographs on Theoretical Computer Science*. Springer, Berlin, 1987.

[Wei91a]   Klaus Weihrauch. The lowest Wadge-degrees of subsets of the Cantor space. Informatik Berichte 107, FernUniversität Hagen, Hagen, April 1991.

[Wei91b]   Klaus Weihrauch. On the complexity of online computations of real functions. *Journal of Complexity*, 7:380–394, 1991.

[Wei92a]   Klaus Weihrauch. The degrees of discontinuity of some translators between representations of the real numbers. Informatik Berichte 129, FernUniversität Hagen, Hagen, July 1992.

[Wei92b]   Klaus Weihrauch. The degrees of discontinuity of some translators between representations of the real numbers. Technical Report TR-92-050, International Computer Science Institute, Berkeley, July 1992.

[Wei92c]   Klaus Weihrauch. The TTE-interpretation of three hierarchies of omniscience principles. Informatik Berichte 130, FernUniversität Hagen, Hagen, September 1992.

[Wei93]   Klaus Weihrauch. Computability on computable metric spaces. *Theoretical Computer Science*, 113:191–210, 1993. Fundamental Study.

[Wei94]   Klaus Weihrauch. Effektive Analysis. Correspondence course 1681, FernUniversität Hagen, 1994.

[Wei95]   Klaus Weihrauch. A simple introduction to computable analysis. Informatik Berichte 171, FernUniversität Hagen, Hagen, July 1995. 2nd edition.

[Wei98]   Klaus Weihrauch. A refined model of computation for continuous problems. *Journal of Complexity*, 14:102–121, 1998.

[Wie80]   E. Wiedmer. Computing with infinite objects. *Theoretical Computer Science*, 10:133–155, 1980.

[WK91]   Klaus Weihrauch and Christoph Kreitz. Type 2 computational complexity of functions on Cantor's space. *Theoretical Computer Science*, 82:1–18, 1991. Fundamental Study.

[Woź99]   Henryk Woźniakowski. Why does information-based complexity use the real number model? *Theoretical Computer Science*, 219:451–465, 1999.

[WS81]   Klaus Weihrauch and Ulrich Schreiber. Embedding metric spaces into cpo's. *Theoretical Computer Science*, 16:5–24, 1981.

[WS83]   Klaus Weihrauch and G. Schäfer. Admissible representations of effective cpo's. *Theoretical Computer Science*, 26:131–147, 1983.

[WZ97]   Klaus Weihrauch and Xizhong Zheng. Effectiveness of the global modulus of continuity on metric spaces. In Eugenio Moggi and Giuseppe Rosolini, editors, *Category Theory and Computer Science*, volume 1290 of *Lecture Notes in Computer Science*, pages 210–219, Berlin, 1997. Springer. 7th International Conference, CTCS'97, Santa Margherita Ligure, Italy, September 4–6, 1997.

[WZ98a]   Klaus Weihrauch and Xizhong Zheng. A finite hierarchy of the recursively enumerable real numbers. In Luboš Brim, Jozef Gruska, and Jiří Zlatuška, editors, *Mathematical Foundations of Computer Science 1998*, volume 1450 of *Lecture Notes in Computer Science*, pages 798–806, Berlin, 1998. Springer. 23rd International Symposium, MFCS'98, Brno, Czech Republic, August, 1998.

[WZ98b]   Klaus Weihrauch and Ning Zhong. The wave propagator is Turing computable. In Ker-I Ko, Anil Nerode, Marian B. Pour-El, Klaus Weihrauch, and Jiří Wiedermann, editors, *Computability and Complexity in Analysis*, volume 235 of *Informatik Berichte*, pages 127–155. FernUniversität Hagen, August 1998. CCA Workshop, Brno, Czech Republik, August, 1998.

[YMT99] Mariko Yasugi, Takakazu Mori, and Yoshiki Tsujii. Effective properties of sets and functions in metric spaces with computability structure. *Theoretical Computer Science*, 219:467–486, 1999.

[Zho96] Qing Zhou. Computable real-valued functions on recursive open and closed subsets of Euclidean space. *Mathematical Logic Quarterly*, 42:379–409, 1996.

[Zho98] Ning Zhong. Recursively enumerable subsets of R^q in two computing models: Blum-Shub-Smale machine and Turing machine. *Theoretical Computer Science*, 197:79–94, 1998.

[ZW99] Xizhong Zheng and Klaus Weihrauch. The arithmetical hierarchy of real numbers. In Mirosław Kutyłowski, Leszek Pacholski, and Tomasz Wierzbicki, editors, *Mathematical Foundations of Computer Science 1999*, volume 1672 of *Lecture Notes in Computer Science*, pages 23–33, Berlin, 1999. Springer. 24th International Symposium, MFCS'99, Szklarska Poręba, Poland, September, 1999.

Index

Former volumes appeared as
EATCS Monographs on Theoretical Computer Science

Printing (Computer to Film): Saladruck, Berlin
Binding: Lüderitz & Bauer, Berlin